Three Lectures on Commutative Algebra

University
LECTURE
Series

Volume 42

Three Lectures on Commutative Algebra

Holger Brenner
Jürgen Herzog
Orlando Villamayor

Edited by

Gemma Colomé-Nin
Teresa Cortadellas Benítez
Juan Elias
Santiago Zarzuela

American Mathematical Society
Providence, Rhode Island USA

Real Sociedad Matemática Española
Madrid, Spain

2000 *Mathematics Subject Classification.* Primary 13–02, 14–02, 05Cxx, 13Dxx, 13Fxx, 13Hxx, 14Bxx, 14Exx, 14Jxx.

For additional information and updates on this book, visit
www.ams.org/bookpages/ulect-42

Library of Congress Cataloging-in-Publication Data

Winter School on Commutative Algebra and Applications (2006 : Institut de Matemàtica de la Universitat de Barcelona)

Three lectures on commutative algebra / Holger Brenner, Jürgen Herzog, Orlando Villamayor ; edited by Gemma Colomé-Nin... [et al.].

p. cm. — (University lecture series ; 42)

Includes bibliographical references.

ISBN-13: 978-0-8218-4434-2 (alk. paper)

ISBN-10: 0-8218-4434-2 (alk. paper)

1. Commutative algebra—Congresses. I. Brenner, Holger, 1966– II. Herzog, Jürgen, 1941– III. Villamayor, Orlando E. IV. Colomé-Nin, Gemma, 1978– V. Title.

QA251.3.W56 2006
512′.4—dc22

2007060568

∞ The paper used in this book is acid-free and falls within the guidelines
established to ensure permanence and durability.
Visit the AMS home page at `http://www.ams.org/`

10 9 8 7 6 5 4 3 2 1   13 12 11 10 09 08

Contents

Preface

The **Winter School on Commutative Algebra and Applications** was held from January 30 to February 3, 2006, at the *Institut de Matemàtica de la Universitat de Barcelona* (IMUB). The Winter School was a scientific activity of the (RTACA) *Red Temática de Algebra Conmutativa y Aplicaciones*[1]. The school consisted of three series of lectures surveying recent developments in commutative algebra and applications and some additional talks and communications. The series of lectures were: Tight Closure and Vector Bundles (Holger Brenner, University of Sheffield), Combinatorics and Commutative Algebra (Jürgen Herzog, University of Essen), Constructive Desingularization (Orlando Villamayor, Universidad Autónoma de Madrid). This volume collects an enlarged version of these series of lectures.

The School was organized by G. Colomé-Nin, T. Cortadellas-Benítez, J. Elias, and S. Zarzuela. It was sponsored by Institut de Matemàtica de la Universitat de Barcelona, Societat Catalana de Matemàtiques and Ministerio de Educación y Ciencia.

Gemma Colomé-Nin
Teresa Cortadellas-Benítez
Juan Elias
Santiago Zarzuela

Barcelona, July 2007

[1]Spanish Net of Commutative Algebra and Applications

Tight Closure and Vector Bundles

by Holger Brenner

Contents

Introduction

These notes on tight closure and vector bundles are based on a series of four one-and-a-half hour lectures which I gave at a winter school in Barcelona in 2006.

Tight closure is a technique in positive characteristic ring theory was introduced by M. Hochster and C. Huneke 20 years ago. It assigns to an ideal I another ideal $I^* \supseteq I$, called the tight closure of the ideal. This theory is now a standard technique in commutative algebra in positive characteristic and, by reduction to positive characteristic, also in characteristic zero. It has many applications in invariant theory, homological algebra, birational geometry and algebraic geometry; see [**49**], [**54**], [**55**], [**97**].

In these notes we present a geometric interpretation of tight closure in terms of vector bundles. This approach has brought some new insights into the theory of tight closure, and it was particularly successful for graded rings of dimension two. In this setting one can obtain a fairly detailed understanding of tight closure in terms of positivity properties (ampleness) of vector bundles on the corresponding projective curve. The bundle of interest here is the syzygy bundle given by homogeneous ideal generators of the ideal. With this tool one can solve some of the basic open problems of tight closure theory in graded dimension two, in particular that tight closure equals plus closure if the base field is finite and that the Hilbert-Kunz multiplicity is a rational number ([**8**], [**12**], [**13**], [**14**], [**17**] [**99**]). Moreover, in a recent preprint (see [**19**]), which was written after these notes were finished, P. Monsky and the author give an example that shows that tight closure does not commute with localization and that the identity of tight closure and plus closure does not hold in dimension two when the base field contains transcendental elements.

The most important geometric notion for this is that of a strongly semistable vector bundle. A vector bundle on a projective variety is called strongly semistable if the pullbacks under every iteration of the absolute Frobenius are again semistable. This notion has gained much attention in previous years from various seemingly unrelated perspectives. Besides the relationship with tight closure and Hilbert-Kunz theory, it played an important role in the work of A. Langer to construct moduli spaces for μ-semistable bundles in positive characteristic in every dimension; see [**69**]. C. Deninger and A. Werner construct representations of the algebraic fundamental group induced by certain vector bundles on a p-adic curve, using a model such that the reduction of the extended bundle to positive characteristic can be characterized in terms of strong semistability; see [**21**], [**22**]. H. Lange, Y. Laszlo, V. B. Mehta, K. Joshi, B. Osserman, C. Pauly, S. Ramanan, E. Z. Xia and J.-K. Yu study the action of the Frobenius on the moduli spaces of bundles on a projective curve, and in particular try to understand where this rational mapping is defined; see [**58**], [**59**], [**67**], [**70**], [**71**], [**73**], [**87**], [**88**].

We now give a short overview of the content of these notes. We will recall the definition of tight closure and some of its applications in Chapter 1. Chapter 2 deals with forcing algebras, solid closure and the affine-linear bundles which arise naturally from tight closure data. The syzygy bundle for ideal generators acts on these affine-linear bundles in such a way that they are a torsor for the vector bundle (Chapter 3). The global cohomological and geometric properties of these torsors are crucial for the study of tight closure. In Chapter 4 we deal with ample vector bundles, and we deal with (strongly) semistable bundles and the (strong) Harder-Narasimhan filtration in Chapter 5. Chapter 6 describes inclusion criteria for tight closure in terms of slope properties of the syzygy bundle. In the strongly semistable case we get an exact degree bound for tight closure. Chapter 7 gives a numerical characterization of when a torsor is an affine scheme in terms of positivity properties of the strong Harder-Narasimhan filtration. In Chapter 8 we establish the same numerical criterion for the non-existence of projective curves inside the torsor. From this, we deduce that tight closure is plus closure in graded dimension two under the condition that the base field is finite. The last two chapters were not covered in the lectures given in Barcelona. Chapter 9 deals with the Hilbert-Kunz function and the Hilbert-Kunz multiplicity, which is intimately related to tight closure. The vector bundle techniques allow us among other things to prove that the Hilbert-Kunz multiplicity is a rational number in graded dimension two, which is in general an open problem stated by P. Monsky. The final chapter deals with arithmetic and geometric deformations of tight closure (and of strongly semistable bundles and of the Hilbert-Kunz multiplicity). The question here is how these notions behave in a family in which we vary the prime number or another parameter. In this context we also give an example of a geometric deformation which provides a counterexample to the localization problem in tight closure theory. Here we only present the example and describe how the problem reduces to an elementary ideal membership problem for a series of ideals. For the proof of all properties we refer to [**19**].

As mentioned above, these notes are based on the lectures given in Barcelona. Some of the material presented here is based on the Habilitationsschrift of the author [**6**]. I think these notes give a fairly complete account on the geometric interpretation of tight closure, though I did not include complete proofs for every statement. An extra problem arises from the fact that the problems we are dealing with have their origins in commutative algebra, while the techniques to attack them come from algebraic geometry, in particular from the theory of vector bundles. The reader will find it helpful to have working experience with the basics of algebraic geometry, including sheaf theory and cohomology, projective varieties and line bundles (as covered in [**45**]).

I thank the organizers of the Winter School on Commutative Algebra and Applications in Barcelona for the excellent job they did and for their hospitality. I thank the participating students for their interests and for suggestions. I thank the referees for their careful reading and helpful comments. For further remarks I am indebted to U. Hackstein, C. Roitzheim, S. Takagi and in particular A. Kaid. For computer-based computations related to the counterexample to the localization problem I am very grateful to D. Heldt, A. Kaid and M. Kreuzer.

CHAPTER 1

Tight closure

We start with a short overview on tight closure, which is a theory initiated by M. Hochster and C. Huneke in 1986; see **[49]**, **[51]**, **[54]**, **[55]**, **[97]**.

Let R denote a commutative ring which contains a field K of positive characteristic $p > 0$. Then the mapping

$$F : R \longrightarrow R, \qquad f \longmapsto f^p,$$

is a ring-endomorphism, the so-called *Frobenius*. The point is that

$$(f+g)^p = \sum_{i=0}^{p} \binom{p}{i} f^i g^{p-i} = f^p + g^p,$$

since the binominal coefficients in between vanish. Then also its e-fold iterates

$$F^e : R \longrightarrow R, \qquad f \longmapsto f^q = f^{p^e},$$

are ring homomorphisms.

DEFINITION 1.1. Let I be an ideal in R. We denote the extended ideal under the Frobenius power $F^e : R \to R$ by $I^{[q]}$, where $q = p^e$. If the ideal I is given by generators $I = (f_1, \ldots, f_n)$, then $I^{[q]} = (f_1^q, \ldots, f_n^q)$.

We assign now to such an ideal I in a ring of positive characteristic two ideals: its Frobenius closure and its tight closure.

DEFINITION 1.2. The *Frobenius closure* of an ideal is

$$I^F = \{f \in R : f^q \in I^{[q]} \text{ for some } q = p^e\}.$$

DEFINITION 1.3. The *tight closure* of an ideal is given by

$$I^* = \{f \in R : \text{ there exists } t \in R, t \notin \text{minimal prime}, tf^q \in I^{[q]} \text{ for almost all } q = p^e\}.$$

If R is reduced and $f \in I^*$, then we may skip the word 'almost' in the definition; i.e., we find then also a t (not in any minimal prime of R) such that $tf^q \in I^{[q]}$ holds for all $q = p^e$. It turns out that I^* is again an ideal which contains I and which has the property that $(I^*)^* = I^*$. For $I \subseteq J$ we get the inclusion $I^* \subseteq J^*$. Hence the name "closure" is justified. The tight closure of an ideal is difficult to compute. Ideals which coincide with their tight closure are called *tightly closed*.

We will restrict ourselves mainly to domains. Then the condition that t is not in any minimal prime just means that $t \neq 0$. For domains, we can also define immediately a third closure operation, the plus closure. This definition makes sense in every characteristic; however it is most important in positive characteristic.

DEFINITION 1.4. For an ideal I in a Noetherian domain we set

$$I^+ = \{f \in R : \text{ there exists a finite extension of domains } R \subseteq S \text{ such that } f \in IS\}$$

and call it the *plus closure* of the ideal.

If R is a domain of positive characteristic, then we have $I^F \subseteq I^+$: the Frobenius morphism $F^e : R \to R$ is not necessarily finite, but it is integral, so if $f^q \in I^{[q]}$, say $f^q = a_1 f_1^q + \ldots + a_n f_n^q$, then also $f^q \in (f_1^q, \ldots, f_n^q)R[a_1, \ldots, a_n]$, and $S = R[a_1, \ldots, a_n]$ is now finite over R. More importantly, the plus closure (and so also the Frobenius closure) lies inside the tight closure. This is the first main source for elements to belong to the tight closure: "tight closure by *contraction*".

THEOREM 1.5. *Let $R \subseteq S$ be a finite extension of commutative Noetherian domains over a field of characteristic $p > 0$. Let $I \subseteq R$ denote an ideal and $f \in R$ an element such that $f \in IS$. Then $f \in I^*$ holds.*

PROOF. It is rather easy to show (see [**97**, Key Lemma] or [**46**, Corollary 2.3]) that there exists an R-linear mapping $\varphi : S \to R$ such that $\varphi(1) = t \neq 0$. Suppose that $f \in IS = (f_1, \ldots, f_n)S$. Then also $f^q \in (f_1^q, \ldots, f_n^q) = I^{[q]}S$ holds, say $f^q = s_1 f_1^q + \ldots + s_n f_n^q$. Applying the R-linear mapping φ to this equation we get $tf^q = \varphi(s_1)f_1^q + \ldots + \varphi(s_n)f_n^q$; hence $tf^q \in (f_1^q, \ldots, f_n^q)$ holds in R. □

Another way to define the plus closure is to consider the absolute integral closure of R. For a Noetherian domain R let $Q = Q(R)$ denote its quotient field and let $\overline{Q}$ denote an algebraic closure. The integral closure of R in $\overline{Q}$ is denoted by R^+ and is called the *absolute integral closure* of R. With this we may also write $I^+ = IR^+ \cap R$.

One of the most difficult questions of tight closure theory is whether Theorem 1.5 has a converse; i.e., does there exist for every $f \in I^*$ in a Noetherian domain R over a field of positive characteristic a finite extension $R \subseteq S$ such that $f \in IS$. Hochster calls this a "*tantalizing question*" [**46**]. We will deal with this question in these notes.

PROBLEM 1.6. Let R denote a Noetherian domain over a field of positive characteristic. Is it true for every ideal $I \subseteq R$ that the tight closure is the same as the plus closure, $I^* = I^+$?

An important result of K. Smith settles the case of parameters. Recall that elements $x_1, \ldots, x_d$ in a local Noetherian ring $(R, \mathfrak{m})$ of dimension d are called *parameters* if $V(x_1, \ldots, x_d) = V(\mathfrak{m})$ holds. An ideal which is generated by a system of parameters is called a *parameter ideal*. From the viewpoint of tight closure, parameter ideals are much better understood than arbitrary ideals.

THEOREM 1.7. *Let R be an excellent local Noetherian domain of dimension d over a field of positive characteristic. Let $x_1, \ldots, x_d$ denote parameters in R. Then*

$$(x_1, \ldots, x_d)^+ = (x_1, \ldots, x_d)^* .$$

PROOF. See [**55**, Theorem 7.1] or [**94**]. □

In these notes we will prove that tight closure is plus closure for graded R_+-primary ideals in a two-dimensional standard-graded normal domain over a finite field (Theorem 8.10). Problem 1.6 is connected with another problem in tight closure theory, the *localization problem.*

PROBLEM 1.8. Let R denote a Noetherian domain over a field of positive characteristic and let M denote a multiplicatively closed system. Does $(IR_M)^* = I^*R_M$ hold?

Here the inclusion $\supseteq$ is a special case of the persistence property (see Theorem 1.12 below) and $\subseteq$ is the problem. A positive answer to Problem 1.6 would imply a positive answer to Problem 1.8, since it is easy to see that the plus closure commutes with localization. Localization in dimension two is fairly easy to prove, at least under some mild technical conditions.

After these notes were written, P. Monsky and the author gave an example showing that Problem 1.6 and Problem 1.8 have a negative answer in general. The example is presented here (Example 10.8), but not the proofs of all its properties. The example, which is a geometric deformation of a two-dimensional tight closure problem over the affine line, shows that in graded dimension two tight closure is not necessarily plus closure if the base field contains transcendental elements and that tight closure does not commute with localization in a three-dimensional hypersurface domain. Problem 1.6 might still have a positive answer for domains of finite type over a finite field.

We look now at a standard example of tight closure theory. We emphasize here already the approach to look at the induced cohomology classes.

EXAMPLE 1.9. Let $R = K[X,Y,Z]/(X^3+Y^3+Z^3)$, where K is a field of positive characteristic $p \neq 3$. We denote the images of X, etc. , in R by small letters. We have $z^2 \notin (x,y)$: if we mod out (x,y), the factor ring is $K[Z]/(Z^3)$, and in this ring z^2 is not zero. We claim that $z^2 \in (x,y)^*$ and even $\in (x,y)^F$ for $p = 2 \mod 3$. Let's look at some small characteristics.

$p = 2$.

$$z^4 = zz^3 = -z(x^3+y^3) = -zxx^2 - zyy^2 \in (x^2,y^2) = (x,y)^{[2]}.$$

$p = 5$.

$$z^{10} = z(z^3)^3 = -z(x^3+y^3)^3 = -z(x^9+3x^6y^3+3x^3y^6+y^9) \in (x^5,y^5).$$

$p = 7$.

$$z^{14} = z^2(z^3)^4 = z^2(x^3+y^3)^4 = z^2(x^{12}+4x^9y^3+6x^6y^6+4x^3y^9+y^{12}) \notin (x^7,y^7).$$

So for $p = 7$ the first Frobenius power of z^2 is not in the Frobenius power of the ideal (and also the higher powers are not in the higher powers). It is however in the tight closure: Take $t = z$. Then (since $q = 1 \bmod 3$)

$$z(z^2)^q = z^{2q+1} = (-x^3-y^3)^{\frac{2q+1}{3}} = \sum_{i+j=\frac{2q+1}{3}} a_{ij}x^{3i}y^{3j}.$$

Here either i or $j \geq (q+1/2)/3$ and so the exponent of x (or y) is at least $q+1/2 > q$, and so this element belongs to (x^q,y^q).

Another and better way is to look at the Koszul resolution (x and y are regular parameters)

$$0 \longrightarrow R \stackrel{(y,-x)}{\longrightarrow} R^2 \stackrel{x,y}{\longrightarrow} R \to R/(x,y) \longrightarrow 0,$$

which is exact. This induces on the open subset $D(x,y) \subseteq \operatorname{Spec} R = X$ a short exact sequence

$$0 \longrightarrow \mathcal{O}_X \stackrel{(y,-x)}{\longrightarrow} \mathcal{O}_X^2 \stackrel{x,y}{\longrightarrow} \mathcal{O}_X \longrightarrow 0$$

of locally free sheaves. The corresponding long exact sequence in sheaf cohomology is (since $\Gamma(D(x,y),\mathcal{O}_X) = R$, because R is normal)

$$0 \longrightarrow R \stackrel{(y,-x)}{\longrightarrow} R^2 \stackrel{x,y}{\longrightarrow} R \stackrel{\delta}{\longrightarrow} H^1(D(x,y),\mathcal{O}_X) \longrightarrow \dots .$$

The element $z^2 \in R = \Gamma(D(x,y), \mathscr{O}_X)$ yields via the connecting homomorphism the Čech cohomology class $c = \delta(z^2) = \frac{z^2}{xy} \in H^1(D(x,y), \mathscr{O}_X)$. The e-th Frobenius pull-back of this short exact sequence is again exact, yielding

$$0 \longrightarrow \mathscr{O}_X \stackrel{(y^q,-x^q)}{\longrightarrow} \mathscr{O}_X^2 \stackrel{x^q,y^q}{\longrightarrow} \mathscr{O}_X \longrightarrow 0,$$

and the corresponding cohomology class is now $\frac{z^{2q}}{x^q y^q}$, which is also the pull-back of z^2/xy under the e-th Frobenius. Now $z^{2q} \in (x^q, y^q)$ if and only if this class $\frac{z^{2q}}{x^q y^q} = 0$ in $H^1(D(x,y), \mathscr{O}_X)$ (the class is zero if and only if z^{2q} comes from the left) , and $tz^{2q} \in (x^q, y^q)$ if and only if t annihilates the cohomology class, i.e., $t\frac{z^{2q}}{x^q y^q} = 0$.

Since R is given by a cubic equation, $C = \operatorname{Proj} R$ is an elliptic curve, and on C we have (since everything is graded) the short exact sequence

$$0 \longrightarrow \mathscr{O}_C(-2) \stackrel{(y,-x)}{\longrightarrow} \mathscr{O}_C(-1)^2 \stackrel{x,y}{\longrightarrow} \mathscr{O}_C \longrightarrow 0$$

of locally free sheaves on C or (by twisting=tensoring with $\mathscr{O}_C(2)$)

$$0 \longrightarrow \mathscr{O}_C \stackrel{(y,-x)}{\longrightarrow} \mathscr{O}_C(1)^2 \stackrel{x,y}{\longrightarrow} \mathscr{O}_C(2) \longrightarrow 0.$$

The global element $z^2 \in \Gamma(C, \mathscr{O}_C(2)) = R_2$ yields now the class

$$\frac{z^2}{xy} \in H^1(C, \mathscr{O}_C) \cong K,$$

since $g(C) = 1$. The Frobenius morphism $H^1(C, \mathscr{O}_C) \to H^1(C, \mathscr{O}_C)$ induced by the *absolute Frobenius morphism* on C sends $z^2/xy \mapsto z^{2p}/x^p y^p$. On an elliptic curve, this mapping can be either bijective (and one says that the elliptic curve has *Hasse invariant* one or is *ordinary*; see [**45**, Proposition IV.4.21]) or the zero map (Hasse invariant zero or *supersingular*). It is known (and not difficult) that the Fermat cubic is supersingular exactly for $p = 2 \bmod 3$. In this case z^2 lies in the Frobenius closure (and already in the first Frobenius power), otherwise not.

With respect to tight closure, one can also make the following observation. For an elliptic curve we have $H^1(C, \mathscr{O}_C(m)) = 0$ for $m \geq 1$ (by Serre duality). Hence, since $z^{2q}/x^q y^q \in H^1(C, \mathscr{O}_C)$, we have $z \cdot z^{2q}/x^q y^q \in H^1(C, \mathscr{O}_C(1))$, so it must be zero and therefore $z^2 \in (x,y)^*$. For $p = 1 \bmod 3$ the Frobenius does not annihilate the cohomology class $\frac{z^2}{xy}$. Can we find also in this case a finite extension $R \subseteq S$ where $z^2 \in (x,y)S$? We do that in the case $p = 7$. We look at

$$T = R[i][u,v]/(u^7 - ux^6 + iz^2(4x^3y^2 + y^5), v^7 - vy^6 - iz^2(x^5 + 4x^2y^3)).$$

The ring T is integral over R, and it is a two-dimensional complete intersection ring. Hence it is Cohen-Macaulay. In T we have

$$\begin{aligned}
(yu - xv - iz^2)^7 &= y^7u^7 - x^7v^7 + iz^2z^{12} \\
&= y^7u^7 - x^7v^7 + iz^2(x^3 + y^3)^4 \\
&= y^7(ux^6 - iz^2(4x^3y^2 + y^5)) - x^7(vy^6 + iz^2(x^5 + 4x^2y^3)) \\
&\quad + iz^2(x^{12} + 4x^9y^3 + 6x^6y^6 + 4x^3y^9 + y^{12}) \\
&= x^6y^6(yu - xv) + iz^2(6x^6y^6) \\
&= x^6y^6(yu - xv - iz^2).
\end{aligned}$$

Therefore the equation $(yu - xv - iz^2)((yu - xv - iz^2)^6 - x^6y^6) = 0$ holds in T. Since T is equidimensional [**25**, Corollary 18.11], every component of $\operatorname{Spec} T$ is two-dimensional and every component dominates the base, since the mapping is finite. Therefore there exists a factor domain $S = T/\mathfrak{p}$ with $yu - xv - iz^2 \in \mathfrak{p}$ and such that $R \subset S$ is finite. But then $z^2 \in (x,y)S$. We will see in Remark 8.6 how to get the equations which define T.

EXAMPLE 1.10. Let $R = K[X,Y,Z]/(X^3+Y^3+Z^3)$ and ask whether $xyz \in (x^2,y^2,z^2)^*$ holds. This is not a parameter ideal anymore, and this was for some time an open question in tight closure theory. The answer is yes, as was first shown by Singh [**93**]; for another proof see the appendix of G. Leuschke to [**49**]. We will come back to this example in Example 5.5 and Example 6.5. This question is also interesting for other Fermat equations $x^d + y^d + z^d = 0$. The answer depends in general on d and p. For $d \geq 5$ the answer is not always known. More generally, we may ask what is the tight closure of (x^a, y^a, z^a) in $K[X,Y,Z]/(F)$, where F is homogeneous of a certain degree?

We come to some further important properties and applications of tight closure.

THEOREM 1.11. *Let R denote a regular ring over a field of characteristic p. Then every ideal in R is tightly closed, $I^* = I$.*

PROOF. See [**54**, Theorem 1.3(e) and Theorem 1.5(3)]. □

This proposition is due to the fact that for regular rings the Frobenius morphism is flat. Rings with the property that every ideal is tightly closed are called *weakly F-regular* (F is for Frobenius) and *F-regular* if this property holds for every localization (it is not known whether this is the same). Hence Theorem 1.11 tells us that regular rings are F-regular. Theorem 1.11 gives rise to the question whether there exist other F-regular rings and what kinds of properties may be deduced from this property.

The following property is called the *persistence* of tight closure.

THEOREM 1.12. *Let R and S denote Noetherian commutative rings over a field of characteristic p and let $\varphi : R \to S$ denote a ring homomorphism. Suppose that R is essentially of finite type over an excellent local ring. Then $I^*S \subseteq (IS)^*$.*

PROOF. See [**54**, Theorem 2.3]. □

REMARK 1.13. The proof of this statement is rather intricate (for inclusions of domains it is trivial). It involves the theory of *test ideals*. The tight closure test ideal of a ring is the ideal generated by all *test elements*. These are elements t with the property that whenever $f \in I^*$, then $tf^q \in I^{[q]}$ for all $q = p^e$. It is a non-trivial fact that such elements exist. The theory of test ideals has strong ties with the theory of *multiplier ideals*, which arose in the context of birational geometry. For this interesting relationship see [**36**], [**37**], [**38**], [**39**], [**96**], [**98**].

A local Noetherian ring is called *Cohen-Macaulay* if one (and then every) system of parameters $x_1, \ldots, x_d$ forms a regular sequence; i.e., x_k is not a zero divisor in $R/(x_1, \ldots, x_{k-1})$ for $k = 1, \ldots, d$. This is equivalent with the property that

$$((x_1, \ldots, x_{k-1}) : x_k) \subseteq (x_1, \ldots, x_{k-1})$$

holds in R for $k = 1, \ldots, d$. The point is that this statement holds without further conditions if we replace the ideal on the right by its tight closure.

THEOREM 1.14. *Let $x_1, \ldots, x_d$ denote parameters in a d-dimensional local Noetherian ring R over a field of characteristic p. Suppose that R is the homomorphic image of a local Cohen-Macaulay ring. Then we have the inclusion*

$$((x_1, \ldots, x_{k-1}) : x_k) \subseteq (x_1, \ldots, x_{k-1})^*$$

for $k = 1, \ldots, d$.

PROOF. See [**54**, Theorem 3.1]. □

This property of tight closure is called "*colon-capturing*". The condition that the ring is the image of a Cohen-Macaulay ring is of a technical nature and holds for complete local rings and for rings which are essentially of finite type over a field. Hence tight closure provides a measure for the deviation from the Cohen-Macaulay property. Formulated in another way, we get the following.

COROLLARY 1.15. *A local weakly F-regular ring which is the homomorphic image of a local Cohen-Macaulay ring is Cohen-Macaulay.*

PROOF. This follows directly from Theorem 1.14. □

The properties of tight closure developed so far allow us to prove interesting statements in which tight closure is only a tool for proving but not a part of the statement itself.

THEOREM 1.16. *A Noetherian ring S which is a direct summand of a regular ring R over a field of characteristic p is (F-regular and in particular) Cohen-Macaulay.*

PROOF. Let $S \hookrightarrow R$ be a direct summand and $I \subseteq S$ an ideal. From $f \in I^*$ it follows due to the persistence Theorem 1.12 (in fact we need here only a trivial case of persistence) that $f \in (IR)^*$ and therefore $f \in IR$ due to the (F-)regularity of R (Theorem 1.11). Since S is a direct summand it follows that $f \in I$. (Since we may pass to the completion, the technical conditions in the cited statements are granted.) □

An important special case of this statement is the following theorem of Hochster and Roberts. Its clear and easy proof using the theory of tight closure is already a justification for the effort! The original proof in [**53**] was rather complicated.

THEOREM 1.17. *Let G denote a linearly reductive group (every finite dimensional representation splits into a direct sum of irreducible representations) which acts on a polynomial ring $K[X_1,\ldots,X_n]$ over a field K of characteristic p as a group of linear automorphisms. Then the ring of invariants $K[X_1,\ldots,X_n]^G$ is Cohen-Macaulay.*

PROOF. The Reynolds operator shows that the ring of invariants is a direct summand of the polynomial ring. Furthermore it is a finitely generated graded K-algebra. Hence this follows from Theorem 1.16. □

Note that the statement of Theorem 1.17 is formulated only for positive characteristic, and therefore it applies to rather few groups, namely to finite groups, if their order is not divisible by the characteristic, and to tori (the operations of tori correspond to multigraduations on the polynomial ring). With the theory of tight closure in characteristic zero, which is defined by reduction to positive characteristic, one can also prove this in zero characteristic, and then it applies also to the classical groups, the general linear group, the special linear group, the orthogonal group and the symplectic group.

We describe briefly the setting to define tight closure in a finitely generated $\mathbb{Q}$-algebra R (this can be done in greater generality; see [**48**]). There exists a finitely generated $\mathbb{Z}$-algebra S (a 'model') such that $S_{\mathbb{Q}} = S \otimes_{\mathbb{Z}} \mathbb{Q} = R$. For an ideal I in R there exists also an ideal J in S with the property that $J_{\mathbb{Q}} = I$, and also an element $f \in R$ extends up to a unit to S. Then for every prime number p we get characteristic p data $R_p = S \otimes_{\mathbb{Z}} \mathbb{Z}/(p)$, the extended ideal $I_p = JR_p$ and the element $f_p = f \otimes 1$. Then one declares that $f \in I^*$, the tight closure in characteristic zero, if and only if $f_p \in (I_p)^*$ holds in R_p for almost all prime numbers p.

It can be shown that this definition is independent of the choice of the model S and the ideal J. Also, since most of the algebraic relevant properties pass from $S_{\mathbb{Q}}$ to R_p for almost all p, most results of tight closure in positive characteristic pass to characteristic zero. For a thorough treatment of tight closure in characteristic zero; see [**50**]. We will come back to this setting in Remark 4.16 and in the last chapter.

CHAPTER 2

Forcing algebras and affine-linear bundles

Starting with this chapter we develop a more geometric view on tight closure.

DEFINITION 2.1. Let R denote a commutative ring and let $f_1,\ldots,f_n \in R$ and $f \in R$ be elements. The R-algebra

$$A = R[T_1,\ldots,T_n]/(f_1T_1+\ldots+f_nT_n+f)$$

is called the *forcing algebra* for the elements (the *forcing data*) $f_1,\ldots,f_n,f$.

REMARK 2.2. This notion was introduced by M. Hochster in [**46**]. The forcing algebra forces that $f \in (f_1,\ldots,f_n)A$. Every other R-algebra S with the property that $f \in (f_1,\ldots,f_n)S$ factors (not uniquely) through A; if $-f = s_1f_1+\ldots+s_nf_n$ with $s_i \in S$, then we just send $T_i \mapsto s_i$. The forcing algebra induces a morphism $\operatorname{Spec} A \longrightarrow \operatorname{Spec} R$ with many geometric features, which we will study. It has a section $\operatorname{Spec} R \rightarrow \operatorname{Spec} A$ if and only if $f \in (f_1,\ldots,f_n)$. If A is the forcing algebra for elements $f_1,\ldots,f_n,f \in R$ and if $\varphi : R \rightarrow R'$ is a ring homomorphism, then $A' = A \otimes_R R'$ is the forcing algebra for the elements $\varphi(f_1),\ldots,\varphi(f_n),\varphi(f)$.

The connection to tight closure in positive characteristic is given by the following theorem, which can also be considered as an alternative definition for tight closure.

THEOREM 2.3 (Hochster). *Suppose that R is a normal excellent domain containing a field of characteristic $p > 0$. Suppose, furthermore, that R is essentially of finite type over an excellent local ring or that the Frobenius endomorphism is finite. Let $I = (f_1,\ldots,f_n)$ denote an ideal which is primary to a maximal ideal $\mathfrak{m}$ of height d. Then $f \in I^*$ if and only if $H^d_{\mathfrak{m}}(A) \neq 0$, where $A = R[T_1,\ldots,T_n]/(f_1T_1+\ldots+f_nT_n+f)$ is the forcing algebra for these data.*

PROOF. See [**46**, Theorem 8.6]. □

One can define a closure operation for ideals in an arbitrary commutative ring by taking the right-hand side of Theorem 2.3 as a definition.

DEFINITION 2.4. Let R be a Noetherian ring and let $I \subseteq R$ be an ideal, $f \in R$. Then f belongs to the *solid closure*, written $f \in I^\star$, if and only if the following hold: for every local complete domain $R' = \hat{R}_{\mathfrak{m}}/\mathfrak{q}$ (here $\mathfrak{m}$ is a maximal ideal of R and $\mathfrak{q}$ is a minimal prime of $\hat{R}_{\mathfrak{m}}$) we have that $H^d_{\mathfrak{m}'}(A') \neq 0$, where $d = \dim R'$ and A' is the forcing algebra over R'.

REMARK 2.5. In an excellent ring of positive characteristic, solid closure is the same as tight closure by Theorem 2.3. In mixed characteristic not much is known. However, in equal characteristic zero and if the dimension of the ring is at least 3, then this closure operation does not have all properties one expects of a tight closure type operation. A computation of Paul Roberts shows that the three-dimensional polynomial ring over a field of characteristic zero is not solid-regular, i.e., not every ideal coincides with its solid closure. Roberts showed in **[89]** that

$$H^3_{(x,y,z)}(K[x,y,z][T_1,T_2,T_3]/(x^3T_1+y^3T_2+z^3T_3+x^2y^2z^2)) \neq 0,$$

and therefore $(xyz)^2 \in (x^3,y^3,z^3)^\star$, though $(xyz)^2 \notin (x^3,y^3,z^3)$.

In characteristic zero and dimension two, solid closure has all the properties which one expects from a tight closure theory, and we will sometimes use this in these notes. In higher dimensions there is a variant of solid closure, called *parasolid closure*, with the right properties, and which is also defined without any reduction to positive characteristic; see **[5]**.

In the characterization of tight closure with local cohomology we have

$$H^d_{\mathfrak{m}}(A) = H^d_{\mathfrak{m}A}(A) = H^{d-1}(D(\mathfrak{m}A),\mathcal{O})$$

for $d \geq 2$ by the long exact sequence relating local cohomology and sheaf cohomology (see **[45**, Exercise III.2.3]). Therefore our main task will be to understand the cohomological properties of the scheme $D(\mathfrak{m}A)$. This is an open subset inside the affine scheme $\operatorname{Spec} A$; such schemes are called *quasiaffine*. $D(\mathfrak{m}A)$ is the complement of the closed subset $V(\mathfrak{m}A)$, which is the fiber over the maximal ideal $\mathfrak{m}$ under the scheme morphism $\operatorname{Spec} A \to \operatorname{Spec} R$. What structure does this quasiaffine scheme have that might help in computing its cohomology?

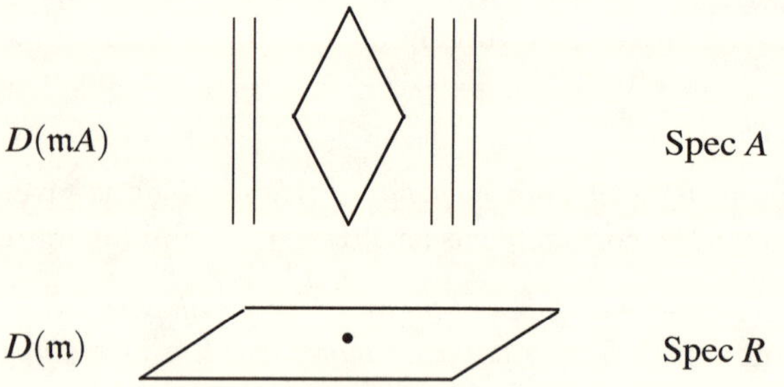

The picture shows an instance where R has dimension two and we are in the parameter case.

What do the fibers over $P \in \operatorname{Spec} R$ look like? If you are not that acquainted with schemes, restrict R to be a K-algebra of finite type, say

$$R = K[X_1,\ldots,X_m]/(G_1,\ldots,G_r)$$

and think of $\operatorname{Spec} R$ as the affine variety

$$\{(x_1,\ldots,x_m) \in \mathbb{A}^m(K) = K^m : G_1(x_1,\ldots,x_m) = 0,\ldots,G_r(x_1,\ldots,x_m) = 0\}$$

given as the common zero set of polynomial equations. The elements f and $f_i \in R$ are then polynomials which are defined on this affine variety. The K-spectrum of the forcing algebra is then

$$\{(x_1,\ldots,x_m,t_1,\ldots,t_n) \in \mathbb{A}^{m+n}(K) : G_j(x) = 0,\, f_1(x)t_1 + \ldots + f_n(x)t_n + f(x) = 0\}.$$

Thus the fiber over a point $x = (x_1,\ldots,x_m)$ is the solution space to the inhomogeneous linear equation $f_1(x)t_1 + \ldots + f_n(x)t_n + f(x) = 0$. As long as this equation does not degenerate, i.e., as long as at least one $f_i(x) \neq 0$, this is an $(n-1)$-dimensional affine space. If, however, all $f_i(x) = 0$ and $f(x) = 0$, then the solution has dimension n, and in the case that all $f_i(x) = 0$ but $f(x) \neq 0$, the solution set is empty.

LEMMA 2.6. *Let R denote a commutative ring and let $f_1, \dots, f_n \in R$ be elements, setting $U = D(I) = D(f_1, \dots, f_n) \subseteq \operatorname{Spec} R$. Let $f \in R$ denote another element and let $A = R[T_1, \dots, T_n]/(f_1T_1 + \dots + f_nT_n + f)$ denote the forcing algebra. Set $\tilde{B} = \operatorname{Spec} A$ and $B = \operatorname{Spec} A|_U$. Then B is locally an affine space of dimension $n-1$ over U, and the transition mappings are affine-linear.*

PROOF. Consider the mapping $\operatorname{Spec} A \longrightarrow \operatorname{Spec} R$ over the open subset $U = D(I) = D(f_1, \dots, f_n)$. On $D(f_i)$, $i \geq 1$, one can identify

$$A_{f_i} = (R[T_1, \dots, T_n]/(f_1T_1 + \dots + f_nT_n + f))_{f_i} \cong R_{f_i}[T_1, \dots, T_{i-1}, T_{i+1}, \dots, T_n].$$

So this mapping looks locally like the projection $D(f_i) \times \mathbb{A}^{n-1} \longrightarrow D(f_i)$. The transition mapping on $D(f_if_j)$ is given by

$$R_{f_if_j}[T_1, \dots, T_{i-1}, T_{i+1}, \dots, T_n] \longrightarrow R_{f_if_j}[T_1, \dots, T_{j-1}, T_{j+1}, \dots, T_n],$$

where $T_k \longmapsto T_k$ for $k \neq i, j$ and $T_i \longmapsto \frac{\sum_{j \neq i} f_jT_j + f}{f_i}$. This is an affine-linear mapping. □

This lemma says that $B \to U$ is an affine-linear bundle of rank $n-1$ in the sense of the following definition. We will call this bundle the *forcing bundle* for the data $f_1, \dots, f_n, f$.

DEFINITION 2.7. A scheme $p : B \to X$ is called an *affine-linear bundle* of rank r if there exist an open affine covering $X = \bigcup_i U_i$ and U_i-isomorphisms

$$\psi_i : B|_{U_i} = p^{-1}(U_i) \longrightarrow U_i \times \mathbb{A}^r = \mathbb{A}^r_{U_i},$$

such that for every open affine subset $U \subset U_i \cap U_j$ the transition mappings $\psi_j \circ \psi_i^{-1} : \mathbb{A}^r_U \to \mathbb{A}^r_U$ are affine-linear automorphisms, i.e., they are induced by an automorphism of the polynomial ring $\Gamma(U, \mathcal{O}_X)[T_1, \dots, T_r]$ given by $T_i \mapsto \sum_{j=1}^r a_{ij}T_j + v_i$.

REMARK 2.8. The difference of two sections in $\tilde{B}$ (over some open subset) is a syzygy for $f_1, \dots, f_n$. Therefore the syzygies act as translations on the affine-linear bundle given by the forcing algebra.

Recall from linear algebra that every solution to an inhomogeneous system of linear equations can be obtained by adding to a particular solution a uniquely determined solution to the corresponding homogeneous system of linear equations. This means that the homogeneous solutions act by addition simply transitive on the inhomogeneous solutions. The same thing holds if the data depend on a parameter space, which in our situation is $\operatorname{Spec} R$ (or the open subset $U = D(I)$). The inhomogeneous solutions form an affine-linear bundle and the solutions to the homogeneous equations form a vector bundle over the base. We recall the notion of vector bundles and locally free sheaves.

DEFINITION 2.9. Let X denote a scheme. A scheme $p : V \to X$ is called a *(geometric) vector bundle* of rank r if there exist an open affine covering $X = \bigcup_i U_i$ and U_i-isomorphisms

$$\psi_i : V|_{U_i} = p^{-1}(U_i) \to U_i \times \mathbb{A}^r = \mathbb{A}^r_{U_i},$$

such that for every open affine subset $U \subset U_i \cap U_j$ the transition mappings $\psi_j \circ \psi_i^{-1} : \mathbb{A}^r_U \to \mathbb{A}^r_U$ are linear automorphisms, i.e., they are induced by an automorphism of the polynomial ring $\Gamma(U, \mathcal{O}_X)[T_1, \dots, T_r]$ given by $T_i \mapsto \sum_{j=1}^r a_{ij}T_j$.

The affine space $\mathbb{A}_X^r$ together with its natural structure of a vector bundle is called the *trivial bundle* over X of rank r. A vector bundle V over X is in particular a commutative *group scheme* over X, where the addition $V \times_X V \to V$ is locally given by the identifications with $\mathbb{A}_U^r$. Due to the compatibility conditions this gives a well-defined addition.

REMARK 2.10. A vector bundle $p : V \to X$ comes along with two locally free sheaves. The *sheaf of sections* $\mathscr{S}_V$ is given by

$$\Gamma(U, \mathscr{S}_V) = \{s : U \to V|_U : p \circ s = \mathrm{id}_U\}.$$

The linear structure on the vector bundle makes this into a quasicoherent sheaf. Since locally $V \cong \mathbb{A}^r$, the sheaf of sections is locally isomorphic to $\mathscr{O}_X^r$, and hence $\mathscr{S}_V$ is a *locally free sheaf* of rank r on X. The group scheme $V \to X$ represents the group functor $X' \mapsto \Gamma(X', \varphi^*(\mathscr{S}_V))$, where $\varphi : X' \to X$ is any scheme over X.

The *sheaf of linear forms* $\mathscr{F}_V$ for a vector bundle $V \to X$ is given by

$$\Gamma(U, \mathscr{F}_V) = \{f : V|_U \to \mathbb{A}|_U \text{ is a linear morphism}\}.$$

Here a morphism $f : V \to W$ of two vector bundles over X is called *linear* if for an open covering $X = \bigcup_i U_i$ with simultaneous identifications $\psi_i : V|_{U_i} \to \mathbb{A}_{U_i}^r$ and $\varphi_i : W|_{U_i} \to \mathbb{A}_{U_i}^s$ (which are compatible with the identifications which define the linear structures on the bundles), the induced mappings

$$\varphi_i \circ \psi_i^{-1} : \mathbb{A}_{U_i}^r \longrightarrow \mathbb{A}_{U_i}^s$$

are linear. This is again a locally free sheaf of rank r, and we have the relationship $\mathscr{S}_V^\vee = \mathrm{Hom}_{\mathscr{O}_X}(\mathscr{S}_V, \mathscr{O}_X) = \mathscr{F}_V$ ($\vee$ denotes the dual).

Vector bundles and locally free sheaves are essentially the same objects. For a locally free sheaf $\mathscr{F}$ of rank r on X, the *symmetric algebra* $S(\mathscr{F}) = \oplus_{k \geq 0} S^k(\mathscr{F})$, is a graded $\mathscr{O}_X$-algebra. The associated spectrum of this algebra (see [**30**, 9.4] or [**29**, 1.7]),

$$V = \operatorname{Spec} S(\mathscr{F}) = \operatorname{Spec} \oplus_{k \geq 0} S^k(\mathscr{F})$$

is a geometric vector bundle over X with the sheaf of linear forms $\mathscr{F}$. This gives a correspondence of vector bundles and locally free sheaves.

A vector bundle of rank one is called a *line bundle*, and the corresponding sheaves are called *invertible shaves*; see [**45**, Section II.6].

In our situation the elements $f_1, \ldots, f_n \in R$ define a vector bundle over the open subset $D(f_1, \ldots, f_n) \subseteq \operatorname{Spec} R$ which acts on the forcing bundle. The next proposition describes the action of the homogeneous solutions on the solutions of an inhomogeneous linear system.

PROPOSITION 2.11. *Let R denote a commutative ring and let $I \subseteq R$ denote an ideal with ideal generators $f_1, \ldots, f_n$. Set $U = D(I) = D(f_1, \ldots, f_n) \subseteq \operatorname{Spec} R$. Then the following hold.*

(i) $\tilde{V} = \operatorname{Spec} R[T_1, \ldots, T_n]/(f_1T_1 + \ldots + f_nT_n)$ *is a commutative group scheme over* $\operatorname{Spec} R$.
(ii) *The sheaf of sections in $\tilde{V}$ is the quasicoherent sheaf of syzygies for the elements $f_1, \ldots, f_n$.*
(iii) *The restriction $V = \tilde{V}|_U$ is a vector bundle on U of rank $n-1$.*

Let $f \in R$ denote another element with forcing algebra $A = R[T_1, \ldots, T_n]/(f_1T_1 + \ldots + f_nT_n + f)$ and set $\tilde{B} = \operatorname{Spec} A$ and $B = \operatorname{Spec} A|_U$. Then the following also hold.

(iv) *The group scheme $\tilde{V}$ acts on $\tilde{B}$.*

(v) *This action is simply transitive; i.e., the morphism*

$$\mu \times p_2 : \tilde{V} \times_X \tilde{B} \longrightarrow \tilde{B} \times_X \tilde{B}$$

is an isomorphism.

(vi) *On $D(I)$, this is the action of the syzygy bundle V on the affine-linear bundle B.*

PROOF. (i). The coaddition is given by

$$R[T_1,\dots,T_n]/(\sum_{i=1}^n f_iT_i) \longrightarrow R[T_1,\dots,T_n,S_1,\dots,S_n]/(\sum_{i=1}^n f_iT_i,\ \sum_{i=1}^n f_iS_i),$$

where $T_i \mapsto T_i + S_i$. It is easy to see that this fulfills the axioms of a group scheme (see **[32]** or **[84]** for the basic notions of group schemes).

(ii). A section

$$s : \operatorname{Spec} S \to \operatorname{Spec} S[T_1,\dots,T_n]/(f_1T_1+\dots+f_nT_n)$$

over an affine subset $W = \operatorname{Spec} S \subseteq \operatorname{Spec} R$ is just a tupel $(s_1,\dots,s_n) \in S^n$ satisfying $f_1s_1 + \dots + f_ns_n = 0$.

(iii). The bundle V is on $D(f_i)$, $i = 1,\dots,n$, isomorphic to

$$\operatorname{Spec} R_{f_i}[T_1,\dots,T_{i-1},T_{i+1},\dots,T_n],$$

and the transition functions send

$$T_i \mapsto -\frac{1}{f_i}(f_1T_1+\dots+f_{i-1}T_{i-1}+f_{i+1}T_{i+1}+\dots+f_nT_n);$$

thus they are linear and V is a vector bundle on $U = \bigcup_{i=1}^n D(f_i)$.

(iv). The action of the group scheme $\tilde{V} = \operatorname{Spec} R[S_1,\dots,S_n]/(\sum f_iS_i)$ on the scheme $\tilde{B} = \operatorname{Spec} R[T_1,\dots,T_n]/(\sum f_iT_i + f)$ is given by the coaction

$$R[T_1,\dots,T_n]/(\sum_{i=1}^n f_iT_i + f) \longrightarrow R[T_1,\dots,T_n,S_1,\dots,S_n]/(\sum_{i=1}^n f_iT_i + f,\ \sum_{i=1}^n f_iS_i)$$

by sending $T_i \mapsto T_i + S_i$. It is easy to verify that this is indeed a group coaction.

(v). The morphism $\mu \times p_2$ corresponds to the ring homomorphism

$$R[S_1,\dots,S_n,T_1,\dots,T_n]/(\sum_{i=1}^n f_iS_i + f,\ \sum_{i=1}^n f_iT_i + f)$$

$$\longrightarrow R[S_1,\dots,S_n,T_1,\dots,T_n]/(\sum_{i=1}^n f_iS_i,\ \sum_{i=1}^n f_iT_i + f)$$

given by $S_i \mapsto S_i + T_i$ and $T_i \mapsto T_i$. This is clearly an isomorphism. □

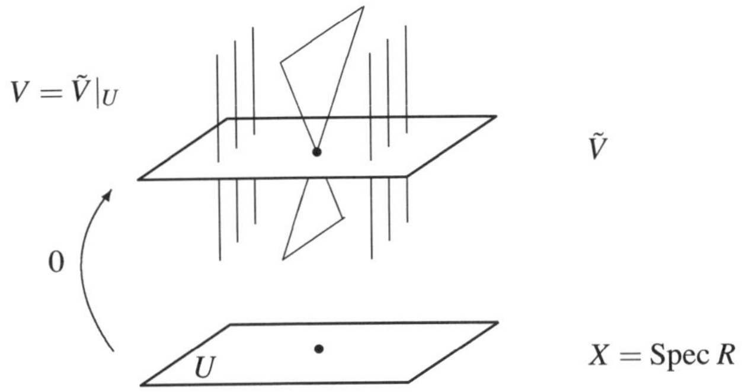

DEFINITION 2.12. We call $\tilde{V} = \operatorname{Spec} R[T_1,\ldots,T_n]/(f_1T_1+\ldots+f_nT_n)$ the *syzygy group scheme* and its restriction $V = \tilde{V}|_U$ the *syzygy bundle* corresponding to the ideal generators $f_1,\ldots,f_n$. We will denote the sheaf of syzygies by $\operatorname{Syz}(f_1,\ldots,f_n)$. It sits inside the short exact sequences of sheaves on U,

$$0 \longrightarrow \operatorname{Syz}(f_1,\ldots,f_n) \longrightarrow \mathcal{O}_U^n \xrightarrow{f_1,\ldots,f_n} \mathcal{O}_U \longrightarrow 0,$$

which we call the *presenting sequence*.

We come back to the cohomological properties of $B = \operatorname{Spec} A|_U$. The property that $H^{d-1}(B,\mathcal{O}) = 0$ may also be expressed with the *cohomological dimension* of B. Recall that the cohomological dimension $\operatorname{cd} X$ of a scheme X is the maximal number i such that there exists a (quasi-)coherent sheaf $\mathcal{F}$ with $H^i(X,\mathcal{F}) \neq 0$. This notion was developed by R. Hartshorne; see [**42**]. In the following proposition we gather some results on cohomological dimension which we will need in the sequel.

PROPOSITION 2.13. *The cohomological dimension has the following properties.*

(i) *Let X denote a Noetherian scheme. Then* $\operatorname{cd}(X) \leq \dim(X)$.
(ii) *If $X' \to X$ is an affine morphism* (*meaning that the preimage of every open affine subset is again affine*) *of separated Noetherian schemes, then* $\operatorname{cd}(X') \leq \operatorname{cd}(X)$.
(iii) *A Noetherian scheme X is an affine scheme* (*that is, isomorphic to* $\operatorname{Spec} S$) *if and only if* $\operatorname{cd}(X) = 0$.
(iv) *Suppose that $(R,\mathfrak{m})$ is a local Noetherian ring of dimension d. Then the cohomological dimension of $D(\mathfrak{m})$ is $d-1$.*
(v) *Suppose that A is a graded K-algebra of finite type over a field K and let $\mathfrak{a}$ denote a homogeneous ideal. Then* $\operatorname{cd}(D(\mathfrak{a})) = \operatorname{cd}(D_+(\mathfrak{a}))$ (*$D_+(\mathfrak{a})$ denotes the corresponding open subset in* $\operatorname{Proj} A$).

PROOF. (i) is the vanishing theorem of Grothendieck ([**45**, Theorem III.2.7]), (ii) follows from Čech cohomology, (iii) is the cohomological characterization of affine schemes due to Serre ([**45**, Theorem III.3.7]) and (iv) is a theorem of Grothendieck, [**20**, Theorem 3.5.7]. For (v) see [**6**, Proposition 1.3.1]. □

PROPOSITION 2.14. *Let R be a normal excellent domain. Let $f_1,\ldots,f_n \in R$ be elements which generate an ideal primary to a maximal ideal $\mathfrak{m}$ of height $d \geq 2$, let $f \in R$ and let $A = R[T_1,\ldots,T_n]/(f_1T_1+\ldots+f_nT_n+f)$ denote the forcing algebra for this data. Then $f \in (f_1,\ldots,f_n)^\star$ if and only if the cohomological dimension of $B = D(\mathfrak{m}A) \subset \operatorname{Spec} A$ is $d-1$.*

PROOF. The cohomological dimension of B is at most $d-1$ by Proposition 2.13(ii). If $H^{d-1}(B,\mathcal{O}) \neq 0$, then the cohomological dimension is $d-1$. If $H^{d-1}(B,\mathcal{O}) = 0$, then one can resolve every coherent sheaf $\mathcal{F}$ by $\mathcal{O}^r \to \mathcal{F} \to 0$, which shows that the cohomological dimension is $< d-1$. □

REMARK 2.15. We will mostly deal with the situation where R is two-dimensional. Then the question whether $f \in (f_1,\ldots,f_n)^\star$ for an $\mathfrak{m}$-primary ideal is the question whether the (forcing) affine-linear bundles is an affine scheme. This is still a difficult question. An open subset $U = D(\mathfrak{a}) \subseteq \operatorname{Spec} A$, $\mathfrak{a} = (a_1,\ldots,a_k)$ is affine if and only if there exist elements $q_1,\ldots,q_k \in \Gamma(U,\mathcal{O}_U)$ such that $\sum_{i=1}^k a_iq_i = 1$. The functions q_i define then the closed embedding of $D(\mathfrak{a})$ into an affine space.

EXAMPLE 2.16. We come back to Example 1.9 from the first chapter. We know already that $z^2 \in (x,y)^*$ on $x^3+y^3+z^3$, and therefore

$$D(x,y) \subset \operatorname{Spec} K[X,Y,Z,S,T]/(X^3+Y^3+Z^3, XT+YS+Z^2)$$

is not affine. This variety is also not affine for a field of characteristic 0. It is interesting to note that for $K=\mathbb{C}$ it is a *Stein space* (meaning that it is isomorphic as a complex space to a closed subset of $\mathbb{C}^k$ given as the zero set of holomorphic functions), and in fact it is the cone over the first example of this phenomenon given by Serre.

Now we want to show that $z \notin (x,y)^*$ by showing that

$$D(x,y) \subset \operatorname{Spec} K[X,Y,Z,S,T]/(X^3+Y^3+Z^3, XT+YS+Z)$$

is affine. We can eliminate Z and get the equation $X^3+Y^3-(XT+YS)^3=0$ or that

$$(1-S^3)X^3 - 3S^2TX^2Y - 3ST^2XY^2 + (1-T^3)Y^3 = 0.$$

In characteristic $\neq 3$ the four coefficients (considering this expression as a polynomial in X,Y over $K[S,T]$) generate the unit ideal and the subset is affine by the following proposition.

PROPOSITION 2.17. *Let R denote a commutative ring, and let*

$$A = R[X,Y]/(\sum_{ij} a_{ij}X^iY^j), \ a_{ij} \in R.$$

If the open subset $D(x,y) \subseteq \operatorname{Spec} A$ is affine, then the a_{ij} generate the unit ideal. On the other hand, if the set $\{a_{ij} : (i,j)$ is minimal (in the product order) among the $a_{ij} \neq 0\}$ generates the unit ideal, then $D(x,y)$ is affine.

PROOF. In the first case, let $\mathfrak{m}$ be a maximal ideal containing all a_{ij} and let K be the residue class field. Then there exists a ring homomorphism $A \to K[X,Y]$ (respectively, $\mathbb{A}^2 \to \operatorname{Spec} A$). If $D(x,y)$ would be affine, so would be (the preimage under the scheme morphism) $D(X,Y) \subseteq \operatorname{Spec} K[X,Y]$, which is, however, not affine.

In the other case we may assume, since affineness is a local property, that one of the a_{ij} is a unit ((i,j) being a minimal index pair). We can then rewrite the equation as $X^iY^j + X^{i+1}P + Y^{j+1}Q$. Then we can build the functions

$$\frac{P}{Y^j} = -\frac{X^i+YQ}{X^{i+1}} \quad \text{and} \quad \frac{Q}{X^i} = -\frac{Y^j+XP}{Y^{j+1}},$$

which are defined on $D(x,y)$. We also get from the equation by dividing through X^iY^j the equation

$$1 = X\frac{P}{Y^j} + Y\frac{Q}{X^i},$$

which shows that $(X,Y)\Gamma(D(X,Y),\mathcal{O})$ is the unit ideal. □

CHAPTER 3

Principal fiber bundles (torsors) and their projective realization

Let $f_1,\dots,f_n,f \in R$ be given. In the last chapter we saw that the syzygy bundle $\operatorname{Syz}(f_1,\dots,f_n)$ on $U = D(f_1,\dots,f_n)$ acts simply transitively on the forcing bundle B, which is given by

$$B = (\operatorname{Spec} R[T_1,\dots,T_n]/(f_1T_1+\dots+f_nT_n+f)|_U),$$

and that the tight closure question $f \in (f_1,\dots,f_n)^\star$ is equivalent to a question concerning the cohomological dimension of B. The bundle B is a torsor for the syzygy bundle in the sense of the following definition.

DEFINITION 3.1. Let G denote a group scheme over a scheme X. A scheme $B \to X$ together with an operation $\sigma : G \times_X B \to B$ is called a *geometric torsor* for G (or a G-*principal fiber bundle* or a *principal homogeneous space*) if there exist an open covering $X = \bigcup_i U_i$ and isomorphisms $\varphi_i : B|_{U_i} \to G|_{U_i}$ such that the diagrams (with $U = U_i$, $\varphi = \varphi_i$)

$$\begin{array}{ccc} G|_U \times_U B|_U & \xrightarrow{\ \sigma\ } & B|_U \\ {\scriptstyle id\times\varphi}\downarrow & & \downarrow{\scriptstyle \varphi} \\ G|_U \times_U G|_U & \xrightarrow{\ \mu\ } & G|_U \end{array}$$

commute.

PROPOSITION 3.2. *Let X denote a Noetherian separated scheme and let $p : V \to X$ denote a vector bundle on X with a sheaf of sections $\mathcal{S}$. Then there exists a correspondence between cohomology classes $c \in H^1(X,\mathcal{S})$ and geometric V-torsors.*

PROOF. We will describe this correspondence; see also [**75**, Proposition III.4.6]. Let B denote a V-torsor. Then there exists by definition an open covering $X = \bigcup U_i$ such that $B|_{U_i}$ together with the action of $V|_{U_i}$ is isomorphic to $V|_{U_i}$ acting on itself. These isomorphisms $\varphi_i : B|_{U_i} \to V|_{U_i}$ induce isomorphisms

$$\psi_{ij} = \varphi_j \circ \varphi_i^{-1} : V|_{U_i \cap U_j} \longrightarrow V|_{U_i\cap U_j}.$$

These isomorphisms are compatible with the operation of V on itself, and this means that they are of the form $\psi_{ij} = \mathrm{id}_{V|_{U_i\cap U_j}} + s_{ij}$ with suitable sections $s_{ij} \in \Gamma(U_i \cap U_j, \mathcal{S})$. This family defines a Čech cocycle for the covering and gives therefore a cohomology class in $H^1(X,\mathcal{S})$.

For the reverse procedure, suppose that the cohomology class $c \in H^1(X,\mathcal{S})$ is represented by a Čech cocycle $s_{ij} \in \Gamma(U_i \cap U_j, \mathcal{S})$ for an open covering $X = \bigcup_i U_i$. Set $B_i := V|_{U_i}$. We take the morphisms

$$B_i|_{U_i\cap U_j} = V|_{U_i\cap U_j} \xrightarrow{\psi_{ij}} V|_{U_i\cap U_j} = B_j|_{U_i\cap U_j}$$

given by $\psi_{ij} := \mathrm{id}_{V|_{U_i \cap U_j}} + s_{ij}$ to glue the B_i together to a scheme B over X. This is possible since the cocycle condition guarantees the gluing condition [**30**, 0, 4.1.7]. The action of V_i on B_i by itself also glues together to give an action on B. □

We now consider some functorial properties of torsors.

PROPOSITION 3.3. *Let X denote a Noetherian separated scheme, let $\mathscr{S}$ denote a locally free sheaf on X, and let $c \in H^1(X, \mathscr{S})$ with corresponding geometric torsor $B = B(c) \to X$.*

(i) *Let Y denote another Noetherian separated scheme and let $\varphi : Y \to X$ denote a morphism. Then $\varphi^*(B) = Y \times_X B$ is the torsor corresponding to $\varphi^*(c) \in H^1(Y, \varphi^*(\mathscr{S}))$.*
(ii) *A morphism $\varphi : Y \to X$ factors through B if and only if $0 = \varphi^*(c) \in H^1(Y, \varphi^*(\mathscr{S}))$. In particular, B is trivial ($c = 0$) if and only if there exists a section $s : X \to B$.*
(iii) *If $\mathscr{T}$ is another locally free sheaf and $\psi : \mathscr{S} \to \mathscr{T}$ is a sheaf homomorphism, then there exists an affine morphism $B(c) \to B(\psi(c))$.*

PROOF. See [**6**, Lemma 2.3.8, Proposition 2.3.9 and Lemma 2.3.7]. The relevant diagram for (ii) is

$$\begin{array}{ccc} Y \times_X B & \longrightarrow & B \\ \downarrow & & \downarrow \\ Y & \xrightarrow{\varphi} & X. \end{array}$$

□

We now describe a more projective realization of a torsor. For a vector bundle $V \to X$ the corresponding *projective bundle* is denoted by $\mathbb{P}(V)$. Over every point $x \in X$, the fiber $V_{\kappa(x)}$ of V over x is an affine space over $\kappa(x)$ and the corresponding projective space $\mathbb{P}(V_{\kappa(x)})$ forms the fiber over x of this projective bundle. A more technical way to describe this follows: let $\mathscr{F}$ be the sheaf of linear forms of V. Then

$$V = \mathrm{Spec}(\bigoplus_{k \in \mathbb{N}} S^k(\mathscr{F})) \quad \text{and} \quad \mathbb{P}(V) = \mathrm{Proj}(\bigoplus_{k \in \mathbb{N}} S^k(\mathscr{F}))$$

(see [**45**, Exercises II.5.16 and II.5.17, Proposition II.7.11]). The invertible sheaves $\mathscr{O}(1)$ on the fibers glue together to give an invertible *relatively ample* sheaf $\mathscr{O}_{\mathbb{P}(V)}(1)$ on the projective bundle. It has the property that $\Gamma(\mathbb{P}(\mathscr{F}), \mathscr{O}_{\mathbb{P}(\mathscr{F})}(k)) = \Gamma(X, S^k(\mathscr{F}))$ and in particular $\Gamma(\mathbb{P}(\mathscr{F}), \mathscr{O}_{\mathbb{P}(\mathscr{F})}(1)) = \Gamma(X, \mathscr{F})$. The cone mappings $V_{\kappa(x)} \dashrightarrow \mathbb{P}(V_\kappa(x))$ glue together to give a mapping $V \dashrightarrow \mathbb{P}(V)$ which is defined outside the zero-section of V. A subbundle $W \subseteq V$ (corresponding to a surjection $\mathscr{F} \to \mathscr{G}$) induces a closed embedding $\mathbb{P}(W) \subseteq \mathbb{P}(V)$.

PROPOSITION 3.4. *Let X denote a Noetherian separated scheme and let V be a vector bundle on X with sheaf of sections $\mathscr{S}$. Let $c \in H^1(X, \mathscr{S})$ be a cohomology class and let*

$$0 \longrightarrow V \longrightarrow V' \longrightarrow \mathbb{A}_X \longrightarrow 0 \ (\textit{resp.}\ 0 \longrightarrow \mathscr{S} \longrightarrow \mathscr{S}' \longrightarrow \mathscr{O}_X \longrightarrow 0)$$

denote the extension of vector bundles corresponding to $c \in H^1(X, \mathscr{S}) \cong \mathrm{Ext}^1(\mathscr{O}_X, \mathscr{S})$. Then the geometric torsor $B \to X$ corresponding to c (as described in Proposition 3.2) *is isomorphic to $\mathbb{P}(V') - \mathbb{P}(V)$.*

PROOF. The inclusion $e : V \hookrightarrow V'$ induces a projective subbundle $\mathbb{P}(V) \subset \mathbb{P}(V')$. Let $T : V' \to \mathbb{A}_X$ denote the linear form on V' in the sequence. Hence V is the zero set of T. We consider the cone mapping $V' \dashrightarrow \mathbb{P}(V')$. We claim that the zero set $Z = V(T-1) \subset V'$ maps isomorphically onto $\mathbb{P}(V') - \mathbb{P}(V)$. This may be checked locally in X and is therefore clear. Hence we shall show that $Z = V(T-1) \subset V'$ is isomorphic to the torsor given by c. First we claim that Z is indeed a V-torsor by the action $V \times_X Z \to Z$ given by the addition inside V'. This is a well-defined action on Z since $T(v+z) = T(v) + T(z) = 1$ (where $v \in V$ and $z \in Z$).

Let U_i denote a covering of X such that there exist sections $s_i' \in \Gamma(U_i, \mathscr{S}')$ that map to $1 \in \Gamma(U_i, \mathscr{O}_X)$. Consider the mappings $\psi_i := e + s_i' \circ p : V|_{U_i} \to V'|_{U_i}$ (let p denote the projections). The image of this mapping lies on Z, since $T(s_i') = 1$, and hence we have mappings $\psi_i : V|_{U_i} \to Z|_{U_i}$, which are isomorphisms. These identifications commute with the operation, since

$$\psi_i(v_1+v_2) = v_1 + v_2 + (s_i' \circ p)(v_1+v_2) = v_1 + (v_2 + (s_i' \circ p)(v_2)) = v_1 + \psi_i(v_2).$$

This shows that Z is a V-torsor. We can compute the corresponding cohomology class of this torsor by using these identifications. The transition mappings are given by

$$\psi_j^{-1} \circ \psi_i : V|_{U_i \cap U_j} \longrightarrow V|_{U_i \cap U_j},$$

which sends $v \mapsto v + (s_i' - s_j')(p(v))$. Therefore the corresponding cohomology class is given by $s_{ij} = s_i' - s_j' \in \Gamma(U_i \cap U_j, \mathscr{S})$. But this is also the image of 1 under the connecting homomorphism for the short exact sequence $0 \to \mathscr{S} \to \mathscr{S}' \to \mathscr{O}_X \to 0$. □

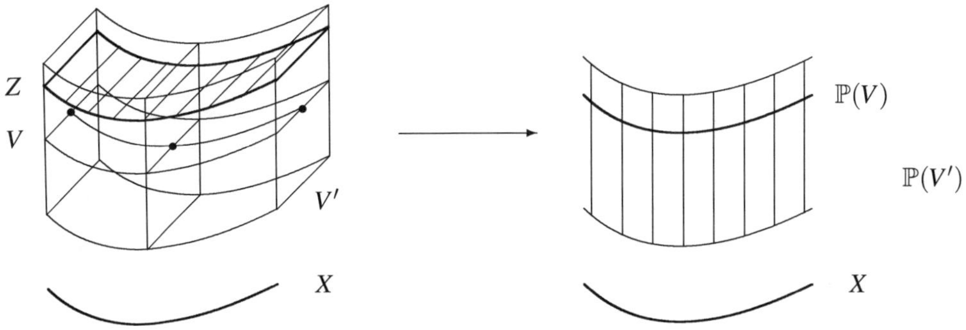

PROPOSITION 3.5. *Let R denote a commutative ring and let $f_1, \ldots, f_n \in R$ be elements. Let V denote the corresponding syzygy bundle on $U = D(f_1, \ldots, f_n)$ and let $\mathrm{Syz} = \mathrm{Syz}(f_1, \ldots, f_n)$ be the sheaf of syzygies. Let $f \in R$ denote another element. Then the affine-linear bundle*

$$B = \operatorname{Spec} R[T_1, \ldots, T_n]/(f_1T_1 + \ldots + f_nT_n + f)|_U$$

given by these forcing data is a V-torsor, which is isomorphic to the V-torsor given by the cohomology class $c = \delta(f) \in H^1(U, \mathrm{Syz})$, where δ is the connecting homomorphism for the presenting sequence on U (Definition 2.12).

PROOF. Set $B_i = \operatorname{Spec} R_{f_i}[T_1, \ldots, T_n]/(f_1T_1 + \ldots + f_nT_n + f)$ and

$$V_i = R_{f_i}[S_1, \ldots, S_n]/(f_1S_1 + \ldots + f_nS_n)$$

for $i = 1, \ldots, n$. Then the isomorphisms $\varphi_i : B_i \to V_i$ given by $T_j \mapsto S_j$ for $j \neq i$ and by $T_i \mapsto S_i - f/f_i$ are compatible with the action of V on B defined in Lemma 2.6(ii). Therefore the forcing bundle is a V-torsor. We shall compute the cohomology class represented by the forcing bundle B, and thus we have to look at the mappings

$$\varphi_j \circ \varphi_i^{-1} : V|_{U_i \cap U_j} \to V|_{U_i \cap U_j}.$$

This mapping is given by $S_k \mapsto S_k$ for $k \neq i, j$, $S_i \mapsto S_i - f/f_i$ and $S_j \mapsto S_j + f/f_j$. The corresponding syzygies are

$$(0,\ldots,0,f/f_i,0,\ldots,0,-f/f_j,0,\ldots,0) \in \Gamma(D(f_if_j),\mathscr{S}).$$

On the other hand, look at the connecting homomorphism for the presenting sequence. The element $f \in \Gamma(U,\mathscr{O}_U)$ (coming from $f \in R$ by the restriction map) is locally given by $(0,\ldots,0,f/f_i,0,\ldots,0) \in \Gamma(D(f_i),\mathscr{O}_U^n)$. The difference of two such local representatives gives $\delta(f)$ as a Čech cocycle. □

Therefore the problem of determining whether an element belongs to the tight closure of an ideal is a special case of the following problem (or, rather, a research program).

PROBLEM 3.6. Let X denote a scheme and let V denote a vector bundle on X with sheaf of sections $\mathscr{S}$. Study the cohomological and geometric properties of the geometric torsor $B \to X$ corresponding to a cohomology class $c \in H^1(X,\mathscr{S})$ in dependence of X, $\mathscr{S}$, and c.

We now fix the following homogeneous situation.

SITUATION 3.7. Let R be a standard-graded normal K-domain of dimension $d \geq 2$ over an algebraically closed field K, and let a homogeneous R_+-primary ideal $(f_1,\ldots,f_n)$ be given. Let $d_i = \deg(f_i)$. Let $Y = \operatorname{Proj} R$ denote the corresponding projective variety with tautological invertible sheaf $\mathscr{O}_Y(1)$. Then these data yield on Y the short exact sequence

$$0 \longrightarrow \operatorname{Syz}(f_1,\ldots,f_n)(m) \longrightarrow \bigoplus_{i=1}^{n} \mathscr{O}_Y(m-d_i) \stackrel{f_1,\ldots,f_n}{\longrightarrow} \mathscr{O}_Y(m) \longrightarrow 0$$

for every $m \in \mathbb{Z}$. If f is another homogeneous element of degree m, then this yields via $\delta : R_m \stackrel{\sim}{\to} \Gamma(Y,\mathscr{O}_Y(m)) \to H^1(Y,\operatorname{Syz}(m))$ the cohomology class $c = \delta(f) \in H^1(Y,\operatorname{Syz}(m))$. Let $B(c) = B \to Y$ be the corresponding torsor.

REMARK 3.8. The extension of $\operatorname{Syz}(f_1,\ldots,f_n)(m)$ given by the cohomology class $\delta(f)$ (as in Proposition 3.4) is just

$$0 \longrightarrow \operatorname{Syz}(f_1,\ldots,f_n)(m) \longrightarrow \operatorname{Syz}(f_1,\ldots,f_n,f)(m) \longrightarrow \mathscr{O}_Y \longrightarrow 0$$

with the natural inclusion and the projection onto the last component.

If $s = (s_1,\ldots,s_n) \in \Gamma(Y,\operatorname{Syz}(f_1,\ldots,f_n)(m))$, then we say that s is a syzygy of *total degree m*. The next results show that we can express the properties that are of interest for tight closure in terms of projective bundles on Y.

PROPOSITION 3.9. *Suppose the homogeneous situation 3.7 and let* $\varphi : D(R_+) \to Y$ *denote the cone mapping. Then*

$$\varphi^*(\operatorname{Syz}(f_1,\ldots,f_n)(m)) = \operatorname{Syz}(f_1,\ldots,f_n)$$

on $D(R_+)$. *Under this pull-back the class* $c = \delta(f) \in H^1(Y,\operatorname{Syz}(f_1,\ldots,f_n)(m))$ *goes to* $c' = \delta(f) \in H^1(D(R_+),\operatorname{Syz}(f_1,\ldots,f_n))$. *Moreover, the following are equivalent.*

(i) $f \in (f_1,\ldots,f_n)^\star$ *(solid closure).*

(ii) *The cohomological dimension of* $B(c) \to Y$ *is* $d-1 = \dim Y$.

PROOF. We have the commutative diagram

$$\begin{array}{ccc} \operatorname{Spec}(A)|_{D(R_+)} = B(c') & \longrightarrow & B(c) \\ \downarrow & & \downarrow \\ \operatorname{Spec} R \supset D(R_+) & \longrightarrow & \operatorname{Proj} R\,. \end{array}$$

The cohomological characterization holds by Proposition 2.14 for $B(c')$. The main point is that now $B(c)$ is the quotient of $B(c')$ under the action of the multiplicative group G_m. In such a situation the cohomological dimension of the quotient and of the total space coincide. (One may also introduce a not necessarily positive grading on the forcing algebra and realize $B(c)$ as an open subset of its Proj.) □

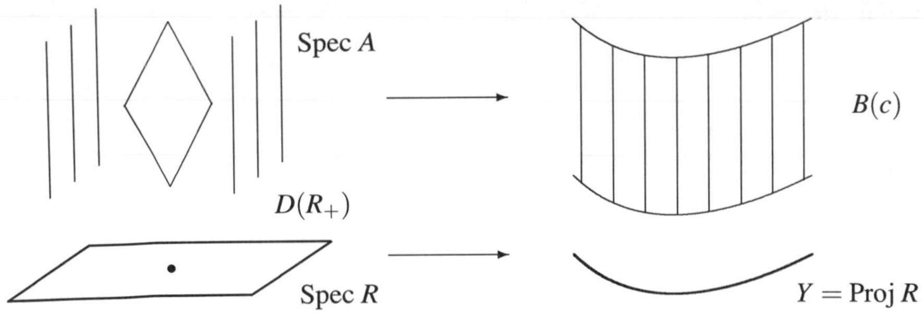

REMARK 3.10. This means that we can check the containment inside the tight closure of a homogeneous primary ideal in a standard-graded normal domain by studying cohomological properties of a torsor over the corresponding projective variety. Moreover, this torsor can be realized as the complement of a projective subbundle of codimension one (the *forcing divisor*) inside a projective bundle. Therefore we can work in an entirely projective and, if the graded ring R has an isolated singularity, smooth setting. So in the first interesting case, where R is a normal standard-graded domain of dimension two, Y is a smooth projective curve, we have a torsor $B \to Y$ and we want to know whether B is an affine scheme. If n is the number of ideal generators, then the rank of the projective bundle $\mathbb{P}(V')$ is $n-1$. For the parameter case, $n = 2$, we are led to a ruled surface together with a section (see [**45**, Section V.2]); for $n = 3$ we have a projective bundle of rank two together with a subbundle of rank one, which is itself a ruled surface.

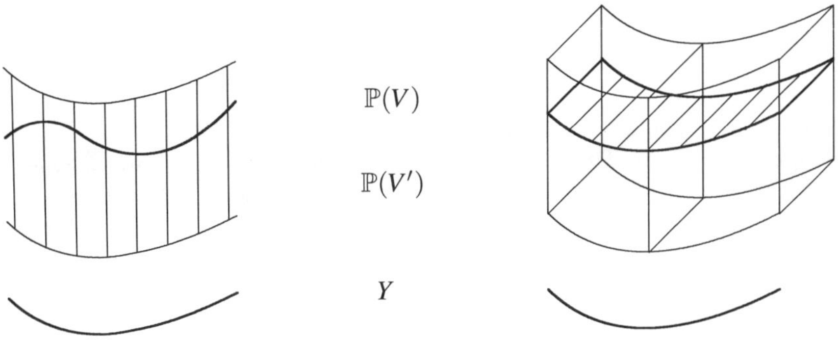

One of the goals of these notes is to show cases where tight closure equals plus closure. So we want to understand plus closure in our geometric picture of forcing algebras and tight closure. In the presence of a grading it is more natural to compare tight closure with the *graded plus closure*, which is defined in the following way.

DEFINITION 3.11. Let R denote an $\mathbb{N}$-graded Noetherian K-domain and let's denote by $I = (f_1, \ldots, f_n)$ a homogeneous ideal. Then

$$I^{+\mathrm{gr}} := \{f \in R : f \in IR' \text{ for a finite graded extension } R \subseteq R'\}.$$

A ring homomorphism $\varphi : R \to R'$ is graded if there exists $a \in \mathbb{N}$ such that $\varphi(R_d) \subseteq R'_{ad}$. If $R \subseteq R'$ is graded and finite, then this induces a morphism $\operatorname{Proj} R' \to \operatorname{Proj} R$. The Frobenius morphism of a graded ring over a field of positive characteristic is therefore always graded. If $R \subseteq R'$ is a finite graded extension such that $f \in IR'$, then this holds also in $R'' = \sum_k R'_{ak}$, and by regrading we may also find a finite graded extension $R \subseteq R''$ such that $f \in R''$ holds, but without any degree shift. We may characterize the containment $f \in (f_1, \ldots, f_n)^{+\text{gr}}$ in our geometric setting in the following way.

PROPOSITION 3.12. *Suppose the homogeneous situation 3.7. Then the following are equivalent.*

(i) $f \in (f_1, \ldots, f_n)^{+\text{gr}}$.
(ii) *There exists a finite surjective morphism $g : Y' \to Y$, where Y' is another projective variety, such that $g^*(c) = 0$.*
(iii) *There exists a finite surjective morphism $g : Y' \to Y$ such that the pull-back $g^*(B)$ is trivial.*
(iv) *There exists a projective subvariety $Y'' \subset B$ of dimension $d-1$ (or a closed subvariety $Y'' \subset \mathbb{P}(V')$ of dimension $d-1$ not intersecting $\mathbb{P}(V)$).*

PROOF. (i) $\Rightarrow$ (ii). Suppose first that $R \subseteq R'$ is a finite extension such that $f \in (f_1, \ldots, f_n)R'$ holds. Then on $Y' = \operatorname{Proj} R'$ there exists due to Proposition 3.3(ii) a section $s : Y' \to \mathbb{P}(V'_{Y'})$ which does not meet $\mathbb{P}(V_{Y'})$, where $V_{Y'}$ denotes the syzygy bundle on Y' corresponding to the elements considered in R'. But $\mathbb{P}(V_{Y'}) = g^*\mathbb{P}(V) = \mathbb{P}(V) \times_Y Y'$, where $g : Y' \to Y$ is the finite morphism corresponding to the ring inclusion $R \subseteq R'$.

(ii) $\Rightarrow$ (i). Let $g : Y' \to Y$ be such a morphism. Then $\mathcal{O}_{Y'}(1) = g^*\mathcal{O}_Y(1)$ is ample and we consider $R' = \oplus_{k \geq 0} \Gamma(Y', g^*\mathcal{O}_Y(k))$, which is finitely generated and finite over R'. We have $Y' \cong \operatorname{Proj} R'$ and we have the commutative diagram

$$\begin{array}{ccc} D(R'_+) & \longrightarrow & D(R_+) \\ \downarrow & & \downarrow \\ Y' & \xrightarrow{g} & Y\,. \end{array}$$

Let $B = \mathbb{P}(V') - \mathbb{P}(V)$. The existence of a section $s : Y' \to \mathbb{P}(V') \times_Y Y'$ not meeting $\mathbb{P}(V) \times_Y Y'$ implies that $g^*(B)$ is trivial. Then the pull-back of B to $D(R'_+)$ is trivial as well, but this is the same as the pull-back of the forcing bundle $\operatorname{Spec} R(T_1, \ldots, T_n]/(\sum f_i T_i + f)|_{D(R_+)}$. It follows that $f \in (f_1, \ldots, f_n)R''$, where R'' is the normalization of R'.

Suppose that (ii) holds. Then the image of the section gives the closed subvariety Y'' finite over Y. This gives (iii) and then (iv). Suppose that (iv) holds. The mapping $Y'' \hookrightarrow \mathbb{P}(V') \to Y$ is projective and the fibers are zero-dimensional, since $\mathbb{P}(V_y) \subset \mathbb{P}(V'_y)$ meets every curve, but $Y'' \cap \mathbb{P}(V) = \emptyset$. Hence this mapping is finite and due to the assumption on the dimension it is surjective. So suppose that (iii) holds. The mapping $g : Y' = Y'' \stackrel{i}{\hookrightarrow} \mathbb{P}(V') \to Y$ is finite and surjective, and the image of the section $i \times id_{Y'} : Y' \to \mathbb{P}(V') \times_Y Y' = g^*\mathbb{P}(V')$ is disjoined to $g^*\mathbb{P}(V) = \mathbb{P}(V) \times_Y Y'$. □

PROBLEM 3.13. For the case $d = 2$ we can rephrase the question as to whether tight closure is graded plus closure in the following way: if $B \to C$ is an affine-linear bundle (a torsor for a vector bundle) on a smooth projective curve over a field of positive characteristic, which is not an affine variety, is it then true that there exists a projective curve C' inside B?

REMARK 3.14. If $K = \mathbb{C}$, then we may consider the torsor corresponding to a cohomology class also as a complex manifold (or complex space), as a real smooth manifold, and as a topological space. It follows from partition of unity that every torsor is trivial (isomorphic to the vector bundle) in the smooth and in the topological category.

CHAPTER 4

Ample vector bundles

In this chapter we will deal with ample vector bundles on smooth projective curves, which is a certain notion of positivity. This positivity is related (though not equivalent) to the positivity of the degree of the bundle, which is, beside the rank, the most important numerical invariant of a vector bundle.

DEFINITION 4.1. Let $\mathscr{S}$ be a locally free sheaf of rank r on a smooth projective curve over an algebraically closed field. Then the degree of $\mathscr{S}$ is defined as

$$\deg(\mathscr{S}) = \deg(\bigwedge^{r}(\mathscr{S})).$$

Here $\bigwedge^r(\mathscr{S})$ is a line bundle, and the degree of a line bundle is the degree of the corresponding Weil divisor (see [**45**, Section II.6]).

REMARK 4.2. This means that if we have a meromorphic section s (which is a rational morphism, defined on an open subset of C) in the line bundle L, then we have to count the zeros and the poles of this section with their multiplicities and sum up. The resulting number is independent of the chosen section.

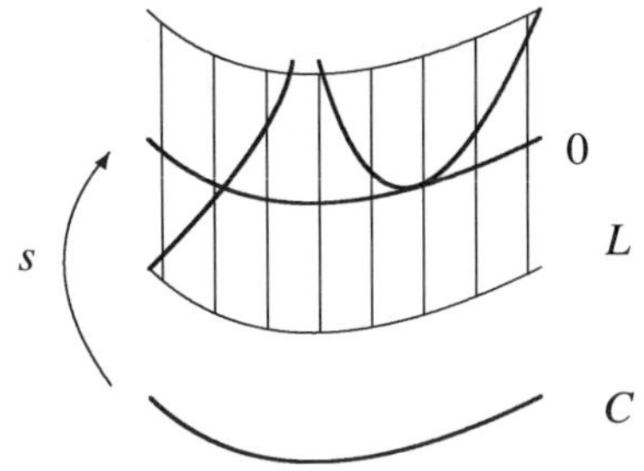

REMARK 4.3. If $\mathbb{P}(\mathscr{S}) = \operatorname{Proj} \oplus_{k\geq 0} S^k(\mathscr{S})$ is the corresponding projective bundle of rank $r-1$ and total dimension r and if ξ denotes the divisor class corresponding to the relatively very ample invertible sheaf $\mathscr{O}_{\mathbb{P}(\mathscr{S})}(1)$, then $\deg(\mathscr{S}) = \xi^r$ also equals the top self-intersection number; see [**72**, Lemma 6.4.10].

REMARK 4.4. If $C = \operatorname{Proj} R$ is a curve, then the degree of $\mathscr{O}_C(1)$ is also called the *degree of the curve*, denoted by $\deg C$. If $C \subset \mathbb{P}^2$ is a plane algebraic curve, then the degree of the curve equation equals the degree of $\mathscr{O}_C(1)$. The degree is additive on short exact sequences. We apply this to our short exact sequence (Situation 3.7)

$$0 \longrightarrow \operatorname{Syz}(f_1,\ldots,f_n)(m) \longrightarrow \bigoplus_{i=1}^{n} \mathscr{O}_C(m-d_i) \stackrel{f_1,\ldots,f_n}{\longrightarrow} \mathscr{O}_C(m) \longrightarrow 0$$

and get

$$\deg(\mathrm{Syz}(f_1,\dots,f_n)(m)) = ((n-1)m - \sum_{i=1}^{n} d_i)\deg C.$$

In particular, the degree of this syzygy bundle is nonnegative if and only if

$$m \geq (\sum_{i=1}^{n} d_i)/(n-1).$$

This number (sometimes called the *degree bound*) will play a crucial role.

DEFINITION 4.5. An invertible sheaf $\mathscr{L}$ (or a line bundle) on a variety X is called *ample* if there exists a positive number k such that $\mathscr{L}^k$ is a *very ample* line bundle. This means that the global sections of $\mathscr{L}^k$ define an embedding $\varphi : X \to \mathbb{P}^N$ into a projective space such that $\mathscr{L}^k = \varphi^* \mathscr{O}_{\mathbb{P}^N}(1)$.

For other equivalent conditions, see [**45**, Section II.7 and Proposition III.5.3]. For line bundles on curves we have a simple numerical criterion for ampleness, which is a special case of the general (Moishezon-Nakai) intersection criterion [**72**, Theorem 1.2.19] for ample divisors.

PROPOSITION 4.6. *An invertible sheaf on a smooth projective curve is ample if and only if its degree is positive.*

PROOF. [**45**, Corollary IV.3.3]. □

REMARK 4.7. If s is a non-zero section of an ample invertible sheaf $\mathscr{L}$ on a projective variety X, then the complement of the corresponding zero-divisor $V(s) = \{x \in X : s(x) = 0\}$ is affine, since $X - V(s) = D(s) = \varphi^{-1}(D_+(x))$ for some linear form $x \in \Gamma(\mathbb{P}^N, \mathscr{O}_{\mathbb{P}^N}(1))$ and since the closed embedding $\varphi : X \to \mathbb{P}^N$ is an affine morphism.

In the situation of a locally free sheaf $\mathscr{F}$ and a (dualized) extension $0 \to \mathscr{O}_C \to \mathscr{F}' \to \mathscr{F} \to 0$ (representing $c \in H^1(C, \mathscr{F}^\vee)$), we are interested in whether $\mathbb{P}(\mathscr{F}') - \mathbb{P}(\mathscr{F})$ is an affine scheme (Proposition 3.4). Now $\Gamma(C, \mathscr{F}') = \Gamma(\mathbb{P}(\mathscr{F}'), \mathscr{O}_{\mathbb{P}(\mathscr{F}')}(1))$, so the section $\mathscr{O}_C \to \mathscr{F}'$ corresponds to a section in $\mathscr{O}_{\mathbb{P}(\mathscr{F}')}(1)$, and the corresponding divisor is just $\mathbb{P}(\mathscr{F})$. So if $\mathscr{O}_{\mathbb{P}(\mathscr{F}')}(1)$ is ample on $\mathbb{P}(\mathscr{F}')$, then the complement of $\mathbb{P}(\mathscr{F})$ is affine. Hence we are interested in the following definition.

DEFINITION 4.8. A locally free sheaf $\mathscr{G}$ on a variety is called *ample* if the invertible sheaf $\mathscr{O}_{\mathbb{P}(\mathscr{G})}(1)$ on $\mathbb{P}(\mathscr{G})$ is ample.

We quote some general results.

PROPOSITION 4.9. *Let X be a variety and let $\mathscr{G}$ be a locally free sheaf. Then the following hold.*

(i) *The quotient $\mathscr{G} \to \mathscr{Q} \to 0$ of an ample sheaf is again ample.*
(ii) *An extension of ample bundles is ample. In particular, a direct sum of ample bundles is ample.*
(iii) *Let $\varphi : X' \to X$ be a finite surjective morphism. Then $\mathscr{G}$ is ample if and only if $\varphi^*(\mathscr{G})$ is ample.*

PROOF. (i) The quotient corresponds to an embedding $\mathbb{P}(\mathscr{H}) \hookrightarrow \mathbb{P}(\mathscr{G})$, and the restriction of $\mathscr{O}_{\mathbb{P}(\mathscr{G})}(1)$ is $\mathscr{O}_{\mathbb{P}(\mathscr{H})}(1)$; hence this is also an ample invertible sheaf. For (ii) see [**72**, Proposition 6.1.12] and for (iii) see [**72**, Proposition 6.1.2 and 6.1.7]. □

We will mainly be interested in ample bundles on curves.

PROPOSITION 4.10. *Let $\mathscr{G}$ denote a locally free sheaf of rank r on a smooth projective curve C of genus g. Then the following hold.*

(i) *An ample sheaf $\mathscr{G}$ has positive degree.*
(ii) *Every quotient of an ample sheaf has positive degree.*
(iii) *Suppose $\deg(\mathscr{G}) > rg$ and that every quotient bundle $\mathscr{Q} \neq \mathscr{G}$ of $\mathscr{G}$ is ample. Then $\mathscr{G}$ is also ample.*
(iv) *(Hartshorne) If the characteristic is zero, then a locally free sheaf $\mathscr{G}$ is ample if and only if every quotient bundle $\mathscr{G} \rightarrow \mathscr{Q} \rightarrow 0$ has positive degree.*

PROOF. (i). By the intersection criterion [**72**, Theorem 1.2.19], the maximal (or top-dimensional) self-intersection number of an ample divisor is positive [**72**, Lemma 6.4.10]. In the case of $\mathscr{O}_{\mathbb{P}(\mathscr{G})}(1)$, the top-dimensional self-intersection equals the degree of the sheaf. Hence $\mathscr{G}$ can only be ample if its degree is positive. (ii) follows from (i) and Proposition 4.9(i). For (iii) see [**27**, Proposition 2.2]. For (iv) see [**44**, Theorem 2.4] or [**72**, 6.4.15]. □

In the following we study under what conditions syzygy bundles, their duals, and their extensions are ample. First we obtain exclusion results for solid closure and tight closure.

PROPOSITION 4.11. *Suppose the homogeneous situation (Situation 3.7) in dimension two and let $0 \rightarrow \mathrm{Syz}(f_1,\ldots,f_n)(m) \rightarrow \mathscr{S}' \rightarrow \mathscr{O}_C \rightarrow 0$ be the extension corresponding to $c \in H^1(C, \mathrm{Syz}(f_1,\ldots,f_n)(m))$. If the dual $(\mathscr{S}')^\vee$ of this extension is ample, then $f \notin (f_1,\ldots,f_n)^\star$. Two necessary conditions for this is that the dual of $\mathrm{Syz}(f_1,\ldots,f_n)(m)$ is ample and that $m < (\sum_{i=1}^n d_i)/(n-1)$.*

PROOF. This follows from Remark 4.7, Proposition 4.10(ii), and the degree computation above. □

If we want to apply this we need sufficient conditions for the ampleness of $\mathscr{S}^\vee$ and of $(\mathscr{S}')^\vee$.

PROPOSITION 4.12. *(Hartshorne-Gieseker) Let $\mathscr{F}$ be an ample locally free sheaf on a smooth projective curve C over an algebraically closed field K. Let $0 \rightarrow \mathscr{O} \rightarrow \mathscr{F}' \rightarrow \mathscr{F} \rightarrow 0$ be an extension given by $c \in H^1(C, \mathscr{F}^\vee)$. Then the following hold.*

(i) *If $c \neq 0$, then all quotients of $\mathscr{F}'$ have positive degree.*
(ii) *If the characteristic is zero and $c \neq 0$, then $\mathscr{F}'$ is also ample.*

PROOF. See [**27**, Proposition 2.2] and [**44**]. □

To apply Proposition 4.12 and also Proposition 4.11 we first have to know criteria for the ampleness of $\mathscr{F}$. In characteristic zero, a useful numerical criterion is provided by Proposition 4.10(iv). We will come back to this in the next chapter in terms of semistability. For now we can only apply it in the parameter case in dimension two. We have the presenting sequence

$$0 \longrightarrow \mathscr{O}(m-d_1-d_2) \cong \mathrm{Syz}(f_1,f_2) \longrightarrow \mathscr{O}(m-d_1) \oplus \mathscr{O}(m-d_2) \stackrel{f_1,f_2}{\longrightarrow} \mathscr{O}(m) \longrightarrow 0$$

and the cohomology class given by $f \in R_m$ is $f/f_1f_2 \in H^1(C, \mathscr{O}(m-d_1-d_2))$. We write the corresponding dual extension as

$$0 \longrightarrow \mathscr{O}_C \longrightarrow \mathscr{G} \longrightarrow \mathscr{O}_C(d_1+d_2-m) \longrightarrow 0.$$

The invertible bundle on the right is ample if and only if $m < d_1+d_2$ (by Proposition 4.6). The torsor is the complement of a section Z inside the ruled surface $\mathbb{P}(\mathscr{G})$.

PROPOSITION 4.13. *Suppose the homogeneous situation in dimension two in the parameter case* ($n=2$). *Then the following are equivalent.*

(i) $f \notin (f_1,f_2)^\star$.
(ii) $\mathbb{P}(\mathscr{G})-Z$ *is affine.*
(iii) *The forcing divisor* Z *on* $\mathbb{P}(\mathscr{G})$ *is ample.*

PROOF. We know the equivalence (i)$\Leftrightarrow$ (ii) from Proposition 3.9 so we have to show the equivalence (ii)$\Leftrightarrow$ (iii). If Z is ample, then its complement is affine as mentioned in Remark 4.7. If $\mathbb{P}(\mathscr{G})-Z$ is affine, then it does not contain any projective curves. Furthermore, there exist global functions on $\mathbb{P}(\mathscr{G})-Z$ that are not constant. Thus aZ, $a\geq 1$, is linearly equivalent with an effective Weil divisor not containing Z. Hence the self-intersection number is positive and the intersection criterion for surfaces [**45**, Theorem V.1.10] yields that Z is ample. □

COROLLARY 4.14. (*Huneke-Smith*) *Let* R *be a normal standard-graded domain of dimension two over an algebraically closed field* K, *and let* f_1,f_2 *be homogeneous parameters of degree* d_1 *and* d_2. *Let* f *be homogeneous of degree* m. *Then the following hold.*

(i) *If* $m\geq d_1+d_2$, *then* $f\in(f_1,f_2)^\star$.
(ii) *If* $m<d_1+d_2$ *and* $f\in(f_1,f_2)^\star$, *then* $f\in(f_1,f_2)$ *in characteristic zero and* $f^q\in(f_1^q,f_2^q)$ *for* $q=p^e>2g$ *in positive characteristic* p.

PROOF. If $m\geq d_1+d_2$, then the self-intersection number of the forcing divisor $Z=\mathbb{P}(V)$ is not positive, hence it is not ample and by Proposition 4.13 the complement is not affine, hence $f\in(f_1,f_2)^\star$. If $m<d_1+d_2$, then $\mathscr{O}_C(d_1+d_2-m)$ is ample. In characteristic zero, if $f\notin(f_1,f_2)$, then the extension $\mathscr{G}$ (given by f/f_1f_2) is non-trivial and by Proposition 4.12(ii) it is also ample. In characteristic p, if we have $f^q\notin(f_1^q,f_2^q)$ for some $q=p^e>2g$, then $F^{e*}(\mathscr{G})$ is a non-trivial dual extension of degree $>2g$. Therefore by Proposition 4.12(i) every quotient (the non-trivials are invertible) of $F^{e*}(\mathscr{G})$ has positive degree, hence they are ample by Proposition 4.6 (iii). So $F^{e*}\mathscr{G}$ is ample by Proposition 4.10(iii) and therefore $\mathscr{G}$ is ample by Proposition 4.9(iii). In both cases it follows that $f\in(f_1,f_2)^\star$ via Remark 4.7. □

REMARK 4.15. There is a generalization due to Hara of Corollary 4.14 for the parameter case in higher dimension, which is called the strong vanishing theorem, because it is related to Kodaira's vanishing theorem. Here the degree bound is $d_1+\ldots+d_n$, where n is the dimension of a normal Cohen-Macaulay graded K-domain; see [**55**, Theorem 6.1] or [**35**].

REMARK 4.16. Let $C\to\operatorname{Spec}\mathbb{Z}$ be a relatively smooth projective curve. Typical examples are $C=\operatorname{Proj}S$, where $S=\mathbb{Z}[X,Y,Z]/(F)$ and where the coefficients of F are integers. Then $C_0=\operatorname{Proj}S\otimes_{\mathbb{Z}}\mathbb{Q}$ and $C_p=\operatorname{Proj}S\otimes_{\mathbb{Z}}\mathbb{Z}/(p)$ are smooth projective curves over $\mathbb{Q}$ and $\mathbb{Z}/(p)$ respectively.

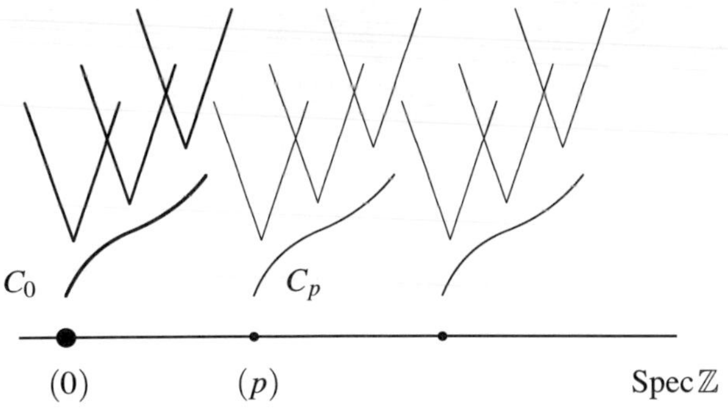

A vector bundle $\mathscr{G}$ on C restricts to a bundle $\mathscr{G}_\mathfrak{p}$ on every fiber $C_\mathfrak{p}$, $\mathfrak{p} \in \operatorname{Spec}\mathbb{Z}$. In such a situation ampleness is an open property [**72**, Theorem 1.2.13]: if $\mathscr{G}_\mathfrak{p}$ is ample on $C_\mathfrak{p}$ for some $\mathfrak{p} \in \operatorname{Spec}\mathbb{Z}$, then there exists an open set $\mathfrak{p} \in D(h) \subseteq \operatorname{Spec}\mathbb{Z}$ such that $\mathscr{G}_\mathfrak{q}$ is ample on $C_\mathfrak{q}$ for $\mathfrak{q} \in D(h)$. In particular, if we can show that the bundle is ample in characteristic zero, then it is ample for almost every prime number. This provides a way to establish tight closure properties in positive characteristic by working in characteristic zero.

The ampleness of the syzygy bundle (not its dual) also has consequences on Frobenius closure and tight closure.

PROPOSITION 4.17. *Suppose the two-dimensional homogeneous situation in positive characteristic. If* $\operatorname{Syz}(f_1,\dots,f_n)(m)$ *is ample, then* $f \in (f_1,\dots,f_n)^F$ *and so* $f \in (f_1,\dots,f_n)^*$.

PROOF. In general, let $\mathscr{G}$ be ample on a curve C. The cohomological characterization of ampleness [**72**, 6.1.9] says that for every coherent sheaf $\mathscr{H}$ we have $H^1(C, S^k(\mathscr{G}) \otimes \mathscr{H}) = 0$ for k big enough. This is equivalent to the notion of being p-cohomologically ample in the sense that $H^1(C, F^{p^k*}(\mathscr{G}) \otimes \mathscr{H}) = 0$ for k big enough (in higher dimension these two notions are not equivalent; see [**61**], [**74**] and [**1**]). So in particular for $c \in H^1(C,\mathscr{G})$ we conclude that $c^{p^k} \in H^1(C, F^{p^k*}(\mathscr{G}))$ must be zero for $k \gg 0$. Applying this to the cohomology class $c = \delta(f) \in H^1(C, \operatorname{Syz}(f_1,\dots,f_n)(m))$, $m = \deg(f)$, yields the result. □

EXAMPLE 4.18. Let $R = K[X_0,\dots,X_d]$ denote the polynomial ring and consider the elements $X_0,\dots,X_d$. Then the presenting sequence for the sheaf of syzygies on the projective space is just

$$0 \longrightarrow \operatorname{Syz}(X_0,\dots,X_d)(m) \longrightarrow \bigoplus_{d+1} \mathscr{O}_{\mathbb{P}^d}(m-1) \overset{X_0,\dots,X_d}{\longrightarrow} \mathscr{O}_{\mathbb{P}^d}(m) \longrightarrow 0.$$

For $m = 0$, this is merely the Euler-sequence for the cotangent sheaf; see [**45**, Theorem II.8.13]. Hence $\operatorname{Syz}(X_0,\dots,X_d) = \Omega_{\mathbb{P}^d}$ and its dual is $\mathscr{F} = \mathscr{T}_{\mathbb{P}^d}$, the (ample) tangent sheaf on $\mathbb{P}^d$.

REMARK 4.19. The syzygy bundle $\operatorname{Syz}(f_1,\dots,f_n)(m)$ can only be ample for $m > (\sum_{i=1}^n d_i)/(n-1)$. The 'big picture' is that $\operatorname{Syz}(f_1,\dots,f_n)(m)$ is ample for m large enough and that $(\operatorname{Syz}(f_1,\dots,f_n)(m))^\vee$ is ample for m small enough, and the degree bound gives a rough idea as to where the behavior changes. The next chapter deals with the notion of strong semistability. Under this condition the degree bound is an exact bound.

CHAPTER 5

Semistable bundles and the Harder-Narasimhan filtration

In this chapter we discuss the notion of the slope of a vector bundle and various related concepts like semistable and strongly semistable sheaves and the Harder-Narasimhan filtration which we will need in the sequel. Our main references are [**40**], [**57**], [**72**, Chapter 6.4], [**76**], and [**91**]. Recall from the last chapter the definition of the degree of a vector bundle on a smooth projective curve over an algebraically closed field K. With the degree one defines the slope of a vector bundle.

DEFINITION 5.1. The *slope* of a locally free sheaf $\mathscr{S}$ on a smooth projective curve is defined by $\mu(\mathscr{S}) = \deg(\mathscr{S})/\operatorname{rk}(\mathscr{S})$.

For two locally free sheaves $\mathscr{S}_1$ and $\mathscr{S}_2$ we have that $\mu(\mathscr{S}_1 \otimes \mathscr{S}_2) = \mu(\mathscr{S}_1) + \mu(\mathscr{S}_2)$.

DEFINITION 5.2. A locally free sheaf $\mathscr{S}$ is called *semistable* if for every locally free subsheaf $\mathscr{T} \subseteq \mathscr{S}$ the inequality $\mu(\mathscr{T}) \leq \mu(\mathscr{S})$ holds (and it is called *stable* if the strict inequality holds for $\mathscr{T} \subset \mathscr{S}$).

It is sufficient in this definition to look only at subbundles $\mathscr{T} \subseteq \mathscr{S}$, i.e., locally free subsheaves such that the quotient $\mathscr{S}/\mathscr{T}$ is also locally free. This notion is also equivalent to the property that for every locally free quotient sheaf of positive rank $\mathscr{S} \to \mathscr{Q} \to 0$ the inequality $\mu(\mathscr{Q}) \geq \mu(\mathscr{S})$ holds.

EXAMPLE 5.3. Every line bundle is semistable. The direct sum $\mathscr{O}(k) \oplus \mathscr{O}(n)$ is not semistable for $k \neq n$: if $k > n$, then $\mathscr{O}(k) \oplus 0 \subset \mathscr{O}(k) \oplus \mathscr{O}(n)$ is a subbundle of degree=slope= $k \deg \mathscr{O}(1)$, which is larger than $\mu(\mathscr{O}(k) \oplus \mathscr{O}(n)) = ((k+n)\deg \mathscr{O}(1))/2$. For $k = n$ it is semistable (but not stable).

REMARK 5.4. The concept of slope and of semistable bundles occurred first in the work of Mumford on the construction of moduli spaces for vector bundles of given rank and degree [**82**]. The question here is: does there exist a variety such that its closed points are in bijective correspondence with isomorphism classes of vector bundles of fixed rank r and fixed degree d? The answer is no in this generality, because such a space can only exist for a family of bundles that has a common bound for the number of its sections for every member. Here the members in the family $\mathscr{O}(k) \oplus \mathscr{O}(-k)$, $k \in \mathbb{N}$, all have degree 0, but for k large enough this bundle has arbitrarily many global sections. In short, the semistable bundles are such that you can deduce from the degree and the rank universal bounds for the number of global sections.

There are a lot of different angles from which one may look at semistable bundles. Beside the algebraic-geometric viewpoint there exist characterizations in terms of representation of the fundamental group (see [**85**]) and in terms of differential geometry, the so-called Kobayashi-Hitchin correspondence; see [**3**], [**24**], [**102**]. The semistability of bundles has not found much attention in commutative algebra, though the Castelnuovo-Mumford regularity has, which arose in the same context and was first an instrument to prove that semistable bundles form a bounded family.

EXAMPLE 5.5. Consider the syzygy bundle $\mathrm{Syz}(x^2,y^2,z^2)$ on the Fermat cubic curve given by $x^3+y^3+z^3=0$. Then this curve equation gives immediately the global syzygy (x,y,z) for x^2,y^2,z^2, which is a global section of $\mathrm{Syz}(x^2,y^2,z^2)(3)$ of total degree 3. This bundle has degree 0 and this section has no zeros. This yields the short exact sequence

$$0 \longrightarrow \mathcal{O}_C \longrightarrow \mathrm{Syz}(x^2,y^2,z^2)(3) \longrightarrow \mathcal{O}_C \longrightarrow 0\,.$$

Here the quotient must also be locally free, and looking at the determinant it must be the structure sheaf. It follows that the bundle is semistable (though not stable): if $\mathcal{L}$ is an invertible sheaf of positive degree and $\mathcal{L} \to \mathrm{Syz}(x^2,y^2,z^2)(3)$ is a sheaf morphism, then the composite morphism to $\mathcal{O}_C$ on the right must be the zero mapping. Then the mapping must factor through $\mathcal{O}_C$ on the left, and it is again zero. In fact this bundle is F_2 in the classification of vector bundles on elliptic curves due to Atiyah [**2**].

PROPOSITION 5.6. *Let C be a smooth projective curve. Then the following hold:*

(i) *If $\mathcal{S}$ is a semistable bundle, then so is $\mathcal{S} \otimes \mathcal{L}$ for every invertible sheaf $\mathcal{L}$.*
(ii) *If $\mathcal{S}$ is semistable, then so is its dual $\mathcal{S}^\vee$.*
(iii) *If $\mathcal{S}$ is semistable of negative degree, then $\Gamma(C,\mathcal{S})=0$.*

PROOF. We only look at the third part. Suppose that there exists a non-trivial sheaf morphism $\mathcal{O}_C \to \mathcal{S}$. This gives a subsheaf $\mathcal{O}_C \subseteq \mathcal{S}$ (not a subbundle in general), which contradicts the semistability of $\mathcal{S}$. □

THEOREM 5.7. *Every locally free sheaf $\mathcal{S}$ has a unique* Harder-Narasimhan filtration. *This is a filtration of locally free subbundles*

$$0 = \mathcal{S}_0 \subset \mathcal{S}_1 \subset \ldots \subset \mathcal{S}_t = \mathcal{S}$$

such that $\mathcal{S}_i/\mathcal{S}_{i-1}$ is semistable for every $i = 1,\ldots,t$ and the slopes $\mu_i = \mu(\mathcal{S}_i/\mathcal{S}_{i-1})$ of these semistable quotients form a decreasing chain $\mu_1 > \ldots > \mu_t$.

PROOF. See [**72**, Proposition 6.4.7] or [**40**, Lemma 1.3.7] for the curve case or [**57**, Theorem 1.3.4] in general (in higher dimension the sheaves are only torsion-free). □

DEFINITION 5.8. $\mathcal{S}_1$ is called the *maximal destabilizing subsheaf* and $\mathcal{S}/\mathcal{S}_{t-1}$ is called the *minimal destabilizing quotient*. The number $\mu_{\max}(\mathcal{S}) = \mu_1(\mathcal{S})$ is called the *maximal slope* and $\mu_{\min}(\mathcal{S}) = \mu_t = \mu(\mathcal{S}/\mathcal{S}_{t-1})$ is called the *minimal slope*.

REMARK 5.9. The minimal slope is the same as $\mu_{\min}(\mathcal{S}) = \min\{\mu(\mathcal{Q}) : \mathcal{S} \to \mathcal{Q} \to 0\}$. For the dual sheaf we have $\mu_{\max}(\mathcal{S}^\vee) = -\mu_{\min}(\mathcal{S})$. If $\mu_{\min}(\mathcal{S}) > \mu_{\max}(\mathcal{F})$, then $\mathrm{Hom}(\mathcal{S},\mathcal{F}) = 0$. In particular, if $\mu_{\max}(\mathcal{F}) < 0$, then $\Gamma(C,\mathcal{F}) = \mathrm{Hom}(\mathcal{O}_C,\mathcal{F}) = 0$. The sheaf $\mathcal{S}$ is semistable if and only if $\mu(\mathcal{S}) = \mu_{\min}(\mathcal{S}) = \mu_{\max}(\mathcal{S})$.

We can rephrase the statement that a bundle in characteristic 0 is ample if and only if every quotient has positive degree (Proposition 4.10(iv)).

THEOREM 5.10. (*Hartshorne*) *Let C denote a smooth projective curve over an algebraically closed field of characteristic* 0. *Then a vector bundle $\mathcal{S}$ is ample if and only if $\mu_{\min}(\mathcal{S}) > 0$. In particular, a semistable vector bundle is ample if and only if its degree is positive.*

PROOF. This is a restatement of Proposition 4.10(iv); see also [**72**, Theorem 6.4.15]. □

In positive characteristic we have to deal first of all with the problem that the Frobenius pull-back may destabilize semistable vector bundles. Recall that the *absolute Frobenius morphism* $F : X \to X$ is the identity on the points of X and on every ring of global sections it is the Frobenius homomorphism. The absolute Frobenius morphism of a K-variety is in general not a K-morphism. The absolute Frobenius on a smooth projective curve has degree p [**45**, Example IV.2.4.3]. The Frobenius pull-back of an invertible sheaf $\mathscr{L}$ is $\mathscr{L}^p$ [**45**, Proposition II.6.9]. To get the degree (and the slope) of the Frobenius pull-back of a bundle we have to multiply by p. The pull-back of a semistable bundle under a separable morphism is again semistable [**76**, Proposition 3.2], but this is not true for the Frobenius.

In the case of a syzygy bundle on a projective variety Y, the Frobenius pull-back of it is easy to describe. If

$$0 \longrightarrow \mathrm{Syz}(f_1,\dots,f_n) \longrightarrow \bigoplus_{i=1}^{n} \mathscr{O}_Y(-d_i) \stackrel{f_1,\dots,f_n}{\longrightarrow} \mathscr{O}_Y \longrightarrow 0$$

is the presenting sequence, then pulling this back by the e-th Frobenius (which is exact, since we are dealing with locally free sheaves) gives ($q = p^e$):

$$0 \longrightarrow \mathrm{Syz}(f_1^q,\dots,f_n^q) \cong F^{e*}(\mathrm{Syz}(f_1,\dots,f_n)) \longrightarrow \bigoplus_{i=1}^{n} \mathscr{O}_Y(-qd_i) \stackrel{f_1^q,\dots,f_n^q}{\longrightarrow} \mathscr{O}_Y \longrightarrow 0\,.$$

The pull-back of the cohomology class $\delta(f) \in H^1(Y,\mathrm{Syz}(f_1,\dots,f_n)(m))$ induced by $f \in \Gamma(Y,\mathscr{O}_Y(m))$ is the cohomology class $\delta(f^q) \in H^1(Y,\mathrm{Syz}(f_1^q,\dots,f_n^q)(qm))$.

EXAMPLE 5.11. Consider the restricted cotangent bundle $\mathrm{Syz}(x,y,z)$ (see Example 4.18) on the smooth projective curve C given by the equation $x^4+y^4+z^4=0$. Then for characteristic $p=3$ we look at the first Frobenius pull-back, which is $\mathrm{Syz}(x^3,y^3,z^3)$. The tuple (x,y,z) gives a global section of $\mathrm{Syz}(x^3,y^3,z^3)(4)$ of total degree 4. This bundle has degree $(2\cdot 4-3\cdot 3)\deg(\mathscr{O}_C(1))$, which is negative, so it cannot be semistable (by Proposition 5.6(iii)). This syzygy does not have a common zero, hence the Harder-Narasimhan filtration is $0 \subset \mathscr{O}_C \subset \mathrm{Syz}(x^3,y^3,z^3)(4)$. See also Example 9.8.

The original bundle $\mathrm{Syz}(x,y,z)$ is, however, semistable. Otherwise there would exist a maximal destabilizing invertible subsheaf $\mathscr{L} \subset \mathrm{Syz}(x,y,z)(0)$, and its pull-back, which is $\mathscr{L}^3$, would equal the maximal destabilizing subsheaf of $\mathrm{Syz}(x^3,y^3,z^3)(0)$ (because of the uniqueness of Harder-Narasimhan filtration). So, after the twist with $\mathscr{O}_C(4)$ we must have $\mathscr{L}^3 \otimes \mathscr{O}_C(4) \cong \mathscr{O}_C \subset \mathrm{Syz}(x^3,y^3,z^3)(4)$. But then the degree of $\mathscr{L}$ would be $-16/3$, which is not possible.

DEFINITION 5.12. A locally free sheaf $\mathscr{S}$ on C is called *strongly semistable* if the pull-back of $\mathscr{S}$ under every iteration of the absolute Frobenius morphism $F : C \to C$ is again semistable.

REMARK 5.13. On the projective line every bundle is the decomposition of invertible sheaves, $\mathscr{S} \cong \mathscr{O}(n_1) \oplus \ldots \oplus \mathscr{O}(n_r)$, by the theorem of Grothendieck. This is semistable only if $n_1 = \ldots = n_r$ (and stable only for $r=1$) and then also strongly semistable.

On an elliptic curve every *indecomposable* sheaf is semistable, and every semistable bundle on an elliptic curve is strongly semistable. This follows from the classification of vector bundles on an elliptic curve due to Atiyah; see [**2**], [**101**]. This is a property which makes the understanding and computation of tight closure over an elliptic curve much easier than in the case of higher genus (see also Theorem 10.1).

DEFINITION 5.14. Let C denote a smooth projective curve over an algebraically closed field of characteristic $p > 0$ and let $\mathscr{S}$ denote a locally free sheaf on C. Then we define

$$\bar{\mu}_{\max}(\mathscr{S}) = \sup_e \{ \frac{\mu_{\max}(F^{e*}\mathscr{S})}{p^e} \}$$

and

$$\bar{\mu}_{\min}(\mathscr{S}) = \inf_e \{ \frac{\mu_{\min}(F^{e*}\mathscr{S})}{p^e} \}.$$

REMARK 5.15. When X is a normal projective variety, then one may define the degree of a locally free (or even torsion-free) sheaf $\mathscr{S}$ with respect to a fixed ample divisor H by $\deg(\mathscr{S}) = \det(\mathscr{S}).H^{\dim(X)-1}$. Then one can define the slope as before and also the notion of semistability and Harder-Narasimhan filtration. We will touch upon this several times, as in the following theorem.

THEOREM 5.16. (*Langer*) *Let $\mathscr{S}$ denote a locally free sheaf on a normal projective variety over an algebraically closed field of positive characteristic p. There exists a number e such that the Harder-Narasimhan filtration of $F^{e*}(\mathscr{S})$ is strong, which means that every semistable quotient $\mathscr{S}_i/\mathscr{S}_{i-1}$ in the filtration is strongly semistable.*

PROOF. See [**69**, Theorem 2.7]. □

In dimension one another proof is given by Trivedi in [**100**, Remark 1.9]. If the Harder-Narasimhan filtration of $\mathscr{S}$ is strong, then its pull-back is the Harder-Narasimhan filtration of $F^*(\mathscr{S})$. Moreover, if $F^{e*}(\mathscr{S})$ has strong Harder-Narasimhan filtration, then $\bar{\mu}_{\max}(\mathscr{S}) = \mu_{\max}(F^{e*}(\mathscr{S}))/p^e$. We can give a numerical ampleness criterion in positive characteristic for bundles on curves.

THEOREM 5.17. *Let C denote a smooth projective curve over an algebraically closed field of positive characteristic p. Then a vector bundle $\mathscr{G}$ is ample if and only if $\bar{\mu}_{\min}(\mathscr{G}) > 0$. This is true if and only if every quotient bundle of $F^{e*}(\mathscr{G})$ has positive degree for every e.*

PROOF. The first equivalence is due to Barton; see [**4**] or [**6**, Theorem 4.2.4]. The second follows from Theorem 5.16 (without Theorem 5.16 it is not clear whether the limit of the positive minimal slopes is positive). □

COROLLARY 5.18. *Let $\mathscr{S}$ be a locally free sheaf of rank r and let $c \in H^1(C,\mathscr{S})$ be given with corresponding dual extension $0 \to \mathscr{O}_C \to \mathscr{F}' \to \mathscr{F} \to 0$ ($\mathscr{F} = \mathscr{S}^\vee$). Suppose that $\mathscr{F}$ is ample. If $0 \neq F^{e*}(c) \in H^1(C, F^{e*}(\mathscr{S}))$ for all e, then $\mathscr{F}'$ is also ample.*

PROOF. By assumption, all Frobenius pull-backs $0 \to \mathscr{O}_C \to F^{e*}(\mathscr{F}') \to F^{e*}(\mathscr{F}) \to 0$ of the dual extension are non-trivial. Therefore by Proposition 4.12(i) every quotient bundle of $F^{e*}(\mathscr{F}')$ has positive degree, and so by Theorem 5.17 $\mathscr{F}'$ is ample. □

It is now rather straightforward to establish slope criteria for tight closure via characteristic zero that hold in the relative situation for $p \gg 0$. But to get results for fixed prime numbers and to bound the Frobenius powers in Theorem 5.17 and Corollary 5.18 we need to do more work.

LEMMA 5.19. *Let C denote a smooth projective curve of genus g over an algebraically closed field K. Suppose that $\mathscr{F}$ is a vector bundle on C and that $\mu_{\min}(\mathscr{F}) > g$. Then $\mathscr{F}$ is ample.*

PROOF. We do induction on the rank; the case $r = 1$ is clear. We know that every quotient bundle $\mathscr{F} \to \mathscr{G} \to 0$ has the property that $\mu_{\min}(\mathscr{G}) \geq \mu_{\min}(\mathscr{F}) > g$. Therefore by the induction hypotheses every quotient $\neq \mathscr{F}$ of $\mathscr{F}$ is ample. Since $\deg(\mathscr{F}) > g\,\mathrm{rk}(\mathscr{F})$, it follows from Proposition 4.10(iii) that $\mathscr{F}$ is ample. □

LEMMA 5.20. *Let $\mathscr{F}$ be a rank r vector bundle on C. Let $0 \to \mathscr{O}_C \to \mathscr{F}' \to \mathscr{F} \to 0$ be a non-trivial (dual) extension and suppose that $\mu_{\min}(\mathscr{F}) > g(r+1)$ (in particular $\mathscr{F}$ is ample by Lemma 5.19). Then $\mathscr{F}'$ is ample.*

PROOF. We will do induction on the rank of $\mathscr{F}$; the case $r = 1$ will also be treated in the induction step. We will show that all quotient bundles of $\mathscr{F}'$ are ample, so by Proposition 4.10(iii) $\mathscr{F}'$ will be ample, since $\deg(\mathscr{F}') = \deg(\mathscr{F}) = r\mu(\mathscr{F}) > gr(r+1)$. Let $\mathscr{F}' \to \mathscr{Q} \to 0$ be a quotient sheaf. If the induced mapping $\mathscr{O}_C \to \mathscr{Q}$ is the zero mapping, then $\mathscr{Q}$ is a quotient of $\mathscr{F}$ and so it is ample. So suppose that the induced map is not zero. Then $\mathscr{O}_C \to \mathscr{Q}$ factors through an invertible sheaf $\mathscr{L}$ of non-negative degree which is a subbundle of $\mathscr{Q}$ and we get a diagram

$$\begin{array}{ccccccccc} 0 & \longrightarrow & \mathscr{O}_C & \longrightarrow & \mathscr{F}' & \longrightarrow & \mathscr{F} & \longrightarrow & 0 \\ \downarrow & & \downarrow & & \downarrow & & \downarrow & & \downarrow \\ 0 & \longrightarrow & \mathscr{L} & \longrightarrow & \mathscr{Q} & \longrightarrow & \mathscr{Q}/\mathscr{L} & \longrightarrow & 0. \end{array}$$

Here the rows are exact, $\mathscr{F} \to \mathscr{Q}/\mathscr{L}$ is surjective, and $\mathscr{Q}/\mathscr{L}$ is locally free. If $\mathscr{Q}$ has rank one, then $\mathscr{Q} = \mathscr{L}$, and $\mathscr{L}$ cannot be $\mathscr{O}_C$, because the given sequence (the first row) is not trivial. Therefore $\mathscr{L}$ has positive degree and is ample. Suppose now that $\mathscr{Q}$ has rank ≥ 2. Then $\mathscr{Q}/\mathscr{L}$ is ample as a quotient of $\mathscr{F}$ and $\mu_{\min}(\mathscr{Q}/\mathscr{L}) \geq \mu_{\min}(\mathscr{F}) > g(r+1) \geq g\,\mathrm{rk}(\mathscr{Q})$. If $\mathscr{L}$ has positive degree, then $\mathscr{Q}$ is ample as an extension of two ample bundles by Proposition 4.9(ii). If $\mathscr{L}$ has degree zero, then it is trivial, since it has a section. Then also the extension in the second row does not split, so we know by the induction hypothesis that $\mathscr{Q}$ is ample. □

LEMMA 5.21. *Let $\mathscr{F}$ denote a locally free sheaf of rank r on a smooth projective curve C of genus g over an algebraically closed field K of positive characteristic p. Suppose that $p \geq 4(g-1)r^3$. Then*

$$\frac{\mu_{\min}(F^*(\mathscr{F}))}{p} \geq \mu_{\min}(\mathscr{F}) - \frac{(r-1)(2g-2)}{p}.$$

PROOF. By a theorem of Shepherd-Barron [**92**, Corollary 2^p] we have for a semistable bundle $\mathscr{F}$ the estimate

$$\mu_{\max}(F^*(\mathscr{F})) - \mu_{\min}(F^*(\mathscr{F})) \leq (r-1)(2g-2)$$

and in particular $\mu(F^*(\mathscr{F})) - \mu_{\min}(F^*(\mathscr{F})) \leq (r-1)(2g-2)$. In general, by a result of Trivedi (under the condition for p) [**100**, Lemma 1.6], the Harder-Narasimhan filtration of the pull-back is a refinement of the pull-back of the Harder-Narasimhan filtration. In particular, the minimal destabilizing quotient of $F^*(\mathscr{F})$ is a quotient of the pull-back of the (semistable) minimal destabilizing quotient of $\mathscr{F}$, that is $\mathscr{F}/\mathscr{F}_{t-1}$. Hence

$$\begin{aligned} p\mu_{\min}(\mathscr{F}) - \mu_{\min}(F^*(\mathscr{F})) &= p\mu(\mathscr{F}/\mathscr{F}_{t-1}) - \mu_{\min}(F^*(\mathscr{F})) \\ &= \mu(F^*(\mathscr{F}/\mathscr{F}_{t-1})) - \mu_{\min}(F^*(\mathscr{F}/\mathscr{F}_{t-1})) \\ &\leq (r-1)(2g-2) \end{aligned}$$

by the aforementioned result of Shepherd-Barron applied to $\mathscr{F}/\mathscr{F}_{t-1}$. Therefore

$$\frac{\mu_{\min}(F^*(\mathscr{F}))}{p} \geq \mu_{\min}(\mathscr{F}) - \frac{(r-1)(2g-2)}{p}.$$

□

PROPOSITION 5.22. *Let $\mathscr{S}$ be a locally free sheaf of rank r and let $c \in H^1(C,\mathscr{S})$ be given with corresponding dual extension $0 \to \mathscr{O}_C \to \mathscr{F}' \to \mathscr{F} \to 0$. Suppose that $\mathscr{F}$ is ample and that $p > 4(g-1)r^3$ and $p \geq 3gr^2$. If $0 \neq c^p \in H^1(C,F^*(\mathscr{S}))$, then $\mathscr{F}'$ is ample.*

PROOF. Since $\mathscr{F}$ is ample, its minimal slope is at least $1/r$. By Lemma 5.21 we have

$$\begin{aligned} \mu_{\min}(F^*(\mathscr{F})) &\geq p\mu_{\min}(\mathscr{F}) - (r-1)(2g-2) \\ &\geq p/r - (r-1)(2g-2) \\ &\geq 3gr - (r-1)(2g-2) \\ &\geq gr + 2g > g(r+1). \end{aligned}$$

Since $c^p \neq 0$ the pulled-back dual extension $0 \to \mathscr{O}_C \to F^*(\mathscr{F}') \to F^*(\mathscr{F}) \to 0$ is non-trivial. Since $\mu_{\min}(F^*(\mathscr{F})) > g(r+1)$ we can apply Lemma 5.20 and deduce that $F^*(\mathscr{F}')$ and so $\mathscr{F}'$ is ample. □

We come to our main numerical exclusion result for tight closure in terms of slopes.

THEOREM 5.23. *Suppose the homogeneous situation (Situation 3.7) in dimension two. If $\bar{\mu}_{\max}(\mathrm{Syz}(f_1,\ldots,f_n)(m)) < 0$ and $f \notin (f_1,\ldots,f_n)$, then also $f \notin (f_1,\ldots,f_n)^\star$. This holds in characteristic 0 or $p \gg 0$. It holds in fixed characteristic $p > 4(g-1)(n-1)^3$ under the condition that $f^p \notin (f_1^p,\ldots,f_n^p)$.*

PROOF. Let $\mathscr{F}$ be the dual of $\mathrm{Syz}(f_1,\ldots,f_n)(m)$. Then $\bar{\mu}_{\min}(\mathscr{F}) > 0$ and so it is ample by Theorem 5.17. If $f \notin (f_1,\ldots,f_n)$, then the corresponding cohomology class $\delta(f)$ is not trivial. Therefore we have a non-trivial dual extension $0 \to \mathscr{O}_C \to \mathscr{F}' \to \mathscr{F} \to 0$, and by the theorem of Hartshorne-Gieseker (Proposition 4.12) the extension $\mathscr{F}'$ is also ample in characteristic zero. Then the corresponding torsor $B = \mathbb{P}(\mathscr{F}') - \mathbb{P}(\mathscr{F})$ is the complement of an ample divisor, hence affine, and so $f \notin (f_1,\ldots,f_n)^\star$ (Proposition 3.9, Remark 3.10, Remark 4.7, Definition 4.8).

In the relative situation, the statement for $p \gg 0$ follows from the characteristic zero result and from the openness of ampleness (Remark 4.16).

Now we look at a fixed characteristic $p > 4(g-1)(n-1)^3$. But then our assumptions ensure that we can apply Proposition 5.22 to $\mathscr{S} = \mathrm{Syz}(f_1,\ldots,f_n)(m)$ of rank $r = n-1$ and $c = \delta(f)$. Hence the dual extension is ample, and $f \notin (f_1,\ldots,f_n)^*$. □

COROLLARY 5.24. *Suppose the relative homogeneous situation in dimension two and suppose that the syzygy bundle is semistable in characteristic zero. Then for $m < (d_1+\ldots+d_n)/(n-1)$ we have $(f_1,\ldots,f_n)^\star \cap R_m \subseteq (f_1,\ldots,f_n)$ in characteristic 0 or $p \gg 0$.*

PROOF. Under the semistability condition we have

$$\mu_{\max}(\mathrm{Syz}(f_1,\ldots,f_n)(m)) = \mu(\mathrm{Syz}(f_1,\ldots,f_n)(m)) = (m - \frac{\sum_{i=1}^n d_i}{n-1})\deg(C) < 0.$$

If f has degree m and does not belong to the ideal, then $f \notin (f_1,\ldots,f_n)^\star$ by Theorem 5.23. □

CHAPTER 6

Numerical inclusion criteria for tight closure

We are coming now to inclusion criteria for tight closure in terms of (minimal) slopes of the corresponding syzygy bundle. We restrict ourselves to the two-dimensional homogeneous situation, but see also Remark 6.10 for the higher-dimensional case.

LEMMA 6.1. *Let C denote a smooth algebraic curve over an algebraically closed field of positive characteristic p, let $\mathscr{S}$ be a locally free sheaf on C, and $c \in H^1(C, \mathscr{S})$. Suppose that $\bar{\mu}_{\min}(\mathscr{S}) \geq 0$. Then c belongs to the tight closure of $0 \subseteq H^1(C, \mathscr{S})$, in the sense that there exists t ($\neq 0$) in some (ample) invertible sheaf $\mathscr{L}$ such that $tF^{e*}(c) = 0$ in $H^1(C, \mathscr{L} \otimes F^{e*}(\mathscr{S}))$ for all $q = p^e$.*

PROOF. Take any invertible sheaf $\mathscr{L}$ of degree $> 2g-2$ (g being the genus of the curve) with a global section $0 \neq t \in \Gamma(C, \mathscr{L})$. For the dual of $\mathscr{S}$ we have $\bar{\mu}_{\max}(\mathscr{S}^\vee) \leq 0$. We will show that $H^1(C, \mathscr{L} \otimes F^{e*}(\mathscr{S})) = 0$. This is by Serre duality [**45**, Corollary II.7.7], dual to $H^0(C, \mathscr{L}^\vee \otimes \omega_C \otimes F^{e*}(\mathscr{S}^\vee))$. Since the canonical bundle ω_C has degree $2g-2$ [**45**, Example IV.3.3], it follows that $\mathscr{L}^\vee \otimes \omega_C$ has negative degree. On the other hand, $\mu_{\max}(F^{e*}(\mathscr{S}^\vee)) \leq \bar{\mu}_{\max}(F^{e*}(\mathscr{S}^\vee)) = q\bar{\mu}_{\max}(\mathscr{S}^\vee) \leq 0$. Hence $\mathscr{L}^\vee \otimes \omega_C \otimes F^{e*}(\mathscr{S}^\vee)$ has negative maximal slope and does not have sections $\neq 0$ by Remark 5.9. □

REMARK 6.2. A variant of Lemma 6.1 that is independent of the characteristic says the following: if $c \in H^1(C, \mathscr{S})$ and $\bar{\mu}_{\min}(\mathscr{S}) \geq 0$, then the corresponding torsor B cannot be affine [**6**, Theorem 4.4.4]. This is proved by showing that under this numerical condition the divisor $\mathbb{P}(V) \subset \mathbb{P}(V')$ is not big. A divisor (or an invertible sheaf) is called *big* [**72**, Definition 2.2.1] if its *Iitaka dimension* is maximal, i.e., if the order of the growth of the global sections of its powers equals the dimension. The complement of an open affine subset is a big divisor, because the transcendence degree of the ring of global functions on an open affine subset equals the dimension. So we have the following implications for an effective divisor D on a projective variety Y: D ample $\Rightarrow$ $Y - D$ affine $\Rightarrow$ D big. These implications are not reversible; see Examples 7.3 and 7.4 below.

We deduce the following numerical inclusion result for tight closure.

COROLLARY 6.3. *Suppose the homogeneous situation in dimension two. If the syzygy bundle fulfills $\bar{\mu}_{\min}(\mathrm{Syz}(f_1, \ldots, f_n)(m)) \geq 0$, then we have $R_m \subseteq (f_1, \ldots, f_n)^*$.*

PROOF. We work in positive characteristic (in characteristic zero this is also true for solid closure by using Remark 6.2). We apply Lemma 6.1 with $\mathscr{S} = \mathrm{Syz}(f_1, \ldots, f_n)(m)$ and $c = \delta(f) \in H^1(C, \mathrm{Syz}(f_1, \ldots, f_n)(m))$, where f is a homogeneous element of degree m. We can take $\mathscr{L} = \mathscr{O}_C(k)$, k big enough, and take $0 \neq t \in \Gamma(C, \mathscr{O}_C(k)) = R_k$. Then $tF^{e*}(c) = 0$ for all e. This is then also true for the corresponding cohomology class $c \in H^1(D(R_+), \mathrm{Syz}(f_1, \ldots, f_n))$. By the exactness of the long cohomology sequence and the normality of R, this says that $tf^q \in (f_1^q, \ldots, f_n^q) = I^{[q]}$ for all q. Hence $f \in (f_1, \ldots, f_n)^*$. □

The results of Corollary 5.24 and Corollary 6.3 together yield an exact degree bound for tight closure in the strongly semistable case.

THEOREM 6.4. *Suppose the homogeneous situation in dimension two. If the syzygy bundle* $\mathrm{Syz}(f_1,\ldots,f_n)$ *is strongly semistable, then the following hold.*

(i) *For* $m \geq \frac{d_1+\ldots+d_n}{n-1}$ *we have* $R_m \subseteq (f_1,\ldots,f_n)^\star$.

(ii) *For* $m < \frac{d_1+\ldots+d_n}{n-1}$ *and* $\deg(f) = m$ *we have* $f \in (f_1,\ldots,f_n)^\star$ *if and only if* $f \in (f_1,\ldots,f_n)$ *already (in characteristic zero) or* $f^p \in (f_1^p,\ldots,f_n^p)$ *(in characteristic* p*) for* $p > 4(g-1)(n-1)^3$ *(or* $f^q \in (f_1^q,\ldots,f_n^q)$ *for a certain* $q = p^e$ *if* p *is too small).*

Therefore the number $(d_1+\ldots+d_n)/(n-1)$ *is in the strongly semistable case the bound which separates tight closure from non-tight closure.*

PROOF. This is a restatement of Corollary 5.24 and Corollary 6.3. In the case $p \leq 4(g-1)(n-1)^3$ the power $q = p^e$ for which we have to check the containment $f^q \in I^{[q]}$ can be computed with Lemma 5.20, because in the strongly semistable case $\mu_{\min}$ behaves fine by pulling the bundle back. □

EXAMPLE 6.5. We come back to Example 1.10 where we have asked whether $xyz \in (x^2,y^2,z^2)^\star$ on $x^3+y^3+z^3=0$ holds. We know by Example 5.5 that $\mathrm{Syz}(x^2,y^2,z^2)$ is semistable on the elliptic curve C and that it is therefore strongly semistable by Remark 5.13. It follows from Theorem 6.4 that $R_{\geq 3} \subseteq (x^2,y^2,z^2)^\star$ and so in particular $xyz \in (x^2,y^2,z^2)^\star$. We also can easily see in positive characteristic that it must belong to the plus closure by looking at the short exact sequence

$$0 \longrightarrow \mathcal{O}_C \longrightarrow \mathrm{Syz}(x^2,y^2,z^2)(3) \longrightarrow \mathcal{O}_C \longrightarrow 0\,.$$

Look at the cohomology class defined by xyz, that is $c = \delta(xyz) \in H^1(C, \mathrm{Syz}(x^2,y^2,z^2)(3))$. Let c' denote its image in $H^1(C,\mathcal{O}_C)$. Then we are in the parameter situation where we know (by the explicit computation in Example 1.9, or by the graded version of the parameter theorem of Smith (1.7), or by Lemma 8.5 below), that there exists a finite morphism $C' \to C$ such that the pull-back of c' is zero. Then looking at the whole situation on C', our class maps to 0 on the right, and therefore it must come from a class $c'' \in H^1(C',\mathcal{O}_{C'})$ from the left. Then we are again in the parameter case, and we find another finite morphism $C'' \to C'$ which annihilates c''. Then $C'' \to C$ annihilates c, so $xyz \in (x^2,y^2,z^2)^+$ by Proposition 3.12.

For genus ≥ 2 it is rather difficult to decide whether a bundle is strongly semistable. But often the semistability in characteristic 0 is sufficient to draw conclusions on tight closure: if we are in the relative situation (as described in Remark 4.16) and the bundle $\mathcal{S}_0$ (characteristic zero) is semistable, then $\mathcal{S}_0(m)$ is ample for $m > (d_1+\ldots+d_n)/(n-1)$ and $(\mathcal{S}_0(m))^\vee$ is ample for $m < (d_1+\ldots+d_n)/(n-1)$. This is then also true for $\mathcal{S}_p$ in positive characteristic $p \gg 0$, and one may deduce tight closure results. If $(d_1+\ldots+d_n)/(n-1)$ is not an integer, then we can deduce results for every degree, otherwise for all degrees with this only exception.

A way to establish semistable bundles on curves in characteristic zero is by using restriction theorems. We describe in the simplest case how this works. Suppose that $S = K[X,Y,Z]$ is a polynomial ring and $R = K[X,Y,Z]/(F)$, F homogeneous. Let $f_1,\ldots,f_n \in K[X,Y,Z]$ be homogeneous and suppose that they are S_+-primary. Then $\mathrm{Syz}(f_1,\ldots,f_n)$ is a vector bundle on the projective plane $\mathbb{P}^2$, and we are interested in the restriction of this bundle to the plane curve $V_+(F) \subset \mathbb{P}^2$. Suppose that the bundle is semistable on $\mathbb{P}^2$.

Then there exist restriction theorems which say that under certain conditions on F the restriction is also semistable (see [**15**]). For example, the theorem of Flenner [**26**] says in characteristic 0 for a bundle $\mathscr{S}$ of rank r and for $\deg(F) \geq \min\{(r^2-3)/2, 2\}$ and F generic (this means that it is true if the coefficients of F are inside a non-empty Zariski-open subset) that the restriction of $\mathscr{S}$ to $V_+(F)$ is also semistable.

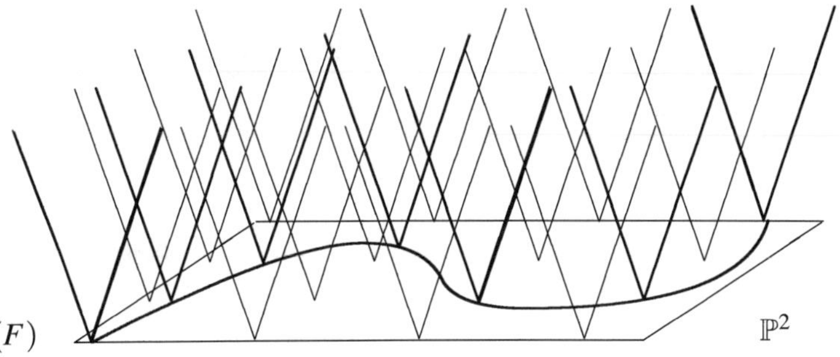

Again, to apply this, we have to look for criteria for syzygy bundles on $\mathbb{P}^N$ to be semistable. This is easier, because the only line bundles on a projective space are $\mathscr{O}_{\mathbb{P}^N}(k)$, $k \in \mathbb{Z}$. The following result gives a combinatorial criterion for semistability in the case when the ideal is generated by monomials.

THEOREM 6.6. *Let $f_i = X^{\sigma_i}$, $i \in I$, denote a set of primary monomials in $K[X_0, \ldots, X_N]$ of degree $d_i = |\sigma_i|$. Suppose that for every subset $J \subseteq I$, $|J| \geq 2$, the inequality*

$$\frac{d_J - \sum_{i \in J} d_i}{|J|-1} \leq \frac{-\sum_{i \in I} d_i}{|I|-1}$$

holds, where d_J is the degree of the highest common factor of $f_i, i \in J$. Then the syzygy bundle $\operatorname{Syz}(f_i, i \in I)$ *is semistable on $\mathbb{P}^N$.*

PROOF. See [**15**, Corollary 6.4]. This is proved with the help of Klyachko's theory of toric bundles; see [**63**], [**62**], [**64**]. □

EXAMPLE 6.7. Look at $\operatorname{Syz}(X^7, Y^8, Z^6, Z^2Y^4, X^3Y^3Z)$. The sum of the degrees is 34, so the crucial number on the right of Theorem 6.6 is $-34/4 = -8.5$. For $|J| = 2$ the worst (biggest) that can happen on the left-hand side of Theorem 6.6 is for the common factor Y^4. This gives $4-8-6 = -10 \leq -8.5$, which is fine. For $|J| = 3$ the worst thing is for Y^3, where we have $(3-8-6-7)/2 = -9 \leq -8.5$, which is also fine. For $|J| = 4$ there are no common factors, so the worst on the left is $(-6-6-7-7)/3 \sim -8.66 \leq -8.5$, which is again fine. So the bundle is semistable on the projective plane and also on the restriction to a generic projective curve of high degree. Therefore on such a generic hypersurface ring we have, for example, $x^2y^2z^4 \notin I^\star$, but $x^2y^2z^5 \in I^\star$. This holds in characteristic 0 and $p \gg 0$.

If we replace the ideal generator Z^6 by Z^3, then we get for $|J| = 4$ the inequality $-23/3 \sim -7.66 > -31/4 = -7.75$, so then the bundle is not semistable anymore. In fact,

$$\operatorname{Syz}(X^7, 0, Z^3, Z^2Y^4, X^3Y^3Z) \subset \operatorname{Syz}(X^7, Y^8, Z^3, Z^2Y^4, X^3Y^3Z)$$

is a destabilizing subbundle.

EXAMPLE 6.8. Consider $I = (x^{100}, y^{100}, z^{100})$ on $K[X,Y,Z]/(F)$. The syzygy bundle $\operatorname{Syz}(X^{100}, Y^{100}, Z^{100})$ is semistable on the projective plane. Its restriction to a generic plane curve $V_+(F)$, F homogeneous of degree $d \geq 2$, is semistable. The same is true for the restriction to the Fermat curves given by $X^d + Y^d + Z^d = 0$ of degree $d \geq 299$; see [**9**, Proposition 6.2]. It follows for these rings that $R_{\geq 151} \subseteq I^*$ (even in the Frobenius closure) and $R_{\leq 149} \cap I^* \subseteq I$, both for $p \gg 0$. So only the critical degree bound 150 is not covered (without looking at semistability, nothing was known before between the degrees 100 and 199).

PROBLEM 6.9. Does there exist for every d and every $n \leq \binom{N+d}{N}$ a family of n monomials in $K[X_0, \ldots, X_N]$ of degree d such that their syzygy bundle is semistable? Due to Theorem 6.6, this is a purely combinatorial problem. The first interesting case is for three variables ($N = 2$), where it has been proved in [**16**].

REMARK 6.10. The inclusion result for tight closure does also hold in higher dimension; see [**10**]. For this suppose that R is a normal Cohen-Macaulay domain with a maximal ideal $\mathfrak{m}$ of height $d \geq 2$ and that $I = (f_1, \ldots, f_n)$ is an $\mathfrak{m}$-primary ideal. Let

$$\ldots \longrightarrow R^{n_2} \longrightarrow R^n \stackrel{f_1,\ldots,f_n}{\longrightarrow} R \longrightarrow R/I \longrightarrow 0$$

be a complex which is exact on $D(\mathfrak{m})$ (so it is not necessarily a resolution of the ideal, but it could be). On $D(\mathfrak{m})$ we have an exact complex

$$\ldots \longrightarrow \mathcal{O}^{n_2} \longrightarrow \mathcal{O}^n \longrightarrow \mathcal{O} \longrightarrow 0$$

of free sheaves. This resolution gives rise to short exact sequences

$$0 \longrightarrow \operatorname{Syz}_1 = \operatorname{Syz}(f_1, \ldots, f_n) \longrightarrow \mathcal{O}^n \longrightarrow \mathcal{O} \longrightarrow 0$$

(as before),

$$0 \longrightarrow \operatorname{Syz}_2 \longrightarrow \mathcal{O}^{n_2} \longrightarrow \operatorname{Syz}_1 \longrightarrow 0$$

etc. until (the last one we need)

$$0 \longrightarrow \operatorname{Syz}_{d-1} \longrightarrow \mathcal{O}^{n_{d-1}} \longrightarrow \operatorname{Syz}_{d-2} \longrightarrow 0 .$$

Here it follows inductively that the Syz_j are also locally free, as they are kernels between surjective morphisms of locally free sheaves.

An element $f \in R$ yields a cohomology class $c = \delta(f) \in H^1(D(\mathfrak{m}), \operatorname{Syz}_1)$. Using the short exact sequences we get inductively by using the connecting homomorphisms again cohomology classes

$$c_j = \delta^j(f) \in H^j(D(\mathfrak{m}), \operatorname{Syz}_j), \ j = 1, \ldots, d-1 .$$

We claim that the 'truth' (at least some of it) about tight closure lies in the cohomology class $c_{d-1} \in H^{d-1}(D(\mathfrak{m}), \operatorname{Syz}_{d-1})$.

Suppose now that we are in positive characteristic $p > 0$. Note that this construction is compatible with the Frobenius, because all the bundles involved are locally free, so the short exact sequences stay exact under the Frobenius. Furthermore, the condition $tf^q \in (f_1^q, \ldots, f_n^q)$ is equivalent to $tF^{e*}(c_j) = 0$ in $H^j(D(\mathfrak{m}), \operatorname{Syz}_j)$ for $j = 1, \ldots, d-1$. Because R is Cohen-Macaulay we have $H^j(D(\mathfrak{m}), \mathcal{O}(m)) = 0$ for $j = 1, \ldots, d-2$ [**20**, Theorem 3.5.7] and therefore $\delta : H^{j-1}(D(\mathfrak{m}), \operatorname{Syz}_{j-1}) \rightarrow H^j(D(\mathfrak{m}), \operatorname{Syz}_j)$ is bijective for $j = 1, \ldots, d-2$ and $\delta : H^{d-2}(D(\mathfrak{m}), \operatorname{Syz}_{d-2}) \rightarrow H^{d-1}(D(\mathfrak{m}), \operatorname{Syz}_{d-1})$ is injective. So $tf^q \in (f_1^q, \ldots, f_n^q)$ is equivalent to $tF^{e*}(c_1) = 0$. This is equivalent to $tF^{e*}(c_2) = 0$ etc., and eventually this is equivalent to $tF^{e*}(c_{d-1}) = 0$.

If we are in a homogeneous situation, then we can shift everything to the corresponding normal projective variety $Y = \operatorname{Proj} R$ of dimension $d-1$. We get locally free sheaves $\operatorname{Syz}_j(m)$ on Y and a homogeneous element f of degree m defines cohomology classes $c_j \in H^j(Y, \operatorname{Syz}_j(m))$. If now $\bar{\mu}_{\min}(\operatorname{Syz}_{d-1}(m)) \geq 0$ and the canonical sheaf on Y is invertible (e.g., if R is a complete intersection or has an isolated singularity), then using Serre duality as in Lemma 6.1 we infer that for $c = c_{d-1} \in H^{d-1}(Y, \operatorname{Syz}_{d-1}(m))$ there exists a section $t \neq 0$ in some ample invertible sheaf $\mathscr{L} = \mathscr{O}_Y(k)$ such that $tF^{e*}(c) = 0$ for all e. This immediately yields the corresponding inclusion result for tight closure.

Which degree bounds for tight closure can we deduce from this consideration? We look at the Koszul resolution of the homogeneous elements $f_1, \ldots, f_n$. This is not globally exact, unless the f_i are a regular sequence, but it is exact on $D(R_+)$ and induces an exact complex on $\operatorname{Proj} R$. Now if the top-dimensional syzygy bundle Syz_{d-1} is strongly semistable (and this is the case if the first syzygy bundle is strongly semistable, because $\operatorname{Syz}_{d-1} \cong \bigwedge^{d-1}(\operatorname{Syz}_1)$, up to certain twists), then we get the following inclusion bound:

$$\text{For } \deg(f) \geq \frac{(d-1)(\sum_{i=1}^n d_i)}{n-1} \text{ we have } f \in (f_1, \ldots, f_n)^* .$$

This number is the degree bound from Theorem 6.4 in dimension two, and in the parameter case ($d = n$) it just gives the sum of the degrees, which is the parameter degree bound; see [**55**, Theorem 2.9 and Theorem 6.1].

CHAPTER 7

Numerical criteria for affineness of torsors

In this and the next chapter we want to prove that tight closure is plus closure for homogeneous primary ideals in a graded domain of dimension two over the algebraic closure of a finite field [**14**]. For this we will formulate a numerical criterion for a torsor to be an affine variety in terms of the behavior of the corresponding cohomology class along the strong Harder-Narasimhan filtration (Theorem 5.16). We will then show in Theorem 8.9 that this numerical criterion also holds in the case of a finite field for the property that one cannot annihilate the cohomology class by a finite mapping. This together shows that tight closure is plus closure (Theorem 8.10).

PROPOSITION 7.1. *Let C denote a smooth projective curve over an algebraically closed field K. Let $\mathscr{S}$ denote a locally free sheaf of rank r and let $c \in H^1(C,\mathscr{S})$ be a cohomology class with corresponding torsor $B \to C$. If $\mathscr{S}$ is strongly semistable, then the following are equivalent.*

(i) *The torsor B is an affine scheme.*
(ii) $\bar{\mu}_{\max}(\mathscr{S}) < 0$ *and $c \neq 0$ (in positive characteristic $F^{e*}(c) \neq 0$ for one e with the property that $p^e > g(r+1)r$).*
(iii) *The dual extension $\mathscr{S}'^{\vee}$ given by c is an ample bundle.*

The implications (ii) $\Leftrightarrow$ (iii) $\Rightarrow$ (i) *hold for any locally free sheaf $\mathscr{S}$.*

PROOF. (i) $\Rightarrow$ (ii). Since B is affine, we have $c \neq 0$; otherwise B would be trivial and would contain projective curves. Because the pull-back $B' = B \times_C C'$ for $C' \to C$ finite is again affine, we also get $F^{e*}(c) \neq 0$ in positive characteristic for all $e \in \mathbb{N}$. The affineness of B implies in general $\bar{\mu}_{\min}(\mathscr{S}) < 0$ (Remark 6.2). But if $\mathscr{S}$ is strongly semistable, then $\bar{\mu}_{\min}(\mathscr{S}) = \bar{\mu}_{\max}(\mathscr{S}) < 0$.

(ii) $\Rightarrow$ (iii). The slope condition yields $\bar{\mu}_{\min}(\mathscr{S}^{\vee}) > 0$ for the dual bundle $\mathscr{S}^{\vee}$. Hence $\mathscr{S}^{\vee}$ is an ample vector bundle by Theorem 5.17. In characteristic zero, since $\mathscr{S}'^{\vee}$ is a non-trivial extension, it follows from Proposition 4.12(ii) that $\mathscr{S}'^{\vee}$ is also ample. In positive characteristic, we may apply the e-th Frobenius and may then assume that $F^{e*}(\mathscr{S}'^{\vee})$ is a non-trivial extension of $F^{e*}(\mathscr{S}^{\vee})$ and that $\bar{\mu}_{\min}(\mathscr{S}^{\vee}) > g(r+1)$. Then the ampleness of $F^{e*}(\mathscr{S}'^{\vee})$ follows from Lemma 5.20 and the ampleness of $\mathscr{S}'^{\vee}$ follows from Proposition 4.9(iii).

(iii) $\Rightarrow$ (i). The ampleness of $\mathscr{S}'^{\vee}$ is by definition the ampleness of the divisor $\mathbb{P}(\mathscr{S}^{\vee}) \subset \mathbb{P}(\mathscr{S}'^{\vee})$. So the complement of $\mathbb{P}(\mathscr{S}^{\vee})$ is an affine scheme as explained in Remark 4.7.

Finally we prove (iii) $\Rightarrow$ (ii) without the condition that $\mathscr{S}$ is strongly semistable. The surjection $\mathscr{S}'^{\vee} \to \mathscr{S}^{\vee} \to 0$ shows that $\mathscr{S}^{\vee}$ is also ample by Proposition 4.9(i); hence $\bar{\mu}_{\min}(\mathscr{S}^{\vee}) > 0$ and therefore $\bar{\mu}_{\max}(\mathscr{S}) < 0$. If $F^{e*}(c) = 0$, then $F^{e*}(\mathscr{S}'^{\vee}) \cong F^{e*}(\mathscr{S})^{\vee} \oplus \mathscr{O}_C$ would be not ample. $\square$

REMARK 7.2. It may indeed happen that a class $0 \neq c \in H^1(C, \mathscr{S})$ for $\mathscr{S}$ of negative degree (even for invertible $\mathscr{S}$) becomes zero under a Frobenius power, but this is an exception. In the case of a relative curve over $\operatorname{Spec} \mathbb{Z}$ (see Remark 4.16), this may happen only for finitely many prime numbers, because the condition $c \neq 0$ implies ampleness over the generic curve of characteristic zero, and this holds then almost everywhere, since ampleness is an open property.

EXAMPLE 7.3. The implication (i) $\Rightarrow$ (iii) in Proposition 7.1 does not hold in general. For example, consider the ideal generators (x^2, y^2, x^2) in $K[x,y]$. The corresponding sequence on $\mathbb{P}^1$ is

$$0 \longrightarrow \mathscr{O}_{\mathbb{P}^1}(-2) \oplus \mathscr{O}_{\mathbb{P}^1} \cong \operatorname{Syz}(x^2, y^2, x^2)(2) \longrightarrow \mathscr{O}_{\mathbb{P}^1}^3 \stackrel{x^2, y^2, x^2}{\longrightarrow} \mathscr{O}_{\mathbb{P}^1}(2) \longrightarrow 0 .$$

The element xy yields the cohomology class

$$(xy/x^2y^2, 0) = (1/xy, 0) \in H^1(\mathbb{P}^1, \mathscr{O}_{\mathbb{P}^1}(-2)) \oplus H^1(\mathbb{P}^1, \mathscr{O}_{\mathbb{P}^1}) .$$

This class is not 0, since xy does not belong to the ideal. It also does not belong to the tight closure, since $K[x,y]$ is regular. Therefore the corresponding torsor is affine. The dual extension $0 \to \mathscr{O}_{\mathbb{P}^1} \to \mathscr{F}' \to \mathscr{O}_{\mathbb{P}^1}(2) \oplus \mathscr{O}_{\mathbb{P}^1} = \mathscr{F} \to 0$ shows however that $\mathscr{F}'$ is not ample, since it has the non-ample quotient $\mathscr{O}_{\mathbb{P}^1}$. So we have constructed a non-ample divisor $\mathbb{P}(\mathscr{F}) \subset \mathbb{P}(\mathscr{F}')$ with affine complement. The first example of this phenomenon is due to Zariski; see [**28**].

EXAMPLE 7.4. We give an example where the forcing divisor is big, but its complement is not affine. For this we can just take $(x,y) \subset K[x,y]$ and the element x. The dual extension is now the trivial extension $\mathscr{F}' = \mathscr{O}_{\mathbb{P}^1} \oplus \mathscr{O}_{\mathbb{P}^1}(1)$. The corresponding forcing divisor $Z \subset \mathbb{P}(\mathscr{F}')$ has positive self-intersection 1. Since it is numerically effective, it is a big divisor [**72**, Theorem 2.2.14]. Its complement is, however, not affine.

We now look at the affineness of the torsor B for an arbitrary locally free sheaf $\mathscr{S}$. The crucial point is to look at the strong Harder-Narasimhan filtration of $\mathscr{S}$ (in characteristic zero, this is just the usual Harder-Narasimhan filtration). Note that the property of being affine is preserved under a finite mapping. So in positive characteristic we often apply an absolute Frobenius morphism $F : C \to C$ to pull back the whole situation.

In order to simplify the formulation of the next criterion we say that the Harder-Narasimhan filtration of a bundle of rank r on a curve of genus g is *very strong* if it is strong and if it has the property that if a quotient $\mathscr{S}_{i+1}/\mathscr{S}_i$ of the filtration has negative slope, then its slope is $< -g(r+1)$. Of course, if $\mathscr{S}$ has strong Harder-Narasimhan filtration, then it is easy by applying the Frobenius to achieve a very strong Harder-Narasimhan filtration. We just have to look for e such that $p^e > g(r+1)r$. In characteristic zero we consider every Harder-Narasimhan filtration as very strong.

THEOREM 7.5. *Let $\mathscr{S}$ denote a locally free sheaf on a smooth projective curve C over an algebraically closed field K. Let $c \in H^1(C, \mathscr{S})$ with corresponding torsor $B \to C$. Let $\mathscr{S}_1 \subset \ldots \subset \mathscr{S}_t$ be a very strong Harder-Narasimhan filtration of $F^{e*}(\mathscr{S})$ on C. Choose i such that $\mathscr{S}_i/\mathscr{S}_{i-1}$ has degree ≥ 0 and that $\mathscr{S}_{i+1}/\mathscr{S}_i$ has degree < 0 (we set $i = -1$ and $\mathscr{S}_{-1} = 0$ or $i = t+1$ and $\mathscr{S}_{t+1} = F^{e*}(\mathscr{S})$ in the cases where all slopes are non-negative or all slopes are negative). Set $0 \to \mathscr{S}_i \to F^{e*}(\mathscr{S}) \to F^{e*}(\mathscr{S})/\mathscr{S}_i = \mathscr{Q} \to 0$. Then B is affine if and only if the image of $F^{e*}(c)$ in $H^1(C, \mathscr{Q})$ is non-zero.*

PROOF. Let $c' \in H^1(C, \mathscr{Q})$ denote the image of $F^{e*}(c)$ and suppose first that $c' \neq 0$. By construction, $\mathscr{Q}$ is a locally free sheaf $\neq 0$ with $\bar{\mu}_{\max}(\mathscr{Q}) < 0$ (and even $< -g(\operatorname{rk}(\mathscr{S})+$

1) in positive characteristic). So its dual sheaf $\mathscr{Q}^\vee$ is an ample vector bundle and the same holds true for the extension $\mathscr{Q}'^\vee$ defined by c' due to Proposition 4.12(ii) (in characteristic 0) or due to Lemma 5.20, since $\mu_{\min}(\mathscr{Q}^\vee) > g(\mathrm{rk}(\mathscr{S})+1)$. Therefore the torsor B' corresponding to c' is affine by Remark 4.7. There exists an affine morphism $F^{e*}(B) \to B'$ induced by $F^{e*}(\mathscr{S}) \to \mathscr{Q}$ (Proposition 3.3(iii)); hence $F^{e*}(B)$ and then also B itself is affine.

On the other hand, suppose that $F^e(c)$ maps to 0 in $H^1(C,\mathscr{Q})$. Then $F^{e*}(c)$ stems from a cohomology class $\tilde{c} \in H^1(C,\mathscr{S}_i)$, where $\bar{\mu}_{\min}(\mathscr{S}_i) \geq 0$ (including the case $\mathscr{S}_i = \mathscr{S}_{-1} = 0$). It is sufficient to show that the affine-linear bundle $\tilde{B}$ corresponding to $\tilde{c}$ is not affine. This was stated in Remark 6.2. □

This criterion becomes easier on an elliptic curve.

THEOREM 7.6. *Let K be an algebraically closed field and let C denote an elliptic curve. Let $\mathscr{S}$ be a locally free sheaf on C and let $\mathscr{S} = \mathscr{S}_1 \oplus \ldots \oplus \mathscr{S}_k$ be the decomposition into indecomposable locally free sheaves. Let $c \in H^1(C,\mathscr{S})$ be a cohomology class with corresponding torsor $B \to C$. Then B is affine if and only if there exists $1 \leq j \leq k$ such that $\deg \mathscr{S}_j < 0$ and $c_j \neq 0$, where c_j denotes the component of c in $H^1(C,\mathscr{S}_j)$.*

PROOF. An indecomposable sheaf on an elliptic curve is strongly semistable, as mentioned in Remark 5.13. So the result follows from the previous theorem. □

We may reformulate Theorem 7.5 as a criterion for tight closure in dimension two.

THEOREM 7.7. *Let R be a two-dimensional standard-graded geometrically normal domain over a field, $C = \mathrm{Proj}\,R$. Let $I = (f_1,\ldots,f_n)$ be an R_+-primary homogeneous ideal. Let f be a homogeneous element of degree m and $\delta(f) = c \in H^1(C,\mathrm{Syz}(f_1,\ldots,f_n)(m))$. Let e ($q = p^e$) be such that the Harder-Narasimhan filtration of $\mathrm{Syz}(f_1^q,\ldots,f_n^q)$ is very strong. Then $f \in I^\star$ if and only if the cohomology class $F^{e*}(c) \in H^1(C,\mathrm{Syz}(f_1^q,\ldots,f_n^q)(qm))$ vanishes in the quotient bundle $Q = \mathrm{Syz}(f_1^q,\ldots,f_n^q)(qm))/S_i$ of negative slope (or, equivalently, comes from the subbundle of non-negative slope).*

PROOF. We may assume that the field is algebraically closed. The result then follows from Theorem 7.5 and Proposition 3.9. □

COROLLARY 7.8. *Let R be a two-dimensional standard-graded normal domain over an algebraically closed field and let x,y be parameters. Let $f_1,\ldots,f_n$ be an R_+-primary family of homogeneous polynomials in x and y. Then the syzygy bundle on $C = \mathrm{Proj}\,R$ is a direct sum*

$$\mathrm{Syz}(f_1,\ldots,f_n) \cong \mathscr{O}_C(-\alpha_1) \oplus \ldots \oplus \mathscr{O}_C(-\alpha_{n-1}),$$

and an element f of degree m belongs to $(f_1,\ldots,f_n)^\star$ if and only if there exists $1 \leq j \leq n-1$ such that $m < \alpha_j$ and $F^{e*}(c_j) \neq 0$, where $c_j \in H^1(C,\mathscr{O}_C(m-\alpha_j))$ is the j-th component of $c = \delta(f)$ and where e is the minimal exponent such that $q = p^e > ng$.

PROOF. The ideal is already defined inside the regular ring $K[x,y] \subseteq R$, so it has finite projective dimension, and the syzygy bundle on $\mathbb{P}^1$ must split in the way written down. This is then also true on C. Since the syzygy bundle splits into line bundles, its Harder-Narasimhan filtration is strong. For $q = p^e > 2g$ the e-th pull-back of it is very strong. Then the statement follows from Theorem 7.7. □

EXAMPLE 7.9. We look at the ideal (x^6,x^3y,x^2y^3,y^4) in $R=K[X,Y,Z]/(G)$, where G is a homogeneous polynomial such that R is normal and such that x and y are parameters (that is, the power $Z^{\deg(G)}$ must occur in G). The syzygies

$$(y,-x^3,0,0),\ (0,y^2,-x,0),\ (0,0,y,-x^2)$$

determine an isomorphism

$$\mathscr{O}_C(-7)\oplus\mathscr{O}_C(-6)\oplus\mathscr{O}_C(-6)\cong \mathrm{Syz}(x^6,x^3y,x^2y^3,y^4).$$

An element f of degree m yields via Čech cohomology the following class. The element f is on $D_+(x)$ represented by $(f/x^6,0,0,0)$ and on $D_+(y)$ by $(0,0,0,f/y^4)$. So their difference is $(f/x^6,0,0,-f/y^4)$, which is a syzygy on $D_+(xy)$. Expressing this syzygy with our three basis syzygies, we get

$$(f/x^6,0,0,-f/y^4)=\frac{f}{x^6y}(y,-x^3,0,0)+\frac{f}{x^3y^3}(0,y^2,-x,0)+\frac{f}{x^2y^4}(0,0,y,-x^2).$$

Therefore f determines the cohomology class $(f/x^6y,f/x^3y^3,f/x^2y^4)$. It follows immediately from Corollary 7.8 that $R_m\subseteq(x^6,x^3y,x^2y^3,y^4)^\star$ for $m\geq 7$, that $R_m\subseteq(x^6,x^3y,x^2y^3,y^4)^F$ for $m\geq 8$ (in positive characteristic) and that $R_m\cap(x^6,x^3y,x^2y^3,y^4)^\star=(x^6,x^3y,x^2y^3,y^4)_m$ for $m\leq 5$ (in characteristic 0 or $p\gg 0$). So the interesting degree is 6, and everything depends on whether the first component of $\delta(f)$, namely the class $f/x^6y\in H^1(C,\mathscr{O}_C(-1))$, is 0, which is equivalent with $f\in(x^6,y)$.

Let's look at $G=X^4+Y^4+Z^4$ and at the monomials $x^ay^bz^c$ of degree 6, $c\leq 3$ (for $c\geq 4$ replace z^c using the curve equation). The monomials of degree 6 that are not in the tight closure are x^5z,x^4z^2,x^3z^3. The monomials of degree 6 that are not in the ideal, but in its tight closure, are $xy^2z^3,xy^3z^2,x^2yz^3,x^2y^2z^2$.

CHAPTER 8

Tight closure and plus closure in graded dimension two

We suppose now that K is the algebraic closure of a finite field of positive characteristic p, $K = \overline{\mathbb{F}}_p$. We shall prove for a smooth projective curve C defined over K that the numerical criterion for the affineness of a torsor (Theorem 7.1) also holds for the property that there does not exist any projective curve inside the torsor. This property itself is equivalent to the fact that the cohomology class does not become trivial under any finite mapping $\varphi : C' \to C$. We first study the case of a strongly semistable bundle. For that we need a theorem of Lange and Stuhler.

THEOREM 8.1. *(Lange-Stuhler) Let C denote a smooth projective curve defined over a finite field. Let $\mathscr{S}$ be a strongly semistable bundle on C of degree 0. Then there exists a finite morphism $C' \to C$ such that the pull-back of $\mathscr{S}$ is trivial.*

PROOF. See [**68**]. The basic observation is the following: the family of semistable bundles of fixed rank and fixed degree is bounded. This means in particular that for a curve defined over a finite field there exist only finitely many semistable bundles of given degree and rank that are defined over this field. If $\mathscr{S}$ is defined on a smooth projective curve over a finite field, then its Frobenius pull-back is also defined over this field. Therefore, if $\mathscr{S}$ is strongly semistable and of degree 0, then all its Frobenius pull-backs are again semistable of degree 0 and are defined over the same finite field. Therefore, looking at the sequence $F^{e*}(\mathscr{S})$, $e \in \mathbb{N}$, we must have a repetition. Say that $F^{e'*}(\mathscr{S}) \cong F^{e*}(\mathscr{S})$, $e' > e$. We may apply the e-th Frobenius $F^{e*} : C \to C$ and achieve a situation where the bundle $\mathscr{T} = F^{e*}(\mathscr{S})$ has the property that it is isomorphic to one of its Frobenius pull-backs. In this situation the next theorem of Lange and Stuhler gives the result. □

THEOREM 8.2. *(Lange-Stuhler) Suppose that C is a smooth projective curve over a (not necessarily finite) field of positive characteristic p, and let $\mathscr{T}$ be a vector bundle of degree 0 on C. Then the following are equivalent.*

(i) *There exists an isomorphism $F^{e*}(\mathscr{T}) \cong \mathscr{T}$ for some $e \in \mathbb{N}$.*
(ii) *There exists a finite étale morphism $C' \to C$ such that the pull-back of $\mathscr{T}$ is trivial.*
(iii) *$\mathscr{T}$ is induced by a representation of the algebraic fundamental group.*

PROOF. See [**68**, Satz 1.4 and Proposition 1.2]; compare also [**86**]. For us the implication (i) $\Rightarrow$ (ii) is most important. □

REMARK 8.3. For the algebraic fundamental group see [**31**] or [**75**, Chapter 1.5]. In relationship with representations of the algebraic fundamental group of p-adic curves there has been renewed interest in strongly semistable bundles,; see [**22**]. The classical result in this respect is the theorem of Narasimhan and Seshadri stating that over the complex numbers $\mathbb{C}$ a polystable bundle of degree zero corresponds to a unitary representation of the (topological) fundamental group [**85**]. It would be interesting to see examples of what the representation of the fundamental group corresponding to certain syzygy bundles of degree 0 looks like, e.g., for $\mathrm{Syz}(x^2,y^2,z^2)(3)$ on the Fermat curve $x^3+y^3+z^3=0$ for $p=1 \bmod 3$ (for $p=2 \bmod 3$ the first Frobenius trivializes the bundle, and it does not correspond to a representation).

We will now deduce from Theorem 8.1 that we can annihilate the cohomology classes of a strongly semistable bundle of non-negative degree.

THEOREM 8.4. *Let C denote a smooth projective curve over the algebraic closure of a finite field. Let $\mathcal{S}$ denote a strongly semistable locally free sheaf of degree ≥ 0 and let $c \in H^1(C,\mathcal{S})$ denote a cohomology class. Then c trivializes under a finite mapping $C' \to C$.*

PROOF. Suppose first that the degree of $\mathcal{S}$ is positive. Then $\mathcal{S}$ is an ample bundle (Theorem 5.17) and so it is also cohomologically p-ample as explained in Proposition 4.17. Therefore a Frobenius power of the cohomology class vanishes.

So suppose that $\deg(\mathcal{S})=0$. Since K is supposed to be the closure of a finite field, every object we encounter is in fact defined already over a finite subfield. Since $\mathcal{S}$ is strongly semistable, there exists by the result of Lange and Stuhler (Theorem 8.1) a finite mapping $\varphi: C' \to C$ (which is an étale mapping followed by a Frobenius power) such that the pull-back is trivial, hence $\varphi^*(\mathcal{S}) = \mathcal{O}_{C'}^r$. Thus we may assume that $\mathcal{S}$ is in fact a trivial bundle. We may deal with the components of $c \in H^1(C,\mathcal{O}_C^r)$ separately, so we may even assume that $\mathcal{S}=\mathcal{O}_C$. But for the structure sheaf this follows from [**95**, Proposition 3.3] or the following lemma. □

LEMMA 8.5. *Let K be an algebraically closed field of characteristic $p>0$. Let C be a smooth projective curve over K and let $c \in H^1(C,\mathcal{O}_C)$. Then there exists a finite morphism $C' \to C$ such that the pull-back of c is 0. Here this morphism may be chosen as the composition of a Frobenius power and an Artin-Schreier mapping.*

PROOF. The Frobenius morphism F acts on $H^1(C,\mathcal{O}_C)$ p-linearly yielding the so-called Fitting decomposition $H^1(C,\mathcal{O}_C) = V_s \oplus V_n$ such that $F|V_s$ is bijective and $F|V_n$ is nilpotent (see [**43**, III §3] or [**75**, Lemma III.4.13]). Thus we may write $c=c_s+c_n$, where c_n becomes zero after applying a certain power of the Frobenius. Thus we may assume that $c=c_s \in V_s$. Consider the Artin-Schreier sequence

$$0 \longrightarrow \mathbb{Z}/(p) \longrightarrow \mathcal{O}_C \stackrel{F-id}{\longrightarrow} \mathcal{O}_C \longrightarrow 0,$$

which is exact in the étale topology. It yields the exact sequence

$$0 \longrightarrow H^1_{et}(C,\mathbb{Z}/(p)) \longrightarrow H^1(C,\mathcal{O}_C) \stackrel{F-id}{\longrightarrow} H^1(C,\mathcal{O}_C) \longrightarrow \dots .$$

There exists a basis c_j of V_s such that $F(c_j)=c_j$; see [**83**, §14]. Thus we may assume that $F(c)-c=0$ and we may consider $c \in H^1_{et}(C,\mathbb{Z}/(p))$. Hence c represents an Artin-Schreier extension C' of C and the cohomology class c vanishes on C'; see [**75**, III. §4]. □

REMARK 8.6. We explicitly describe the Artin-Schreier extension appearing in the last proof. Let $c=f/f_1f_2$, where f_1 and f_2 are homogeneous parameters in R, $C=\operatorname{Proj} R$, and suppose that $c^p-c=0$ in $H^1(C,\mathcal{O}_C)$. This means that $c^p-c=a_2-a_1$, where $a_i \in \Gamma(U_i,\mathcal{O}_C)$, $U_i=D_+(f_i)=\operatorname{Spec} R_i$. Let $U_i' = \operatorname{Spec} R_i[T_i]/(T_i^p-T_i+a_i)$, $i=1,2$. The transition mapping $T_1 \mapsto T_2+c$ is due to

$$(T_2+c)^p-(T_2+c)+a_1 = T_2^p-T_2+c^p-c+a_1 = T_2^p-T_2+a_2$$

being well defined and U_1' and U_2' glue together to give a scheme $C' \to C$. The cohomology class in Y' is $c=f/f_1f_2=T_1-T_2$, $T_i \in \Gamma(U_i',\mathcal{O}_{C'})$, so that $c=0$ in $H^1(C',\mathcal{O}_{C'})$.

THEOREM 8.7. *Let C denote a smooth projective curve over the algebraic closure of a finite field. Let $\mathcal{S}$ denote a locally free sheaf on C and let $c \in H^1(C, \mathcal{S})$ with corresponding torsor $B \to C$. Let $\mathcal{S}_1 \subset \ldots \subset \mathcal{S}_t$ be the strong Harder-Narasimhan filtration of $F^{e*}(\mathcal{S})$ on C. Choose i such that $\mathcal{S}_i/\mathcal{S}_{i-1}$ has degree ≥ 0 and that $\mathcal{S}_{i+1}/\mathcal{S}_i$ has degree < 0. Let $0 \to \mathcal{S}_i \to F^{e*}(\mathcal{S}) \to F^{e*}(\mathcal{S})/\mathcal{S}_i = \mathcal{Q} \to 0$. Then the following are equivalent:*

(i) *The $\mathcal{S}$-torsor B contains a projective curve.*
(ii) *There exists a smooth projective curve C' and a finite mapping $\varphi : C' \to C$ such that $\varphi^*(c) = 0$.*
(iii) *Some Frobenius power of the image of $F^{e*}(c)$ in $H^1(C, \mathcal{Q})$ is zero.*

PROOF. The equivalence of (i) and (ii) is clear, since a torsor is trivial if and only if it has section (Proposition 3.3(ii)).

Suppose that (iii) holds. We may assume that the image of $F^{e*}(c)$ is 0 inside $H^1(C, \mathcal{Q})$. Then $F^{e*}(c)$ stems from a class $c_i \in H^1(C, \mathcal{S}_i)$. The filtration $0 \subset S_1 \subset \ldots \subset S_{i-1} \subset S_i$ is such that all quotients are strongly semistable with $\deg(\mathcal{S}_j/\mathcal{S}_{j-1}) \geq 0$ for $j = 1, \ldots, i$. We look at the short exact sequence $0 \to \mathcal{S}_{i-1} \to \mathcal{S}_i \to \mathcal{S}_i/\mathcal{S}_{i-1} \to 0$. Since the sheaf on the right is strongly semistable of non-negative degree, it follows from Theorem 8.4 that the image c_i' of c_i in $H^1(C, \mathcal{S}_i/\mathcal{S}_{i-1})$ vanishes under a finite mapping $\varphi : C' \to C$. Then $\varphi^*(c_i)$ stems from a class $c_{i-1} \in H^1(C', \varphi^*(\mathcal{S}_{i-1}))$. Going on like this inductively we find a finite mapping such that the pull-back of c is zero.

Condition (i) implies that B is not an affine scheme. So (i) $\Rightarrow$ (iii) follows from Theorem 7.5. □

PROBLEM 8.8. Suppose that $\mathcal{L}$ is an invertible sheaf on a smooth projective curve C over an algebraically closed field of positive characteristic, and let $c \in H^1(C, \mathcal{L})$ be a cohomology class. Does there exist a finite map $C' \to C$ from another curve such that the pull-back of c gets trivial? This is true if $\mathcal{L}$ is torsion in the Picard group and in particular if the base field is finite, but open in general.

We deduce the following corollary, which gives a geometric criterion for the cohomological property of being affine.

THEOREM 8.9. *Let C denote a smooth projective curve over the algebraic closure of a finite field. Let $\mathcal{S}$ denote a locally free sheaf on C and let $c \in H^1(C, \mathcal{S})$ with corresponding geometric torsor $B \to C$. Then B is an affine variety if and only if it does not contain any projective curve.*

PROOF. This follows from Theorem 7.5 and Theorem 8.7, since for both properties the same numerical criterion holds. □

We finally come back to tight closure and its relation to plus closure.

THEOREM 8.10. *Suppose that K is the algebraic closure of a finite field. Let R denote a standard-graded two-dimensional normal domain over K. Then for every homogeneous R_+-primary ideal I we have the identities $I^* = I^{+\mathrm{gr}} = I^+$.*

PROOF. The inclusions $I^{+\mathrm{gr}} \subseteq I^+ \subseteq I^*$ are clear, so it is enough to show that $I^* \subseteq I^{+\mathrm{gr}}$. Due to [**52**, Theorem 4.2] we only have to consider homogeneous elements. Let $I = (f_1, \ldots, f_n)$ be given by homogeneous ideal generators. These give rise to the locally free syzygy bundle $\mathrm{Syz}(f_1, \ldots, f_n)$ on the smooth projective curve $C = \mathrm{Proj}\, R$. A homogeneous element f of degree m induces a cohomology class $c = \delta(f) \in H^1(C, \mathrm{Syz}(f_1, \ldots, f_n)(m))$ with corresponding torsor $B \to C$. The containment of f in the graded plus closure,

$f \in I^{+\mathrm{gr}}$, is equivalent to the existence of a projective curve inside B (Proposition 3.12). The containment in the tight closure, $f \in I^*$, is equivalent to the non-affineness of B (Proposition 3.9). So the result follows from Theorem 8.9. □

REMARK 8.11. If R is a homogeneous coordinate ring over an elliptic curve, then Theorem 8.10 is also true in positive characteristic without the assumption that the base field is the algebraic closure of a finite field [**12**]. In fact it can be shown using Atiyah's classification [**2**] of bundles on an elliptic curve that the relevant cohomology classes are annihilated by certain iterations of multiplication by p (given by the group structure on the elliptic curve). Therefore the only finite extensions needed are the corresponding ring homomorphism $[p]^* : R \to R$ and its iterations. In case of Hasse invariant 0, we even get $I^* = I^F$ for all primary homogeneous ideals, which answers [**97**, Open Problem 4.14] for such ideals.

REMARK 8.12. This result can be generalized to non-normal graded domains and to homogeneous ideals which are not necessarily primary. Moreover, it can be generalized to arbitrary graded submodules inside finitely generated graded modules. Starting from this, G. Dietz has shown [**23**] that one can even get rid of the assumption that the modules are graded: he proved in general that if the equation $N^+ = N^*$ is known for graded coprimary submodules $N \subseteq M$ over a graded ring R, then this is also true for arbitrary coprimary submodules.

CHAPTER 9

Hilbert-Kunz theory

The geometric approach to tight closure that we pursue in these notes is also successful in *Hilbert-Kunz theory*. We will describe here, mostly without proofs, the basics of this theory and the main results that can be obtained by applying vector bundle techniques. Let R be a commutative noetherian ring of positive characteristic p and let I be an ideal that is primary to a maximal ideal $\mathfrak{m}$. We recall the definition of the Hilbert-Kunz function and the Hilbert-Kunz multiplicity.

DEFINITION 9.1. The *Hilbert-Kunz function* of I is the function

$$e \longmapsto \varphi_I(e) = \lg(\mathrm{R}/\mathrm{I}^{[\mathrm{p}^\mathrm{e}]}),$$

where lg denotes the length of a module. The *Hilbert-Kunz multiplicity* is the limit (setting $d = \dim(R)$)

$$e_{HK}(I) = \lim_{e \mapsto \infty} \frac{\varphi_I(e)}{p^{ed}}.$$

The Hilbert-Kunz multiplicity of the maximal ideal in a local ring R is also called the *Hilbert-Kunz multiplicity* of the ring, and is denoted by $e_{HK}(R)$.

This function was first considered by E. Kunz in [**65**] and is often said to be mysterious. P. Monsky showed in [**77**] that the Hilbert-Kunz multiplicity exists and that it is a positive real number. He also showed that the Hilbert-Kunz function has the form (we write $q = p^e$ for the argument, not e)

$$\varphi(q) = \varphi_I(q) = e_{HK}(I)q^d + O(q^{d-1}).$$

Recall that $O(q^{d-1})$ stands for a function $\gamma(q)$ with the property that $\gamma(q)/q^{d-1}$ is bounded. It is conjectured that the Hilbert-Kunz multiplicity is always a rational number.[1] The best general result about the Hilbert-Kunz function to date is due to Huneke, McDermott, and Monsky.

THEOREM 9.2. *Suppose R is a normal excellent local ring of positive characteristic with a perfect residue field. Let I be an ideal that is primary to the maximal ideal. Then*

$$\varphi(q) = e_{HK}(I)q^d + \beta q^{d-1} + O(q^{d-2}),$$

where β is another real number.

PROOF. See [**56**, Theorem 1.12]. □

The geometric approach to Hilbert-Kunz theory in terms of vector bundles shows that for a two-dimensional normal standard-graded domain the Hilbert-Kunz multiplicity is a rational number. This result was obtained independently by Brenner and Trivedi. In fact, the Hilbert-Kunz multiplicity can be expressed by a formula in terms of data coming from

[1] In a recent preprint [**81**], Monsky gives an example of a four-dimensional ring where there is strong numerical evidence that the Hilbert-Kunz multiplicity is not rational.

the strong Harder-Narasimhan filtration (see Theorem 5.16) of the syzygy bundle. One can also show that the second term β is 0 (hence also rational). Moreover, if the base field is finite (or the algebraic closure of a finite field), then the lowest term (the $O(q^0)$-term) is an eventually periodic function. Before we describe these results we recall the relationship between tight closure and Hilbert-Kunz multiplicity.

THEOREM 9.3. (*Hochster, Huneke*) *Let R be an analytically unramified and formally equidimensional local ring with maximal ideal $\mathfrak{m}$ and let I be an $\mathfrak{m}$-primary ideal. Let $f \in R$. Then $f \in I^*$ if and only if $e_{HK}((I,f)) = e_{HK}(I)$.*

PROOF. See [**51**, Theorem 8.17]. □

Suppose now that R is a finitely generated K-algebra over an algebraically closed field K of characteristic $p > 0$. Then $\mathrm{lg}(\mathrm{R}/\mathrm{I}^{[q]}) = \dim_{\mathrm{K}}(\mathrm{R}/\mathrm{I}^{[q]})$. If, moreover, R is standard-graded, normal, and of dimension $\dim(R) \geq 2$ and $I = (f_1, \ldots, f_n)$ is an R_+-primary homogeneous ideal with homogeneous generators f_i of degree d_i, then we use again the short exact sequence (compare Situation 3.7 and the paragraph after Theorem 5.10)

$$0 \longrightarrow \mathrm{Syz}(f_1^q, \ldots, f_n^q)(m) \longrightarrow \bigoplus_{i=1}^{n} \mathscr{O}(m - qd_i) \stackrel{f_1^q, \ldots, f_n^q}{\longrightarrow} \mathscr{O}(m) \longrightarrow 0$$

on the projective variety $Y = \mathrm{Proj}\, R$. The cokernel of the global evaluation of this sequence in degree m yields

$$\dim_K((R/I^{[q]})_m) = \dim_K(\Gamma(Y, \mathscr{O}(m))/(f_1^q, \ldots, f_n^q)).$$

Therefore we get

$$\begin{aligned} \varphi(q) &= \mathrm{lg}(\mathrm{R}/\mathrm{I}^{[q]}) = \sum_{\mathrm{m}=0}^{\infty} \dim_{\mathrm{K}}(\Gamma(\mathrm{Y}, \mathscr{O}(\mathrm{m}))/(\mathrm{f}_1^q, \ldots, \mathrm{f}_n^q)) \\ &= \sum_{m=0}^{\infty} \Big(h^0(\mathscr{O}(m)) - \sum_{i=1}^{n} h^0(\mathscr{O}(m - qd_i)) + h^0(\mathrm{Syz}(f_1^q, \ldots, f_n^q)(m))\Big). \quad (*) \end{aligned}$$

This sum is finite, because for m large enough the alternating sum is 0. This is because $H^1(\mathrm{Syz}(f_1^q, \ldots, f_n^q)(m)) = 0$ for $m \gg 0$, as these twists of the syzygy bundle are getting ample (see [**45**, Proposition III.5.3]). It is comparably easy to control the global sections of the involved line bundles, so the problem is to compute the global sections of $\mathrm{Syz}(f_1^q, \ldots, f_n^q)(m)$ for varying q and m running in certain ranges, where the range of m depends on q. This problem is still difficult in general, so suppose from now on that R has dimension two, so that $C = \mathrm{Proj}\, R$ is a smooth projective curve. In this case one can control the global sections of the syzygy bundle by knowing its stability properties.

LEMMA 9.4. *Suppose that the syzygy bundle $\mathrm{Syz}(f_1, \ldots, f_n)$ on C is strongly semistable. Then the following hold (ω_C denotes the canonical divisor).*
For $m < (q\sum_{i=1}^n d_i)/(n-1)$ we have $H^0(C, \mathrm{Syz}(f_1^q, \ldots, f_n^q)(m)) = 0$.
For $m > (q\sum_{i=1}^n d_i)/(n-1) + \deg(\omega_C)/\deg(C)$ we have $H^1(C, \mathrm{Syz}(f_1^q, \ldots, f_n^q)(m)) = 0$.
In particular, in the formula $(*)$ above we only have to sum up to this bound.

PROOF. (i). The degree of the syzygy bundle $\mathrm{Syz}(f_1^q, \ldots, f_n^q)(m)$ is $((n-1)m - q\sum_{i=1}^n d_i)\deg(C)$; see Remark 4.4. Hence this is negative under the given condition and since the bundle is semistable it has no sections in this range (Proposition 5.6(iii)).

(ii). The first cohomology module is dual to $H^0((\mathrm{Syz}(f_1^q, \ldots, f_n^q)(m))^\vee \otimes \omega_C)$ by Serre duality. Its degree is $-((n-1)m - q\sum_{i=1}^n d_i)\deg(C) + (n-1)\deg(\omega_C)$. So this is again negative under the given condition. □

This means in particular that in the strongly semistable case there is only a finite degree range (which is independent of q) where we have no direct control on the global sections. This turns out to be enough to determine the asymptotic behavior of the Hilbert-Kunz function. In general, when the syzygy bundle is not strongly semistable, then the strong Harder-Narasimhan filtration (Theorem 5.16) contains enough information and yields the rationality of the Hilbert-Kunz multiplicity in graded dimension two. More precisely, there exists a formula for the Hilbert-Kunz multiplicity. To write it down we make the following definition (compare with Theorem 5.7 and Definition 5.14).

DEFINITION 9.5. Let $\mathcal{S}$ be a locally free sheaf on a smooth projective curve C and let

$$0 = \mathcal{S}_0 \subset \mathcal{S}_1 \subset \ldots \subset \mathcal{S}_t = F^{e*}(\mathcal{S})$$

be its strong Harder-Narasimhan filtration. Then for $k = 1, \ldots, t$ we set

$$\bar{\mu}_k(\mathcal{S}) := \frac{\mu_k(F^e(\mathcal{S}))}{p^e} = \frac{\mu(\mathcal{S}_k/\mathcal{S}_{k-1})}{p^e}.$$

Moreover, if R is a two-dimensional normal standard-graded domain and $C = \operatorname{Proj} R$ endowed with $\mathcal{O}_C(1)$, then we set $\nu_k = -\bar{\mu}_k/\deg(C)$.

The length t of the strong Harder-Narasimhan filtration, the ranks $r_k = \operatorname{rk}(\mathcal{S}_k/\mathcal{S}_{k-1})$, and these numbers ν_k are independent for q large enough. Applying this to the syzygy bundle $\mathcal{S} = \operatorname{Syz}(f_1, \ldots, f_n)$ we get numbers that control the global syzygies for varying q. With these rational numbers we can express the Hilbert-Kunz multiplicity in the following way.

THEOREM 9.6. *Let R denote a two-dimensional standard-graded normal domain and let $I = (f_1, \ldots, f_n)$ denote a homogeneous R_+-primary ideal generated by homogeneous elements f_i of degree $d_i, i = 1, \ldots, n$. Then the Hilbert-Kunz multiplicity $e_{HK}(I)$ is given by*

$$e_{HK}(I) = \frac{\deg(C)}{2}\left(\sum_{k=1}^{t} r_k \nu_k^2 - \sum_{i=1}^{n} d_i^2\right).$$

In particular, the Hilbert-Kunz multiplicity is a rational number.

This result was proved independently by H. Brenner in **[13]** and by V. Trivedi in **[99]**. As an easy corollary we get a description for the Hilbert-Kunz multiplicity for cones over plane curves. Note that in this case the maximal ideal is $\mathfrak{m} = (x, y, z)$ and so the syzygy bundle $\operatorname{Syz}(x, y, z)$ is the restriction of the cotangent bundle $\Omega_{\mathbb{P}^2} \cong \operatorname{Syz}(X, Y, Z)$ to the curve (see also Example 4.18).

COROLLARY 9.7. *Let $C = V_+(H) \subset \mathbb{P}^2$ denote a smooth plane projective curve of degree δ, $R = K[X, Y, Z]/(H)$. Then there exists a rational number $\frac{3}{2} \leq \nu_2 \leq 2$ such that the Hilbert-Kunz multiplicity of R is*

$$e_{HK}(R) = \delta(\nu_2^2 - 3\nu_2 + 3).$$

The Hilbert-Kunz multiplicity is $3\delta/4$ if and only if the syzygy bundle $\operatorname{Syz}(x, y, z)$ *(the restriction of the cotangent bundle) is strongly semistable.*

This corollary is independent of the result of Langer, since the existence of the strong Harder-Narasimhan filtration is trivial in this case.

EXAMPLE 9.8. We have seen in Example 5.11 that $\mathrm{Syz}(x,y,z)$ on the Fermat quartic given by $x^4+y^4+z^4=0$ is not strongly semistable in characteristic 3. The Harder-Narasimhan filtration of its first Frobenius pull-back is

$$0 \longrightarrow \mathcal{O}_C(-4) \longrightarrow \mathrm{Syz}(x^3,y^3,z^3),$$

which is strong. Hence $\bar{\mu}_1 = \mu_1(\mathrm{Syz}(x^3,y^3,z^3))/3 = -16/3$ and $\nu_1 = 4/3$ (and $\nu_2 = 5/3$). Therefore the Hilbert-Kunz multiplicity of $R = K[X,Y,Z]/(X^4+Y^4+Z^4)$ in characteristic 3 is $28/9$ (in the strongly semistable case it would be $27/9 = 3$).

The vector bundle approach to Hilbert-Kunz theory also yields results on the lower terms of the Hilbert-Kunz function.

THEOREM 9.9. *Let K denote an algebraically closed field of positive characteristic p. Let R denote a normal two-dimensional standard-graded K-domain and let I denote a homogeneous R_+-primary ideal. Then the Hilbert-Kunz function of I has the form*

$$\varphi(q) = e_{HK}(I)q^2 + \gamma(q),$$

where $e_{HK}(I)$ is a rational number and $\gamma(q)$ is a bounded function. In particular, the linear term in the theorem of Huneke, McDermott, and Monsky is 0 (Theorem 9.2). Moreover, if K is the algebraic closure of a finite field, then $\gamma(q)$ is an eventually periodic function.

PROOF. See [**17**, Theorem 6.1]. □

We want to emphasize that the finiteness of the base field is essential for the proof of this periodicity result (it is open without this assumption), as it was for the proof of Theorem 8.10. The main point here is again that the number of semistable bundles in a bounded degree range and over a finite field is finite. Hence every operator on such a set must have repetitions.

The rationality result for the Hilbert-Kunz multiplicity passes over to the non-normal case (see [**13**, Corollary 3.7]). The behavior of the lower terms in the Hilbert-Kunz function of a two-dimensional standard-graded domain is more difficult to understand. In particular, the 'linear term' need not be zero. Monsky succeeded in proving the following Theorem about plane irreducible curves. It is interesting to note that the singularities of the curve have a contribution to the Hilbert-Kunz function. One should also note that many of the properties that we have used on a smooth curve do not hold anymore. In particular, the Frobenius pull-back is not an exact functor, and not every torsion-free sheaf is locally free.

THEOREM 9.10. *Let $C \subset \mathbb{P}^2$ be a plane irreducible curve, and write $C = \mathrm{Proj}\, R$, $R = K[X,Y,Z]/(H)$ with a homogeneous polynomial H of degree δ. Then the Hilbert-Kunz function of R (i.e., of the maximal homogeneous ideal) has the form*

$$\varphi(p^e) = \left(\frac{3\delta}{4} + \frac{\alpha^2}{4\delta}\right)p^{2e} + b_e^* \cdot \frac{\alpha}{\delta} \cdot p^e + \gamma(p^e).$$

Here $\alpha \in \mathbb{Z}_p$, $0 \leq \alpha < \delta$ and b_e^* is a periodic, integer-valued function, which can be written as $b_e^* = \sum_P \beta_e^*(P)$, where the sum runs over the singular points of the curve and where $\beta_e^*(P)$ is an integer-valued periodic function for each singular point. The function $\gamma(q)$ is bounded and is an eventually periodic function if the base field is finite.

PROOF. See [**79**, Theorems I and II]. □

EXAMPLE 9.11. Let $R = K[X_{ij}, i = 1,2; j = 1,2,3]/I_2$, where I_2 is the ideal generated by the 2×2-minors of the generic 2×3-matrix. This is a four-dimensional normal Cohen-Macaulay domain; its Hilbert-Kunz function was computed by K.-I. Watanabe to be

$$\frac{13}{8}q^4 - \frac{1}{4}q^3 - \frac{1}{8}q^2 - \frac{1}{4}q$$

(see [**66**, Example 3.1 (2)]). In particular, even in the normal higher-dimensional case, the β in the theorem of Huneke, McDermott, and Monsky need not be 0.

A natural question is how the Hilbert-Kunz multiplicity behaves for varying prime numbers. For this we need the relative setting as described at the end of the first chapter where we defined tight closure in characteristic zero. In this setting Trivedi has proved the following Theorem.

THEOREM 9.12. *Let S be a standard-graded $\mathbb{Z}$-domain such that for almost all prime numbers p the fiber ring $R_p = S \otimes_{\mathbb{Z}} \mathbb{Z}/(p)$ is normal and two-dimensional. Then for a homogeneous S_+-primary ideal I with reductions I_p, the limit*

$$\lim_{p \to \infty} e_{HK}(R_p, I_p)$$

exists.

PROOF. See [**100**, Theorem 2.4]. □

REMARK 9.13. One can also characterize this limit directly in characteristic zero. On the curve $C_{\mathbb{Q}} = \operatorname{Proj} S \otimes_{\mathbb{Z}} \mathbb{Q}$ of characteristic zero we also have the syzygy bundle and hence its Harder-Narasimhan filtration. With this we can again (as we did in Definition 9.5 using the strong Harder-Narasimhan filtration) define numbers r_k and ν_k. Then the limit is $\frac{\deg(C)}{2}(\sum_{k=1}^{t} r_k \nu_k^2 - \sum_{i=1}^{n} d_i^2)$, which is also called the *Hilbert-Kunz multiplicity* of I in *characteristic zero* (see [**13**, Remark 4.8 and Proposition 4.9]). It was shown in [**7**] that this characteristic zero Hilbert-Kunz multiplicity behaves to solid closure like the Hilbert-Kunz multiplicity behaves to tight closure (Theorem 9.3). Namely, in the graded two-dimensional situation we have that $f \in I^\star$ if and only if $e_{HK,0}((I,f)) = e_{HK,0}(I)$. Hence in the relative situation we can also say (by the result of Trivedi) that $f \in I^\star$ (in characteristic zero) if and only if $\lim_{p \to \infty} e_{HK}(R_p, (I_p, f)) = \lim_{p \to \infty} e_{HK}(R_p, I_p)$. In this sense solid closure in characteristic zero is the limit of tight closure in positive characteristic.

We will see in the next section that the behavior of the Hilbert-Kunz multiplicity, of strong semistability and of tight closure for varying prime numbers (arithmetic deformations) may be quite complicated.

CHAPTER 10

Deformations of tight closure and the localization problem

In this last chapter we want to discuss arithmetic and geometric deformations of tight closure, of (strong) semistability, and of the Hilbert-Kunz multiplicity. In arithmetic deformations we look at the dependence of tight closure etc. on varying prime numbers, so the base space of the deformation is $\operatorname{Spec}\mathbb{Z}$.[1] The generic case (over the 0-ideal) in such a situation is in characteristic zero and the special cases (over a maximal ideal) are in varying positive characteristics. In geometric deformations, the characteristic is fixed and positive, and the data depend on certain parameters, the easiest case being $\mathbb{A}^1_K = \operatorname{Spec} K[U]$. Generically we have transcendental elements, while the special instances are algebraic over K. We will give below (Example 10.8) an example of a geometric deformation over $\mathbb{A}^1_K$ which gives a negative answer to the localization problem in tight closure theory (Problem 1.8).

We start with the situation of arithmetic deformations, which is the setting where tight closure in characteristic zero is defined (see also the end of Chapter 1). So let a finitely generated $\mathbb{Z}$-domain $\mathbb{Z} \subseteq S$ be given and consider the generic fiber $S_{\mathbb{Q}} = S \otimes_{\mathbb{Z}} \mathbb{Q} = R$ in characteristic zero and the special fibers $R_p = S \otimes_{\mathbb{Z}} \mathbb{Z}/(p)$ in positive characteristic p. Recall that for an ideal I in S and an element $f \in S$ one says that $f \in I^*$ holds in R if and only if the corresponding statement holds for almost all prime reductions, that is, if $f \in I_p^*$ holds in R_p for almost all primes p.

How does the containment $f \in I^*$ depend on the prime characteristic? A natural question is the following (see [**54**, Appendix 1], [**47**, Question 13], and [**60**, §4]): suppose that $f \in I^*$ holds for infinitely many prime reductions. Does it then hold for almost all primes?

A related question is how does the Hilbert-Kunz multiplicity of an $\mathfrak{m}$-primary ideal I vary with the prime number? Here we have already mentioned the result of Trivedi (Theorem 9.12) that in graded dimension two the limit $e_{HK}(I_p)$ exists for $p \longrightarrow \infty$. Is this limit achieved, and if so, for infinitely many prime numbers or even for almost all prime numbers? In terms of semistability, the corresponding question (for projective curves) was first raised by Miyaoka [**76**, Problem 5.4]. Suppose that $C \longrightarrow \operatorname{Spec}\mathbb{Z}$ is a relative generically smooth projective curve, and let a vector bundle $\mathcal{S}$ on C be given. Then $\mathcal{S}$ induces a vector bundle on every fiber of the curve. Suppose that the restriction of $\mathcal{S}$ to the generic fiber (of characteristic zero) is semistable. It is known that $\mathcal{S}_p$ on C_p is semistable for almost all p; see [**76**, §5]. However, what can be said about the behavior of strong semistability? Miyaoka asked whether there exist infinitely many prime numbers such that the reduction is strongly semistable. Shepherd-Barron [**92**] asked the stronger question whether it is strongly semistable for almost all prime reductions.

We first give a result where everything behaves nicely.

THEOREM 10.1. *Let $\mathbb{Z} \subseteq S$ be a standard-graded normal three-dimensional domain such that* $\operatorname{Proj} S \longrightarrow \operatorname{Spec}\mathbb{Z}$ *is a relative elliptic curve. Let I be a homogeneous S_+-primary*

[1] One can also consider more general arithmetic base schemes, like the spectrum of the ring of integers in an algebraic number field.

ideal, and $f \in S$ homogeneous. Then the Hilbert-Kunz multiplicity $e_{HK}(I_p)$ is almost constant (constant except for finitely many prime numbers). Moreover, for a homogeneous element f we have either $f \in I_p^*$ or $f \notin I_p^*$ for almost all prime numbers.

PROOF. Almost all fibers are elliptic curves. On an elliptic curve a semistable bundle is strongly semistable (Remark 5.13). Hence a bundle that is semistable in characteristic zero is strongly semistable for almost all prime numbers. This implies that the Harder-Narasimham filtration of the syzygy bundle in characteristic zero induces the strong Harder-Narasimhan filtration for almost all prime numbers. Hence the result follows from the formula for the Hilbert-Kunz multiplicity (Theorem 9.6) and from the numerical characterization of tight closure (Theorem 7.6 and Theorem 7.7). □

However, this result is restricted to elliptic curves; for curves of higher genus the arithmetical behavior is more complicated. With regard to the Hilbert-Kunz multiplicity, Monsky and Han ([**34**], [**33**], [**80**], [**100**]) gave the following example where the Hilbert-Kunz multiplicity in an arithmetic family is not almost constant.

EXAMPLE 10.2. Let $R_p = \mathbb{Z}/(p)[X,Y,Z]/(X^4+Y^4+Z^4)$. Then the Hilbert-Kunz multiplicity of R_p is

$$e_{HK}(R_p) = \begin{cases} 3 \text{ for } p = \pm 1 \bmod 8 \\ 3 + 1/p^2 \text{ for } p = \pm 3 \bmod 8 . \end{cases}$$

Because of Corollary 9.7 this means that the syzygy bundle $\mathrm{Syz}(x,y,z)$ on the Fermat quartic is strongly semistable for $p = \pm 1 \bmod 8$ (and semistable in characteristic zero), but not strongly semistable for $p = \pm 3 \bmod 8$. Since there are infinitely many primes in every arithmetic progression (Theorem of Dirichlet, see [**90**, Chapitre VI, Section 4, Théorème and Corollaire]), this means that for infinitely many prime reductions the restricted cotangent bundle is not strongly semistable (though it is semistable in characteristic zero). Hence the question of Shepherd-Barron has a negative answer.

The original question of Miyaoka, whether there exist infinitely many prime numbers with strongly semistable reduction, is still open. The following class of examples shows that, under the hypothesis that there exist infinitely many Sophie Germain prime numbers (widely believed to be true), the density of prime numbers with strongly semistable reduction can be arbitrarily small. Recall that a prime number h is called a *Sophie Germain prime* if $2h+1$ is also prime.

EXAMPLE 10.3. Let $h > 5$ be a Sophie Germain prime and set $d = 2h+1$. Consider the syzygy bundle $\mathrm{Syz}(x^2,y^2,z^2)$ on the Fermat curve given by $x^d + y^d + z^d = 0$. It can be shown (see [**11**, Proposition 2]) that, for prime numbers $p \neq \pm 1 \bmod d$, this bundle is not strongly semistable. Hence the density of primes with strongly semistable reduction is $\leq 1/h$, which is arbitrarily small for h large enough.

We now come to the arithmetical behavior of tight closure. The following example given in [**18**] shows that the arithmetical behavior of tight closure is more complicated than first hoped for: an element might belong to the tight closure of an ideal for infinitely many prime numbers, but not for almost all prime numbers. This result also challenges the very definition of tight closure in characteristic zero.

EXAMPLE 10.4. Let $R = K[X,Y,Z]/(X^7+Y^7-Z^7)$, $I = (x^4,y^4,z^4)$, and $f = x^3y^3$. It was proved in [**18**] that $f \in I^*$ holds for $p = 3 \bmod 7$ and does not hold for $p = 2 \bmod 7$.

REMARK 10.5. This example has some interesting consequences. The ideal $\mathfrak{a} = (x,y,z)$ in the forcing algebra

$$A = K[X,Y,Z,U,V,W]/(X^7+Y^7-Z^7, UX^4+VY^4+WZ^4+X^3Y^3)$$

is such that the open subset $D(\mathfrak{a}) \subset \operatorname{Spec} A$ is affine for infinitely many but not for almost all prime reductions. This means that its cohomological dimension fluctuates arithmetically between 0 and 1. In characteristic zero the open subset $D(\mathfrak{a})$ is not affine, so f does belong to the solid closure of I, but it does not belong to the tight closure of I.

We now come to geometric deformations of tight closure. This means basically that we replace $\operatorname{Spec}\mathbb{Z}$, which was our base space for the arithmetic deformations, by an equicharacteristic scheme (of positive characteristic). In fact we will only deal with $\operatorname{Spec} K[U]$ as deformation space. So we have a faithfully flat extension $K[U] \subseteq S$ and are interested in the fibers of the morphism $\operatorname{Spec} S \to \operatorname{Spec} K[U] = \mathbb{A}^1_K$. The (generic or transcendental) fiber over (0) is just $\operatorname{Spec}(S \otimes_{K[U]} K(U))$, which is a ring over the field of rational functions $K(U)$. The (special or algebraic) fibers over a maximal ideal $\mathfrak{m} = (P(U))$ in $K[U]$ are $\operatorname{Spec}(S \otimes_{K[U]} \kappa(\mathfrak{m}))$. We can also identify $\mathfrak{m}$ with a certain value $u \in \bar{K}$ in an algebraic closure. We are again mainly interested in the case where S has relative dimension two over the basis (hence dimension three) and is standard-graded. The corresponding relative projective curve $C = \operatorname{Proj} S$ is then a family over $\mathbb{A}^1$.

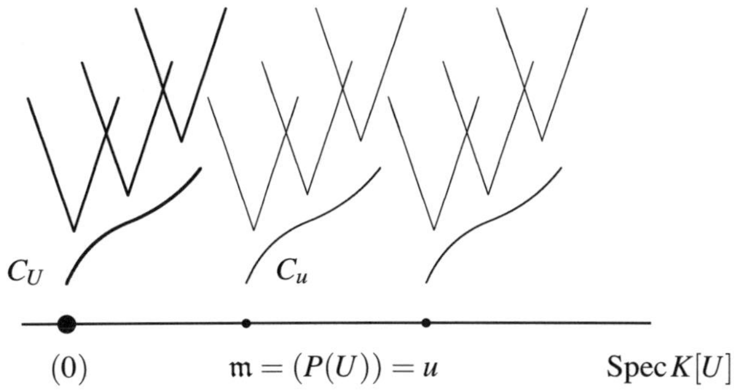

In this situation we ask again how the Hilbert-Kunz multiplicity, strong semistability, and tight closure behave in such a geometric family. The behavior of tight closure under such a geometric deformation is related to the localization problem (Problem 1.8) by the following proposition.

PROPOSITION 10.6. *Let K be a field of positive characteristic, and let $K[U] \subset S$ be a three-dimensional standard-graded domain such that almost all fibers are a normal standard-graded domain of dimension two. Let $M \subseteq K[U] \subset S$ be the multiplicative system of all non-zero polynomials in U. Let $I = (f_1, \dots, f_n)$ be a homogeneous S_+-primary ideal, generated by homogeneous elements of degree $d_i = \deg(f_i)$ with syzygy bundle $\operatorname{Syz}(f_1, \dots, f_n)$. Let $f \in S$ be a homogeneous element of degree $\geq (\sum_{i=1}^n d_i)/(n-1)$. If the syzygy bundle is generically strongly semistable, then $f \in I^*$ in S_M. If tight closure localizes, then $f \in I^*$ also holds for almost all algebraic specializations.*

PROOF. Note that $S_M = S \otimes_{K[U]} (K[U]_M) = S \otimes_{K[U]} K(U)$. So the ring localized at the given multiplicative system is the generic fiber ring in the family. The statement about the containment in the tight closure follows from the degree bound (Theorem 6.4). So suppose that $f \in I^*$ in S_M. If tight closure localizes, then there exists an element $h \in M$ such that $hf \in I^*$ holds in S. By the persistence of tight closure (Theorem 1.12) we have for every algebraic specialization $U \mapsto u \in \overline{K}$ that $hf \in I^*$ in S_u. Since $h(u) = 0$ for only finitely many values $u \in \overline{K}$ it follows that $f \in I^*$ in S_u for almost all algebraic values. □

This means that we can disprove localization if we find a geometric deformation where $f \in I^*$ holds generically, but $f \notin I^*$ holds for infinitely many algebraic values $u \in \overline{K}$. Such a behavior can only happen if the syzygy bundle is strongly semistable in the transcendental case, but not strongly semistable for infinitely many algebraic values (so this cannot happen for a family of elliptic curves). Monsky has given in **[78]** the following example, where the Hilbert-Kunz multiplicity varies with an algebraic parameter. This example can also be interpreted in terms of strong semistability (see also **[99**, Section 5]).

EXAMPLE 10.7. Let $K = \mathbb{F}_2$ and $S = K[U][X,Y,Z]/(g)$ for

$$g = Z^4 + Z^2XY + Z(X^3+Y^3) + (U+U^2)X^2Y^2 .$$

This example was studied by Monsky in the context of Hilbert-Kunz theory. He showed that the Hilbert-Kunz multiplicity of $S_{\kappa(\mathfrak{p})} = S \otimes_{K[U]} \kappa(\mathfrak{p})$, $\mathfrak{p} \in \operatorname{Spec} K[U]$, is

$$e_{HK}(S_{\kappa(\mathfrak{p})}) = \begin{cases} 3 \text{ for } \kappa(\mathfrak{p}) = K(U) \\ 3 + 1/4^m \text{ for } \kappa(\mathfrak{p}) = K(u) \subset \overline{K}, \text{ where } m = \deg(K(u)|K) . \end{cases}$$

By Corollary 9.7 this means that the syzygy bundle $\operatorname{Syz}(x,y,z)$ on $C_{\mathfrak{p}} = \operatorname{Proj} S_{\kappa(\mathfrak{p})}$ is strongly semistable in the transcendental situation and not strongly semistable in all algebraic situations. Monsky shows in fact that the syzygy bundle $\operatorname{Syz}(x^{\tilde q}, y^{\tilde q}, z^{\tilde q})(\frac{3\tilde q}{2} - 1)$ has a global non-zero section on $C_{\mathfrak{p}}$, where $\kappa(\mathfrak{p})$ is algebraic of degree m and $\tilde q = 2^{m+1}$ **[78**, Theorem 3.1]. Thus this pull-back is not semistable anymore. Hence we have a destabilizing short exact sequence

$$0 \longrightarrow \mathscr{L} \longrightarrow \operatorname{Syz}(x^{\tilde q}, y^{\tilde q}, z^{\tilde q}) \longrightarrow \mathscr{M} \longrightarrow 0 .$$

Suppose that $\mathscr{L}$ has degree $6\tilde q + a$ and $\mathscr{M}$ has degree $6\tilde q - a$ (the sum must be $12\tilde q$). Then we have $\nu_1 = -\frac{3}{2} + \frac{a}{4\tilde q}$ and $\nu_2 = -\frac{3}{2} - \frac{a}{4\tilde q}$. Hence by Theorem 9.6 we must have

$$3 + (\frac{1}{4})^m = 2((-\frac{3}{2} + \frac{a}{4\tilde q})^2 + (-\frac{3}{2} - \frac{a}{4\tilde q})^2 - 3) = 2(2\frac{9}{4} - 3 + 2\frac{a^2}{16\tilde q^2}) = 3 + \frac{a^2}{4 \cdot 4^{m+1}} ;$$

therefore $a = 4$. So $\mathscr{L} \otimes \mathscr{O}(\frac{3\tilde q}{2} - 1)$ has degree 0 and it has a section, thus $\mathscr{L} = \mathscr{O}(1 - \frac{3\tilde q}{2})$. It follows that

$$0 \longrightarrow \mathscr{O}(1) \longrightarrow \operatorname{Syz}(x^{\tilde q}, y^{\tilde q}, z^{\tilde q})(\frac{3\tilde q}{2}) \longrightarrow \mathscr{O}(-1) \longrightarrow 0 \quad (*)$$

must be the destabilizing sequence.

Building on Monsky's example we have the following example which gives a negative answer to the localization problem. We describe here how the disproving of this problem reduces it to an elementary ideal membership problem, which was finally settled in **[19]**.

EXAMPLE 10.8. Let $K = \mathbb{F}_2$ and $S = K[U][X,Y,Z]/(g)$ for

$$g = Z^4 + Z^2XY + Z(X^3+Y^3) + (U+U^2)X^2Y^2 .$$

Consider the ideal

$$I = (x^4, y^4, z^4) \text{ and } f = y^3z^3 \text{ in } S_{\kappa(\mathfrak{p})}, \ \mathfrak{p} \in \operatorname{Spec} K[U] .$$

Since $\mathrm{Syz}(x,y,z)$ is strongly semistable for U transcendental, the same is true for $\mathrm{Syz}(x^4,y^4,z^4) = F^{2*}(\mathrm{Syz}(x,y,z))$. Hence by the degree bound Theorem 6.4 we know that $f \in I^*$ holds in the transcendental case (in $S_{K(U)}$). There is however strong numerical evidence, which we present in the following, that $f \notin I^*$ in $S_{\kappa(\mathfrak{p})}$ for all algebraic instances $\mathfrak{p} = (u)$, $u \in \overline{K}$.

In order to show that $f \notin I^*$ in $S_{\kappa(\mathfrak{p})}$ we have to show for a test element t and a certain power $q = 2^e$ (depending on $u \in \overline{K}$) that $tf^q \neq I^{[q]}$. We first need a lemma on test elements.

LEMMA 10.9. *For every point* $\mathfrak{p} \in \operatorname{Spec} K[U]$ *the test ideal of* $S_{\kappa(\mathfrak{p})}$ *is* $(S_{\kappa(\mathfrak{p})})_{\geq 2}$.

PROOF. The test ideal $\tau = \tau_{S_{\kappa(\mathfrak{p})}}$ is the annihilator of the tight closure 0^* inside $H^2_{\mathfrak{m}}(S_{\kappa(\mathfrak{p})}) \cong H^1(D(\mathfrak{m}), \mathscr{O})$ (for the test ideal see [**55**], in particular Proposition 4.1), where $\mathfrak{m}$ is the graded irrelevant ideal. We claim that $0^* = H^2_{\mathfrak{m}}(S_{\kappa(\mathfrak{p})})_{\geq 0}$ (note that the local cohomology module is graded). It is clear that 0^* contains everything of nonnegative degree. An element of negative degree belongs to the tight closure only if it is annihilated by some Frobenius power (Corollary 4.14). We consider a non-zero class of negative degree, $c \in H^1(C, \mathscr{O}(-n))$. This gives a non-trivial extension $0 \to \mathscr{O}_C(-n) \to \mathscr{S} \to \mathscr{O}_C \to 0$ and dually $0 \to \mathscr{O}_C \to \mathscr{F} \to \mathscr{O}_C(n) \to 0$. In such a situation every quotient bundle of $\mathscr{F}$ has positive degree (Proposition 4.12(i)), and we have $\deg(\mathscr{F}) = n \deg(C) \geq 4 > 2 \geq \frac{2}{p}(g(C)-1)$. Hence by [**8**, Theorem 2.7] or directly [**41**, Corollary 7.7] the rank-two bundle $\mathscr{F}$ is ample and every Frobenius pull-back of it stays ample. Hence the Frobenius cannot trivialize the extension.

We have $(H^2_{\mathfrak{m}}(S_{\kappa(\mathfrak{p})}))_m = 0$ for $m \geq 2$, hence $(S_{\kappa(\mathfrak{p})})_{\geq 2}$ annihilates $0^* = (H^2_{\mathfrak{m}}(S_{\kappa(\mathfrak{p})}))_{\geq 0}$. Therefore $(S_{\kappa(\mathfrak{p})})_{\geq 2} \subseteq \tau$ (we only need this inclusion). We also have $(H^2_{\mathfrak{m}}(S_{\kappa(\mathfrak{p})}))_0 = K\langle \frac{z^3}{x^2y}, \frac{z^3}{xy^2}, \frac{z^2}{xy} \rangle$ and $(H^2_{\mathfrak{m}}(S_{\kappa(\mathfrak{p})}))_1 = K\langle \frac{z^3}{xy} \rangle$, and the annihilators of the three cohomology classes of degree zero are (y,z), (x,z), and (x,y). Hence no linear form annihilates $(H^2_{\mathfrak{m}}(S_{\kappa(\mathfrak{p})}))_0$, and we have equality.

We also show that the Frobenius is injective on the degree zero part of the cohomology module. The class z^2/xy is mapped to itself, since

$$\frac{z^4}{x^2y^2} = \frac{z^2xy + z(x^3+y^3) + (u+u^2)x^2y^2}{x^2y^2} = \frac{z^2}{xy}.$$

The class z^3/x^2y is mapped to

$$\begin{aligned}\frac{z^6}{x^4y^2} &= \frac{z^2(z^2xy + z(x^3+y^3) + (u+u^2)x^2y^2)}{x^4y^2} \\ &= \frac{(z^2xy + z(x^3+y^3) + (u+u^2)x^2y^2)xy + z^3x^3}{x^4y^2} = \frac{z^3}{xy^2},\end{aligned}$$

so the other two basis classes swap. In particular this p-linear map is a bijection over a finite field. □

Lemma 10.9 shows particularly that y^2 is a test element. The next lemma helps to simplify the computations.

LEMMA 10.10. *Let* $u \in \overline{K}$, $m = \deg(K(u)|K)$ *and set* $q = 2^m$. *Then* $y^2 f^q \notin I^{[q]}$ *if and only if* $yf^{\frac{q}{2}} \notin I^{[\frac{q}{2}]}$.

PROOF. One direction is clear. For the other direction we tensor the sequence $(*)$ in Example 10.7 with $\mathscr{O}(1)$ and get ($\tilde{q} = 2q$)

$$0 \longrightarrow \mathscr{O}(2) \longrightarrow \mathrm{Syz}(x^{2q}, y^{2q}, z^{2q})(3q+1) \longrightarrow \mathscr{O} \longrightarrow 0 .$$

The element $yf^{\frac{q}{2}} = yy^{3\frac{q}{2}}z^{3\frac{q}{2}}$ determines the cohomology class $\delta(yf^{q/2})$ in $H^1(C, \mathrm{Syz}(x^{2q}, y^{2q}, z^{2q})(3q+1))$, which is 0 if and only if the element belongs to the ideal $(x^{2q}, y^{2q}, z^{2q}) = I^{[q/2]}$. So by assumption this class is not zero. Since $H^1(C, \mathscr{O}(2)) = 0$ we have an isomorphism

$$H^1(C, \mathrm{Syz}(x^{2q}, y^{2q}, z^{2q})(3q+1)) \cong H^1(C, \mathscr{O}) .$$

The Frobenius maps the class $\delta(yf^{\frac{q}{2}})$ to the class $\delta(y^2 f^q)$ in $H^1(C, \mathrm{Syz}(x^{4q}, y^{4q}, z^{4q})(6q+2))$. The Frobenius acts, however, injectively on the zero degree of the cohomology, as shown in the proof of Lemma 10.9. Hence $y^2 f^q \notin (x^{4q}, y^{4q}, z^{4q}) = I^{[q]}$. □

The following numerical observation (Lemma 10.10 makes the computation easier by halving the Frobenius exponent of I) gave important numerical evidence that in Example 10.8 the conclusion of Proposition 10.6 does not hold and hence localization cannot hold.

OBSERVATION 10.11. Let $u \in \overline{K}$ be an algebraic element of degree m, and let $\mathfrak{p} \in \mathrm{Spec}\,\mathbb{F}_2[U]$ be the corresponding closed point. Then the following hold in $S_{\kappa(\mathfrak{p})}$: For $q' = 2^e$, $e < m$, we have

$$y(yz)^{\frac{3q'}{2}} \in (x^{2q'}, y^{2q'}, z^{2q'})$$

and (more importantly) for $q = 2^m$ we have

$$y(yz)^{\frac{3q}{2}} \notin (x^{2q}, y^{2q}, z^{2q}), \quad \text{hence } y^2 f^q \notin I^{[q]} \text{ and } f \notin I^* .$$

This has been computed for all algebraic elements of degree ≤ 8, that is, for all elements in $\mathbb{F}_{256} \subset \overline{K}$ with the help of Almar Kaid and CoCoA. For this just take an irreducible polynomial $h(U)$ of degree m and check the ideal membership in $\mathbb{F}_2[U, X, Y, Z]/(g, h)$. Moreover, for each degree $9, 10, 11, 12$ and for one chosen irreducible polynomial of that degree, the same behavior was observed in computations done by Martin Kreuzer and Daniel Heldt.

After these notes were written, P. Monsky succeeded in proving that this behavior holds for all algebraic values $u \in \overline{K}$, thus completing the proof that tight closure does not commutate with localization; see **[19]**.

Bibliography

[1] D. Arapura. Frobenius amplitude and strong vanishing theorems for vector bundles. With an appendix by Dennis S. Keeler. *Duke Math. J.*, 121(2):231–267, 2004.

[2] M. F. Atiyah. Vector bundles over an elliptic curve. *Proc. London Math. Soc.*, 7:414–452, 1957.

[3] M. F. Atiyah and R. Bott. The Yang-Mills equations over Riemann surfaces. *Philos. Trans. Roy. Soc. London. Ser. A*, 308:523–615, 1982.

[4] C. M. Barton. Tensor products of ample vector bundles in characteristic p. *Amer. J. Math.*, 93:429–438, 1971.

[5] H. Brenner. How to rescue solid closure. *J. Algebra*, 265:579–605, 2003.

[6] H. Brenner. *The theory of tight closure from the viewpoint of vector bundles.* Habilitationsschrift, Ruhr-Universität Bochum, 2003. ArXiv.

[7] H. Brenner. A characteristic zero Hilbert-Kunz criterion for solid closure in dimension two. *Math. Research Letters*, 11:563–574, 2004.

[8] H. Brenner. Slopes of vector bundles and applications to tight closure problems. *Trans. Amer. Math. Soc.*, 356(1):371–392, 2004.

[9] H. Brenner. Computing the tight closure in dimension two. *Mathematics of Computations*, 74(251):1495–1518, 2005.

[10] H. Brenner. A linear bound for Frobenius powers and an inclusion bound for tight closure. *Michigan Math. J.*, 53(3):585–596, 2005.

[11] H. Brenner. On a problem of Miyaoka. In B. Moonen, R. Schoof, G. van der Geer, editors, *Number Fields and Function Fields - Two Parallel Worlds*, volume 239 of *Progress in Mathematics*, pages 51–59. Birkhäuser, 2005.

[12] H. Brenner. Tight closure and plus closure for cones over elliptic curves. *Nagoya Math. J.*, 177:31–45, 2005.

[13] H. Brenner. The rationality of the Hilbert-Kunz multiplicity in graded dimension two. *Math. Ann.*, 334:91–110, 2006.

[14] H. Brenner. Tight closure and plus closure in dimension two. *Amer. J. Math.*, 128(2):531–539, 2006.

[15] H. Brenner. Looking out for stable syzygy bundles. *ArXiv*, 2007.

[16] H. Brenner. Semistable monomial syzygy bundles on $\mathbb{P}^2$. *In preparation*, 2007.

[17] H. Brenner. The Hilbert-Kunz function in graded dimension two. *Comm. in Algebra (to appear)*, 2008.

[18] H. Brenner and M. Katzman. On the arithmetic of tight closure. *J. Amer. Math. Soc.*, 19(3):659–672, 2006.

[19] H. Brenner and P. Monsky. Tight closure does not commute with localization. *arXiv:0710.2913v2*, Oct 2007.

[20] W. Bruns and J. Herzog. *Cohen-Macaulay Rings, Revised edition.* Cambridge University Press, 1998.

[21] C. Deninger and A. Werner. Line bundles and p-adic characters. In B. Moonen, R. Schoof, G. van der Geer, editors, *Number Fields and Function Fields - Two Parallel Worlds*, volume 239 of *Progress in Mathematics*, pages 101–131. Birkhäuser, 2005.

[22] C. Deninger and A. Werner. Vector bundles on p-adic curves and parallel transport. *Ann. Scient. Éc. Norm. Sup.*, 38:535–597, 2005.

[23] G. Dietz. Tight closure of finite length modules in graded rings. *J. Algebra*, 306(2):520–534, 2006.

[24] S. K. Donaldson. Anti self-dual Yang-Mills connections over complex algebraic surfaces and stable vector bundles. *Proc. London Math. Soc. (3)*, 50:1–26, 1985.

[25] D. Eisenbud. *Commutative Algebra with a View Toward Algebraic Geometry.* Springer, 1995.

[26] H. Flenner. Restrictions of semistable bundles on projective varieties. *Comment. Math. Helv.*, 59:635–650, 1984.

[27] D. Gieseker. p-ample bundles and their Chern classes. *Nagoya Math. J.*, 43:91–116, 1971.

[28] J. E. Goodman. Affine open subsets of algebraic varieties and ample divisors. *Ann. of Math.*, 89:160–183, 1969.

[29] A. Grothendieck and J. Dieudonné. *Eléments de géométrie algébrique II*, volume 8. Inst. Hautes Études Sci. Publ. Math., 1961.

[30] A. Grothendieck and J. Dieudonné. *Eléments de géométrie algébrique I*. Springer, 1971.

[31] A. Grothendieck et al. *Revêtements Étales et Groupe Fondamental (=SGA1)*. Springer, 1970.

[32] A. Grothendieck et al. *Schema en groupes (=SGA3) I,II,III*. Springer, 1970.

[33] C. Han. *The Hilbert-Kunz function of a diagonal hypersurface*. PhD thesis, Brandeis University, 1991.

[34] C. Han and P. Monsky. Some surprising Hilbert-Kunz functions. *Math. Z.*, 214:119–135, 1993.

[35] N. Hara. A characterization of rational singularities in terms of injectivity of Frobenius maps. *Amer. J. of Math.*, 120(5):981–996, 1998.

[36] N. Hara. Geometric interpretation of tight closure and test ideals. *Trans. Amer. Math. Soc.*, 353(5):1885–1906, 2001.

[37] N. Hara and S. Takagi. Some remarks on a generalization of test ideals. *Nagoya Math. J.*, 175:59–74, 2004.

[38] N. Hara and K. I. Watanabe. F-regular and F-pure rings vs. log terminal and log canonical singularities. *J. Algebraic Geom.*, 11(2):363–392, 2002.

[39] N. Hara and K. I. Yoshida. A generalization of tight closure and multiplier ideals. *Trans. Amer. Math. Soc.*, 355(8):3143–3174, 2003.

[40] G. Harder and M. S. Narasimhan. On the cohomology groups of moduli spaces of vector bundles on curves. *Math. Ann.*, 212:215–248, 1975.

[41] R. Hartshorne. Ample vector bundles. *Publ. Math. I.H.E.S.*, 29:63–94, 1966.

[42] R. Hartshorne. Cohomological dimension of algebraic varieties. *Ann. of Math.*, 88:403–450, 1968.

[43] R. Hartshorne. *Ample Subvarieties of Algebraic Varieties*. Springer, Berlin, Heidelberg, New York, 1970.

[44] R. Hartshorne. Ample vector bundles on curves. *Nagoya Math. J.*, 43:73–89, 1971.

[45] R. Hartshorne. *Algebraic Geometry*. Springer, New York, 1977.

[46] M. Hochster. Solid closure. *Contemp. Math.*, 159:103–172, 1994.

[47] M. Hochster. Tight closure in equal characteristic, big Cohen-Macaulay algebras, and solid closure. *Contemp. Math.*, 159:173–196, 1994.

[48] M. Hochster. The notion of tight closure in equal characteristic zero, Appendix 1 to : C. Huneke, Tight Closure and Its Applications. AMS, 1996.

[49] M. Hochster. Tight closure theory and characteristic p methods. With an appendix by Graham J. Leuschke. In *Trends in Commutative Algebra*, volume 51 of *Math. Sci. Res. Inst. Publ.*, pages 181–210, Cambridge, 2004. Cambridge Univ. Press.

[50] M. Hochster and C. Huneke. Tight closure in equal zero characteristic. Preprint.

[51] M. Hochster and C. Huneke. Tight closure, invariant theory, and the Briançon-Skoda theorem. *J. Amer. Math. Soc.*, 3:31–116, 1990.

[52] M. Hochster and C. Huneke. Tight closure of parameter ideals and splittings in module finite extensions. *J. Algebraic Geom.*, 3:599–670, 1994.

[53] M. Hochster and J. Roberts. Rings of invariants of reductive groups acting on regular rings are Cohen-Macaulay. *Advances in Math.*, 13:115–175, 1974.

[54] C. Huneke. *Tight Closure and Its Applications*, volume 88 of *CBMS Lecture Notes in Mathematics*. AMS, Providence, 1996.

[55] C. Huneke. Tight closure, parameter ideals, and geometry. In *Six Lectures on Commutative Algebra*, pages 187–329. Birkhäuser, 1998.

[56] C. Huneke, M. McDermott, and P. Monsky. Hilbert-Kunz functions for normal rings. *Math. Res. Letters*, 11(4):539–546, 2004.

[57] D. Huybrechts and M. Lehn. *The Geometry of Moduli Spaces of Sheaves*. Vieweg, 1997.

[58] K. Joshi, S. Ramanan, E. Z. Xia, and J.-K. Yu. On vector bundles destabilized by Frobenius pull-back. *Compositio*, 142:616–630, 2006.

[59] K. Joshi and E. Z. Xia. Moduli of vector bundles on curves in positive characteristic. *Compositio Math.*, 122(3):315–321, 2000.

[60] M. Katzman. Finite criteria for weak F-regularity. *Illinois J. Math.*, 40(3):453–463, 1996.

[61] S. L. Kleiman. Ample vector bundles on surfaces. *Proc. Amer. Math. Soc.*, 21:673–676, 1969.

[62] A. Klyachko. Equivariant bundles on toral varieties. *Math. USSR Izvestiya*, 35(2):337–375, 1990.

[63] A. Klyachko. Stable bundles, representation theory and Hermitian operators. *Selecta Mathematica*, 1998.

[64] A. Klyachko. Vector bundles, linear representations, and spectral problems. In Ta Tsien, editor, *Proceedings of the International Congress of Mathematicians, ICM Beijing, China 2002*, volume II, pages 599–613, 2002.

[65] E. Kunz. Characterizations of regular local rings of characteristic p. *Amer. J. Math.*, 91:772–784, 1969.

[66] K. Kurano. The singular Riemann-Roch theorem and Hilbert-Kunz functions. *J. Algebra*, 304(1):487–499, 2006.

[67] H. Lange and C. Pauly. On Frobenius-destabilized rank-2 vector bundles over curves. *Comm. Math. Helv. (to appear)*, 2007.

[68] H. Lange and U. Stuhler. Vektorbündel auf Kurven und Darstellungen der algebraischen Fundamentalgruppe. *Math. Zeitschrift*, 156:73–83, 1977.

[69] A. Langer. Semistable sheaves in positive characteristic. *Ann. Math.*, 159:251–276, 2004.

[70] Y. Laszlo and C. Pauly. The action of the Frobenius map on rank 2 vector bundles in characteristic 2. *J. Algebraic Geometry*, 11:219–243, 2002.

[71] Y. Laszlo and C. Pauly. The Frobenius map, rank 2 vector bundles and Kummer's quartic surface in characteristic 2 and 3. *Adv. Math.*, 185(2):246–269, 2004.

[72] R. Lazarsfeld. *Positivity in Algebraic Geometry*. Springer, Berlin, 2004.

[73] V. B. Mehta and C. Pauly. Semistability of Frobenius direct images over curves. *ArXiv*, 2006.

[74] L. Migliorini. Some observations on cohomologically p-ample bundles. *Ann. di Math. pura ed appl.*, CLXIV:89–102, 1993.

[75] J. S. Milne. *Étale Cohomology*. Princeton University Press, 1980.

[76] Y. Miyaoka. The Chern class and Kodaira dimension of a minimal variety. In *Algebraic Geometry, Sendai 1985*, volume 10 of *Adv. Stud. Pure Math.*, pages 449–476, 1987.

[77] P. Monsky. The Hilbert-Kunz function. *Math. Ann.*, 263:43–49, 1983.

[78] P. Monsky. Hilbert-Kunz functions in a family: Point S_4-quartics. *J. Algebra*, 208:343–358, 1998.

[79] P. Monsky. Hilbert-Kunz functions for irreducible plane curves. *Preprint*, 2006.

[80] P. Monsky. The Hilbert-Kunz multiplicity of an irreducible trinomial. *J. Algebra*, 304:1101–1107, 2006.

[81] P. Monsky. Are Hilbert-Kunz multiplicities always rational - a likely counterexample. *Preprint*, 2007.

[82] D. Mumford. Projective invariants of projective structures and applications. In *Proc. Intern. Cong. Math. Stockholm*, pages 526–530, 1962.

[83] D. Mumford. *Abelian Varieties*. Oxford University Press, 1970.

[84] D. Mumford. *Geometric Invariant Theory*. Springer, 2nd edition, 1982.

[85] M. S. Narasimhan and C. S. Shesadri. Stable and unitary vector bundles on a compact Riemann surface. *Ann. Math.*, 82:540–567, 1965.

[86] M. V. Nori. On the representation of the fundamental group. *Compos. Math.*, 33:29–41, 1976.

[87] B. Osserman. The generalized Verschiebung map for curves of genus 2. *Math. Ann.*, 336(4):963–986, 2006.

[88] B. Osserman. Frobenius-unstable bundles and p-curvature. *Trans. Amer. Math. Soc. (to appear)*, 2007.

[89] P. Roberts. A computation of local cohomology. *Contemp. Math.*, 159:351–356, 1994.

[90] J. P. Serre. *Cours d' Arithmétique*. Presses Universitaires de France, 1970.

[91] C. S. Seshadri. Fibrés vectoriels sur les courbes algébrique. *Asterisque*, 96, 1982.

[92] N. I. Shepherd-Barron. Semi-stability and reduction mod p. *Topology*, 37(3):659–664, 1997.

[93] A. Singh. A computation of tight closure in diagonal hypersurfaces. *J. Algebra*, 203:579–589, 1998.

[94] K. E. Smith. Tight closure of parameter ideals. *Invent. Math.*, 115:41–60, 1994.

[95] K. E. Smith. Tight closure in graded rings. *J. Math. Kyoto Univ.*, 37(1):35–53, 1997.

[96] K. E. Smith. The multiplier ideal is a universal test ideal. *Commun. in Algebra*, 28(12):5915–5929, 2000.

[97] K. E. Smith. An introduction to tight closure. In *Geometric and combinatorial aspects of commutative algebra*, volume 217 of *Lect. Notes Pure Appl. Math.*, pages 353–377, 2001.

[98] S. Takagi. Formulas for multiplier ideals on singular varieties. *Amer. J. Math.*, 128(6):1345–1362, 2006.

[99] V. Trivedi. Semistability and Hilbert-Kunz multiplicity for curves. *J. Algebra*, 284(2):627–644, 2005.

[100] V. Trivedi. Hilbert-Kunz multiplicity and reduction mod p. *Nagoya Math. J.*, 185:123–141, 2007.

[101] L. W. Tu. Semistable bundles over an elliptic curve. *Adv. Math.*, 98:1–26, 1993.

[102] K. Uhlenbeck and S. T. Yau. On the existence of Hermitian Yang-Mills connections in stable vector bundles. *Commun. Pure. Appl. Math.*, 39:257–293, 1986.

Combinatorics and Commutative Algebra

by Jürgen Herzog

Contents

Introduction

In the following four lectures I give a survey on some combinatorial topics from the viewpoint of an algebraist. The lectures reflect my joint research with Takayuki Hibi over the last 10 years. Among other coauthors, Aneta Aramova and Xinxian Zheng have also contributed substantially to these studies.

The first lecture deals with algebraic shifting theory. We mainly discuss the comparison of the graded Betti-numbers of the Stanley-Reisner ring of a simplicial complex and its shifted complex and discuss the open conjectures related to this topic.

In the second lecture polymatroids are considered. Two algebraic objects are attached to each polymatroid: its base ring and its polymatroidal ideal. We discuss a challenging conjecture of White concerning the relations of the base ring of a polymatroid and study the homological properties of polymatroidal ideals.

The third lecture is devoted to Dirac's theorem on chordal graphs and to an application of this theorem to the study of powers of ideals.

Finally, in the fourth lecture we consider the edge ideal of a graph and show that the edge ideal of a bipartite graph is Cohen-Macaulay. This leads to the investigation of meet-distributive meet-semilattices.

It is assumed that the reader of these lectures is familiar with basic concepts of commutative algebra, including the theory of Stanley–Reisner rings. Even though most of the proofs are omitted, some effort is made to illuminate the relations between the algebraic and combinatorial concepts.

CHAPTER 1

Algebraic shifting theory

Generic initial ideals. We recall a few basic facts about generic initial ideals which are required in shifting theory. Most of these facts and their proofs can be found in the book by Eisenbud [**25**].

Let K be an infinite field, and $S = K[x_1,\dots,x_n]$ the polynomial ring over K. The set of monomials of degree d in S will be denoted by M_d.

We will fix a term order $<$ satisfying $x_1 > x_2 > \ldots > x_n$. Let $I \subset S$ be an ideal. Then we denote by $\mathrm{in}_<(I)$ (or simply by $\mathrm{in}(I)$) the initial ideal of I, that is, the ideal which is generated by all initial terms of I.

Let $GL(n)$ denote the general linear group with coefficients in K. Any $\varphi = (a_{ij}) \in GL(n)$ induces an automorphism of the graded K-algebra S, again denoted by φ, namely

$$\varphi(f(x_1,\dots,x_n)) = f(\sum_{i=1}^{n} a_{i1}x_i,\dots,\sum_{i=1}^{n} a_{in}x_i) \quad \text{for all} \quad f \in S.$$

One basic fact in the theory of generic initial ideals is the following:

THEOREM 1.0.1 (Galligo-Bayer-Stillman). Let $I \subset S$ be a graded ideal. Then there is a nonempty Zariski open set $U \subseteq GL(n)$ such that $\mathrm{in}(\varphi(I))$ does not depend on $\varphi \in U$. Moreover, U meets nontrivially the Borel subgroup of $GL(n)$ consisting of all upper triangular invertible matrices.

For $\varphi \in U$ the monomial ideal $\mathrm{in}(\varphi(I))$ is called the *generic initial ideal of* I and will be denoted $\mathrm{Gin}(I)$.

Another basic result on generic ideals is

THEOREM 1.0.2 (Galligo-Bayer-Stillman). Let $I \subset S$ be a graded ideal. Then $\mathrm{Gin}(I)$ is Borel fixed; that is, $\varphi(\mathrm{Gin}(I)) = \mathrm{Gin}(I)$ for all φ which belong to the Borel group of invertible upper triangular matrices.

Borel fixed ideals can be characterized combinatorially as follows: let p be a prime number, and k and l be non-negative integers with p-adic expansion $k = \sum_i k_i p^i$ and $l = \sum_i l_i p^i$. We set $k \leq_p l$ if $k_i \leq l_i$ for all i. In order to have consistent notation, we also set $k \leq_0 l$ if $k \leq l$ (in the usual sense).

DEFINITION 1.0.3. Let p be a prime number, or $p = 0$. A monomial ideal $I \subset S$ is p-Borel if the following condition holds: for each monomial $u \in I$, $u = \prod_i x_i^{\mu_i}$, one has $(x_i/x_j)^{\nu} u \in I$ for all i,j with $1 \leq i < j \leq n$ and all $\nu \leq_p \mu_j$.

One has

PROPOSITION 1.0.4. Suppose $\mathrm{char}\, K = p \geq 0$, and let $I \subset S$ be a monomial ideal. Then I is Borel-fixed if and only if I is p-Borel.

For the proof of 1.0.4 we refer to [**25**, Theorem 15.23].

For $p > 0$, the p-Borel ideals have a rather complicated combinatorial structure. The reader who is interested in more details about such ideals may consult [**60**], [**2**] and [**47**]. In these notes we will concentrate on 0-Borel ideals, which henceforth will be called *strongly stable ideals*.

For a monomial $u \in S$ we set $m(u) = \max\{i\colon x_i \text{ divides } u\}$.

DEFINITION 1.0.5. A subset $B \subset S$ of monomials is called *strongly stable* if $x_i(u/x_j) \in B$ for all $u \in B$, all x_j that divide u, and all $i < j$. The set B is called *stable* if $x_i(u/x_{m(u)}) \in B$ for all $u \in B$, and all $i < m(u)$.

It follows from 1.0.3 that a strongly stable ideal is a monomial ideal I for which the set of monomials in I is a strongly stable monomial set. If the set of monomials in I is a stable set, then I is called a *stable monomial ideal*. Stable monomial ideals were introduced by Eliahou and Kervaire [**26**].

A monomial $u \in S$, $u = x_1^{a_1} x_2^{a_2} \cdots x_n^{a_n}$ is called *squarefree* if $a_i \leq 1$ for $i = 1, \ldots, n$, and a monomial ideal in S is called *squarefree* if it is generated by squarefree monomials. In combinatorial contexts, squarefree monomial ideals are interesting since they appear as the defining ideals of Stanley-Reisner rings. Unfortunately $\text{Gin}(I)$ of a squarefree monomial ideal I is never squarefree unless I is generated by a subset of the variables. Thus for combinatorial applications one has to find an analogue of the operation Gin which yields a squarefree monomial ideal. The most natural way to define such an operation is to work in the exterior algebra instead of the symmetric algebra.

Let V be an n-dimensional K-vector space with basis $e_1, \ldots, e_n$. The exterior algebra $E = \bigwedge^{\bullet}(V)$ is a finite dimensional graded K-algebra. The ith graded component $\bigwedge^i(V)$ has the K-basis

$$e_{j_1} \wedge e_{j_2} \wedge \cdots \wedge e_{j_i} \quad \text{with} \quad j_1 < j_2 < \ldots < j_i.$$

Let $[n] = \{1, \ldots, n\}$; for a subset $F \subset [n]$, $F = \{j_1 < j_2 < \ldots < j_i\}$, we set $e_F = e_{j_1} \wedge e_{j_2} \wedge \cdots \wedge e_{j_i}$. The elements e_F are called the monomials of E. Term orders, initial terms and initial ideals are defined as in the polynomial ring. For example, the lexicographic or the reverse lexicographic order is defined by restriction to squarefree monomials.

In the exterior algebra the generic initial ideal $\text{Gin}(I)$ of a graded ideal $I \subset E$ is defined similarly as in the case of the polynomial ring. In other words, $\text{Gin}(I) = \text{in}(\varphi(I))$ where φ is a linear automorphism of E. Of course, $\text{Gin}(I)$ is a monomial ideal in E (which is automatically squarefree). The analogues of the theorems of Galligo, Bayer and Stillman hold and are proved similarly in the exterior case. We refer the reader to [**5**] for some general facts on Gröbner basis theory in exterior algebras.

Let $B \subset E$ be a set of monomials in the exterior algebra. Then B is called *(strongly) stable* if B satisfies conditions analogous to those of Definition 1.0.5. Thus, for example, B is stable if for all monomials $u \in B$, $u = e_{i_1} \wedge e_{i_2} \wedge \ldots \wedge e_{i_j}$ with $i_1 < i_2 < \ldots < i_j$ it follows that $e_i \wedge e_{i_1} \wedge \ldots \wedge e_{i_{j-1}} \in B$ for all $i < i_j$ and $i \notin \{i_1, \ldots, i_j\}$.

Let $u = x_{i_1} x_{i_2} \ldots x_{i_j} \in S$ be a squarefree monomial. Then we call $e_{i_1} \wedge e_{i_2} \wedge \ldots \wedge e_{i_j}$ the monomial in E corresponding to u. Let $I \subset S$ be a squarefree monomial ideal, $B \subset I$ the set of squarefree monomials in I, and B' the corresponding set of monomials in E. Notice that the K-subspace J of E spanned by B' is an ideal in E. We call it the monomial ideal in E corresponding to I, and I is called a *squarefree* (*strongly*) *stable* monomial ideal, resp. a *squarefree lexsegment*, ideal if J is a (strongly) stable, resp. lexsegment ideal, in E.

Corresponding to 1.0.4, one has

PROPOSITION 1.0.6. A Borel-fixed ideal $J \subset E$ is strongly stable. In particular, the generic initial ideal of any graded ideal in E is strongly stable.

Graded Betti-numbers. Let M be a finitely generated graded S-module. Then M has a graded free S-resolution of the form

$$\ldots \to \bigoplus_j S(-j)^{\beta_{ij}} \to \ldots \to \bigoplus_j S(-j)^{\beta_{1j}} \to \bigoplus_j S(-j)^{\beta_{0j}} \to M \to 0.$$

The numbers β_{ij} are called the *graded Betti numbers of M*. Note that the Tor-groups $\operatorname{Tor}_i(K,M)$ are finitely generated, graded K-vector spaces and that

$$\beta_{ij}(M) = \dim_K \operatorname{Tor}_i(K,M)_j \quad \text{for all} \quad i,j.$$

The following basic result holds:

THEOREM 1.0.7. Let $I \subset S$ be a graded ideal. Then for any term order $<$ one has

$$\beta_{ij}(S/I) \leq \beta_{ij}(S/\operatorname{in}_<(I)) \quad \text{for all} \quad i,j.$$

The *regularity* of M is defined to be the number $\operatorname{reg}(M) = \max\{j-i\colon \beta_{ij}(M) \neq 0\}$. As an immediate consequence of 1.0.7 we have

COROLLARY 1.0.8. Let $I \subset S$ be a graded ideal. Then for any term order $<$ one has:

(a) $\operatorname{proj}\dim S/I \leq \operatorname{proj}\dim S/\operatorname{in}_<(I)$;
(b) $\operatorname{depth} S/I \geq \operatorname{depth} S/\operatorname{in}_<(I)$;
(c) if $S/\operatorname{in}_<(I)$ is Cohen-Macaulay (Gorenstein), then so is S/I;
(d) $\operatorname{reg} S/I \leq \operatorname{reg} S/\operatorname{in}_<(I)$.

Since the generic initial ideal of graded ideal $I \subset S$ is strongly stable in characteristic 0 and for $J \subset E$ is squarefree strongly stable, it is desirable to have explicit formulas for the graded Betti numbers of stable and squarefree stable ideals.

Let $I \subset S$ be a monomial ideal. We denote by $G(I)$ the unique minimal monomial set of generators of I and by $G(I)_j$ the elements of degree j in $G(I)$. In [**26**] the following result is shown.

THEOREM 1.0.9 (Eliahou-Kervaire). Let $I \subset S$ be a stable ideal. Then

(a) $\beta_{ii+j}(I) = \sum_{u\in G(I)_j} \binom{m(u)-1}{i}$,
(b) $\operatorname{proj}\dim S/I = \max\{m(u)\colon u \in G(I)\}$,
(c) $\operatorname{reg}(I) = \max\{\deg(u)\colon u \in G(I)\}$.

A similar result holds in the squarefree case (cf. [**4**]).

THEOREM 1.0.10 (Aramova-Herzog-Hibi). Let $I \subset S$ be a squarefree stable ideal. Then

(a) $\beta_{ii+j}(I) = \sum_{u\in G(I)_j} \binom{m(u)-j}{i}$,
(b) $\operatorname{proj}\dim S/I = \max\{m(u) - \deg(u) + 1\colon u \in G(I)\}$,
(c) $\operatorname{reg}(I) = \max\{\deg(u)\colon u \in G(I)\}$.

Let M be a finitely generated graded S-module. A Betti number of M is called *extremal* if $\beta_{ii+j} = 0$ for all $(i,j) \neq (k,m)$ with $i \geq k$ and $j \geq m$.

The following theorem [**12**] describes what the Betti-numbers of I and $\operatorname{Gin}(I)$ have in common.

THEOREM 1.0.11 (Bayer-Charalambous-S. Popescu). Let $I \subset S$ be a graded ideal, and let $\operatorname{Gin}(I)$ be the generic initial ideal of I with respect to the reverse lexicographic order. Then for any two integers $i,j \subset \mathbb{N}$ one has

(a) the (i,j)th Betti number of S/I is extremal if and only if the (i,j)th Betti number of $S/\operatorname{Gin}(I)$ is extremal;
(b) the corresponding extremal Betti numbers of S/I and $S/\operatorname{Gin}(I)$ are equal.

This theorem implies the classical result of Bayer and Stillman [**11**], which says that $\operatorname{reg}(I) = \operatorname{reg}(\operatorname{Gin}(I))$. There is a similar theorem for squarefree monomial ideals; see Theorem 1.0.21.

Finally we quote a theorem due to Hochster [**49**] (see also [**15**]). We say that $a = (a_1,\dots,a_n) \in \mathbb{N}^n$ is squarefree if $a_i \leq 1$ for all i.

Let Δ be a simplicial complex on the vertex set $[n]$, K a field and $S = K[x_1,\dots,x_n]$ the polynomial ring in n variables. The ideal $I_\Delta = (x_F)_{F \notin \Delta} \subset S$ is called the *Stanley-Reisner ideal* and $K[\Delta] = S/I_\Delta$ the *Stanley-Reisner ring* of Δ. Here $x_F = x_{i_1} \cdots x_{i_k}$ for $F = \{i_1,\dots,i_k\}$. Similarly one defines the ideal $J_\Delta = (e_F)_{F\in\Delta} \subset E$.

THEOREM 1.0.12. Let Δ be a simplicial complex with vertex set $[n]$, and $a \in \mathbb{N}^n$. Then for all $i \geq 0$, we have

(a) $\beta^S_{ia}(K[\Delta]) = 0$ if a is not squarefree;
(b) $\beta^S_{ia}(K[\Delta]) = \dim_K \tilde{H}_{|a|-i-1}(\Delta_{\operatorname{supp}(a)};K)$ if a is squarefree.

Shifting operations. In this subsection we study shifting operations. They assign to each simplicial complex a shifted simplicial complex which shares basic properties with the original simplicial complex but is combinatorially simpler.

DEFINITION 1.0.13. The simplicial complex Δ is *shifted* if I_Δ is strongly stable. In other words, Δ is shifted if it satisfies the following property: if $F \in \Delta$, $i \in F$ and $j > i$, then $(F \setminus \{i\}) \cup \{j\} \in \Delta$.

Following Kalai [**52**] we define a shifting operation by a list of properties.

DEFINITION 1.0.14. A map which assigns to each simplicial complex Δ on the vertex set $[n]$ a simplicial complex $\operatorname{Shift}(\Delta)$ on the same vertex set $[n]$ is called a *shifting operation* if it satisfies the following conditions:

(S_1) $\operatorname{Shift}(\Delta)$ is shifted;
(S_2) $\operatorname{Shift}(\Delta) = \Delta$ if Δ itself is shifted;
(S_3) the simplicial complexes Δ and $\operatorname{Shift}(\Delta)$ have the same f-vector;
(S_4) if Γ is a subcomplex of Δ, then $\operatorname{Shift}(\Gamma) \subset \operatorname{Shift}(\Delta)$.

Shifting operations were first considered by Erdös, Ko, and Rado (see [**27**]), while algebraic shifting was introduced by Kalai [**51**], [**52**]. In this section we will present and compare the most important shifting operations.

Let us begin with *combinatorial shifting:* In the combinatorics of finite sets one considers the following operation (cf. [**27**]): Let $\mathscr{A}$ be a collection of subsets of $[n]$. For given integers $1 \leq i < j \leq n$ and for all $F \in \mathscr{A}$ one defines:

$$S_{ij}(F) = \begin{cases} (F\setminus\{j\})\cup\{i\}, & \text{if} \quad j \in F, \quad i \notin F, \quad (F\setminus\{j\})\cup\{i\} \notin \mathscr{A}; \\ F, & \text{otherwise.} \end{cases}$$

For $1 \leq i < j \leq n$ and $a \in K$ we define an elementary automorphism $\varphi^a_{ij} \colon V \to V$ as follows: $\varphi^a_{ij}(e_k) = e_k$ if $k \neq j$, and $\varphi^a_{ij}(e_j) = ae_i + e_j$.

The following fact is easily checked.

LEMMA 1.0.15. Let $J \subset E$ be a monomial ideal, and let $a \in K$, $a \neq 0$. Then $\operatorname{in}(\varphi^a_{ij}(J))$ has the K-basis $\{e_{S_{ij}(F)} \colon F \in \mathscr{A}\}$, where $\mathscr{A} = \{F \subset [n] \colon e_F \in J\}$.

It follows in particular that the ideal $\text{in}(\varphi^a_{ij}(J))$ does not depend on the choice of a. If Δ is a simplicial complex, then $\text{Shift}_{ij}(\Delta)$ is the simplicial complex defined by

$$J_{\text{Shift}_{ij}(\Delta)} = \text{in}(\varphi^a_{ij}(J_\Delta)).$$

LEMMA 1.0.16. The operator Shift_{ij} satisfies the conditions (S_2), (S_3) and (S_4).

PROOF. Suppose Δ is shifted; then I_Δ is squarefree strongly stable, and so $J_\Delta \subset E$ is strongly stable. By 1.0.15, $\text{in}(\varphi^a_{ij}(J_\Delta))$ has the K-basis $\{e_{S_{ij}(F)} : F \in \mathscr{A}\}$. As J_Δ is strongly stable it follows that $e_{S_{ij}(F)} \in J_\Delta$ for all $e_F \in J_\Delta$. This proves (S_2).

For the proof of (S_3) we note that J_Δ and $\text{in}(\varphi^a_{ij}(J_\Delta))$ have the same Hilbert function. Condition (S_4) follows from 1.0.15. □

Simple examples show that (S_1) is in general not satisfied for Shift_{ij}. We will see however that a suitable sequence of these operators yields a shifted simplicial complex.

For a monomial $u \in E$ of degree d, $u = e_{j_1} \wedge \cdots \wedge e_{j_d}$, we set $c_d(u) = \sum_{k=1}^d j_k$. Moreover if $J \subset E$ is a monomial ideal, we set $c_d(J) = \sum c_d(u)$ where the sum is taken over all monomials of degree d in J.

The following result was shown in [**4**].

PROPOSITION 1.0.17. Let Δ be a simplicial complex. Then

(a) $c_d(J_{\text{Shift}_{ij}(\Delta)}) \le c_d(J_\Delta)$ for all d;
(b) if Δ is not shifted, then there exist i and j with $i < j$ such that $c_d(J_{\text{Shift}_{ij}(\Delta)}) < c_d(J_\Delta)$ for some d.

PROOF. Assertion (a) follows from the fact that $c_d(e_{S_{ij}(F)}) \le c_d(e_F)$ for all monomials of degree d.

Suppose now that J_Δ is not strongly stable. Then there exists a squarefree monomial $e_F \in J_\Delta$ (of degree d) and integers i and j with $i < j$ such that $e_{S_{ij}(F)} \notin J_\Delta$. Since $c_d(e_{S_{ij}(F)}) < c_d(e_F)$, it follows that $c_d(J_{\text{Shift}_{ij}(\Delta)}) < c_d(J_\Delta)$, as desired. □

COROLLARY 1.0.18. Let Δ be a simplicial complex. Then there exists a sequence of pairs of integers $(i_1, j_1), \ldots, (i_r, j_r)$ with $i_k < j_k$ for $k = 1, \ldots, r$ such that

$$\text{Shift}_{i_r j_r}(\text{Shift}_{i_{r-1} j_{r-1}}(\ldots(\text{Shift}_{i_1 j_1}(\Delta)\ldots))$$

is shifted.

Any simplicial complex which is obtained from Δ by a sequence of operations as described in 1.0.18 will be denoted by Δ^c. It follows from our discussions that $\Delta \mapsto \Delta^c$ is a shifting operator. We call this operator *combinatorial shifting*. Combinatorial shifting is not very natural. In fact, Δ is not even uniquely defined. The only advantage of this operator is that it is easily computable.

It is easy to see that $\beta_{0j}(I_\Delta) \le \beta_{0j}(I_{\Delta^c})$ for all j. A similar inequality holds for all graded Betti-numbers; see Theorem 1.0.35.

Exterior algebraic shifting: Let Δ be simplicial complex, and $J_\Delta \subset E$ its Stanley-Reisner ideal in the exterior algebra. The exterior algebraic shifted complex Δ^e of Δ is defined by the equation $J_{\Delta^e} = \text{Gin}(J_\Delta)$.

PROPOSITION 1.0.19. Exterior algebraic shifting is in fact a shifting operator; that is, it satisfies the conditions (S_1) - (S_4).

PROOF. Condition (S_1) follows from 1.0.6, and (S_3) and (S_4) follow as for combinatorial shifting. In order to prove (S_2) we notice that for any strongly stable ideal $J \subset E$ and any invertible upper triangular matrix φ one has $\varphi(J) = J$. The assertion is clear for elementary upper triangular matrices, as well as for invertible diagonal matrices. Since these matrices generate all invertible upper triangular matrices, we get the desired conclusion. Therefore if $J \subset E$ is strongly stable, then $\mathrm{Gin}(J) = \mathrm{in}(\varphi(J)) = \mathrm{in}(J) = J$. □

CONJECTURE 1.0.20. Let Δ be simplicial complex. Then

$$\beta_{ij}(I_\Delta) \leq \beta_{ij}(I_{\Delta^e}).$$

Note that a result similar to Theorem 1.0.7 holds for ideals in the exterior algebra, so that in particular one has $\beta_{ij}(J) \leq \beta_{ij}(\mathrm{Gin}^E(J))$ for all i and j. Unfortunately this does not imply the conjecture.

Here is the analogue [3] to the Bayer-Charalambous-Popescu, Theorem 1.0.11.

THEOREM 1.0.21. Let Δ be a simplicial complex. Then for any two integers $i, j \in \mathbb{N}$ one has

(a) the (i, j)th Betti number of S/I_Δ is extremal if and only if the (i, j)th Betti number of S/I_{Δ^e} is extremal;
(b) the corresponding extremal Betti numbers of S/I_Δ and S/I_{Δ^e} are equal.

As a consequence of the fact that I_Δ and I_{Δ^e} have the same extremal Betti numbers, we now derive further properties of exterior algebraic shifting which were discovered first (with different methods) by Kalai [51].

PROPOSITION 1.0.22 (Kalai). For all i one has $\tilde{H}^i(\Delta; K) \cong \tilde{H}^i(\Delta^e; K)$.

PROOF. Hochster's formulas (cf. Theorem 1.0.12) imply that

$$\beta_{in}(K[\Delta]) = \dim_K \tilde{H}_{n-i-1}(\Delta; K) \quad \text{for all } i, \tag{1.0.1}$$

and $\beta_{ij}(K[\Delta]) = 0$ for all i and all $j > n$. In particular we see that the Betti numbers $\beta_{in}(K[\Delta])$ are extremal. Thus $\beta_{in}(K[\Delta]) = \beta_{in}(K[\Delta^e])$, by Theorem 1.0.21. Since K is a field, it follows that $\tilde{H}_{n-i-1}(\Delta; K) = \tilde{H}^{n-i-1}(\Delta; K)$ for all i, and the assertion follows. □

The *Alexander dual* of the simplicial complex Δ (on the vertex set $[n]$) is the simplicial complex

$$\Delta^\vee = \{F \subset [n] \colon [n] \setminus F \notin \Delta\}.$$

A simplicial complex Δ is called *Cohen-Macaulay over K* if $K[\Delta]$ is Cohen-Macaulay, and Δ is called *Cohen-Macaulay* if Δ is Cohen-Macaulay over any field.

We shall need the following result ([22]).

THEOREM 1.0.23 (Eagon-Reiner). Let Δ be a simplicial complex. Then the following conditions are equivalent:

(a) I_Δ has a linear resolution;
(b) the dual simplicial complex $\Delta^\vee$ is Cohen-Macaulay over K.

THEOREM 1.0.24 (Kalai). The following conditions are equivalent:

(a) Δ is Cohen-Macaulay over K,
(b) Δ^e is Cohen-Macaulay over K,
(c) Δ^e is pure.

PROOF. We observe the simple fact (see for example [**46**, Lemma 1.1]) that $(\Delta^\vee)^e = (\Delta^e)^\vee$.

(a) $\Longleftrightarrow$ (b): By Theorem 1.0.23, Δ is Cohen-Macaulay over K if and only if $I_{\Delta^\vee}$ has linear resolution. Since, by Theorem 1.0.21, the regularity of $K[\Delta^\vee]$ and $K[(\Delta^\vee)^e]$ is the same, it follows that $K[\Delta^\vee]$ has a linear resolution if and only if $K[(\Delta^\vee)^e] = K[(\Delta^e)^\vee]$ has a linear resolution. This in turn, again by Theorem 1.0.23, is the case if and only if Δ^e is Cohen-Macaulay over K.

The implication (b) $\Rightarrow$ (c) is true for any simplicial complex.

(c) $\Rightarrow$ (b): Since the maximal faces of Δ^e correspond to the minimal non-faces of $(\Delta^e)^\vee$, the purity of Δ^e implies that the minimal generators of the defining ideal of $(\Delta^e)^\vee$ all have the same degree. As $(\Delta^e)^\vee = (\Delta^\vee)^e$, we see that $(\Delta^e)^\vee$ is shifted, and hence its defining ideal is strongly stable. The resolution of a strongly stable ideal which is generated in one degree is linear, as follows from Theorem 1.0.9. This concludes the proof. ☐

Symmetric algebraic shifting: Let $I \subset S = K[x_1,\dots,x_n]$ be a squarefree ideal, where K is a field of characteristic 0. We let $\operatorname{Gin}(I)$ be the generic initial ideal of I with respect to the reverse lexicographic term order. We know from Theorem 1.0.2 and Proposition 1.0.4 that $\operatorname{Gin}(I)$ is a strongly stable ideal in S. But of course it is no longer squarefree.

We will transform $\operatorname{Gin}(I)$ into a squarefree monomial ideal by applying a certain operator: for a monomial $u \in S$, $u = x_{i_1}x_{i_2}\cdots x_{i_j}\cdots x_{i_d}$ with $i_1 \le i_2 \le \cdots \le i_j \le \cdots \le i_d$, we set

$$u^\sigma = x_{i_1}x_{i_2+1}\cdots x_{i_j+(j-1)}\cdots x_{i_d+(d-1)}.$$

One has

$$m(u^\sigma) - \deg u^\sigma = m(u) - 1. \tag{1.0.2}$$

If L is a monomial ideal with $G(L) = \{u_1,\dots,u_s\}$, then we write L^σ for the squarefree monomial ideal generated by $u_1^\sigma,\dots,u_s^\sigma$ in $K[x_1,\dots,x_m]$, where $m = \max\{m(u)+\deg u - 1 : u \in G(L)\}$.

Let Δ be a simplicial complex on the vertex set $[n]$. The *symmetric algebraic shifted complex* of Δ is defined to be the simplicial complex Δ^s with

$$I_{\Delta^s} = (\operatorname{Gin}(I_\Delta))^\sigma.$$

The definition of symmetric algebraic shifting presented here is formally different from that given by Kalai [**52**]. However, it is an easy exercise to see that both notions coincide.

A priori it is not clear from the definition of symmetric algebraic shifting that Δ^s has the same vertex set $[n]$. The next lemma shows that this indeed is the case.

LEMMA 1.0.25. *If I is a squarefree monomial ideal of $S = K[x_1,\dots,x_n]$, then $m(u) + \deg u \le n+1$ for all $u \in G(\operatorname{Gin}(I))$.*

PROOF. Recall from Theorem 1.0.9 that the graded Betti numbers of a strongly stable ideal I are given by

$$\beta_{i,i+j}(I) = \sum_{u \in G(I)_j} \binom{m(u)-1}{i} \tag{1.0.3}$$

for all i and j.

Since $\operatorname{Gin}(I)$ is strongly stable, formula (1.0.3) implies that $\max\{m(u)+\deg u - 1 : u \in G(\operatorname{Gin}(I))\}$ is the highest shift in the resolution of $\operatorname{Gin}(I)$. The monomial ideal I being squarefree, Hochster's formula (see 1.0.12) guarantees that the highest shift in the resolution of I is less than or equal to n. Since the Betti numbers with highest shift in the resolution of I are extremal it follows from Theorem 1.0.11 that the highest shift in the resolution of I and that of $\mathrm{Gin}(I)$ coincide (see also [**46**]). This yields the desired inequalities. $\square$

We want to point out that we defined symmetric algebraic shifting only in a polynomial ring whose base field is of characteristic 0, because otherwise we do not know whether $\mathrm{Gin}(I)$ is strongly stable. It may be possible that symmetric algebraic shifting can be defined in any characteristic, provided the following question can be answered affirmatively.

PROBLEM 1.0.26. Let $I \subset S = K[x_1, \ldots, x_n]$ be a squarefree monomial ideal. Is it true that $\mathrm{Gin}(I)$ is strongly stable in any characteristic? More generally, one may even ask whether $\mathrm{Gin}(I)$ of a monomial ideal is strongly stable if the characteristic of the field is larger than all exponents appearing in the monomial generators of I.

On the other hand, if I is squarefree and $\mathrm{char}\, K > n$, then $\mathrm{Gin}(I)$ is strongly stable. In fact, the highest degree of a generator of $\mathrm{Gin}(I)$ is less than or equal to $\mathrm{reg}(\mathrm{Gin}(I))$, and $\mathrm{reg}\,\mathrm{Gin}(I) = \mathrm{reg}\, I$ by Theorem 1.0.11. Since $\mathrm{reg}(I) \leq n$ by Theorem 1.0.12, we conclude that the highest degree of a generator of $\mathrm{Gin}(I)$ is $\leq n$. Hence the assertion follows from Proposition 1.0.4.

Note that condition (S_1) is satisfied since we have

LEMMA 1.0.27. Let I be a strongly stable ideal with $G(I) = \{u_1, \ldots, u_s\}$. Then the squarefree monomial ideal I^σ is squarefree strongly stable with $G(I^\sigma) = \{u_1^\sigma, \ldots, u_s^\sigma\}$.

PROOF. Suppose that for some $u \in G(I)$ we have $u^\sigma \notin G(I^\sigma)$. Let $u = x_{i_1} \cdots x_{i_d}$ with $i_1 \leq \cdots \leq i_d$. Then, for some proper subset N of $\{1, 2, \ldots, d\}$ and for some $1 \leq q \leq s$, we have $u_q^\sigma = \prod_{j \in N} x_{i_j + (j-1)}$. Hence $u_q = \prod_{j \in N} x_{i_j + h_j}$, where h_j is the number of integers $1 \leq k < j$ with $k \notin N$. Since I is strongly stable, $\prod_{j \in N} x_{i_j}$ must belong to I. This contradicts $u \in G(I)$. Thus we have $G(I^\sigma) = \{u_1^\sigma, \ldots, u_s^\sigma\}$.

Next, to see why I^σ is squarefree strongly stable, let $u = x_{i_1} \cdots x_{i_d} \in G(I)$ and consider the monomial $(x_b u^\sigma)/x_{i_a + (a-1)}$ with $b \notin \mathrm{supp}(u^\sigma)$ and $b < i_a + (a-1)$. Let $i_p + (p-1) < b < i_{p+1} + p$ for some $p < a$ and set

$$v = \Big(\prod_{j=1}^{p} x_{i_j}\Big) x_{b-p} \Big(\prod_{j=p+1}^{a-1} x_{i_j - 1}\Big)\Big(\prod_{j=a+1}^{d} x_{i_j}\Big).$$

Then, since $b - p < i_{p+1} \leq i_a$ and since I is strongly stable, the monomial v belongs to I. Note that $v^\sigma = (x_b u^\sigma)/x_{i_a + (a-1)}$. Say, $v = x_{\ell_1} \cdots x_{\ell_d}$ with $\ell_1 \leq \cdots \leq \ell_d$. Again, since I is strongly stable, it follows that $w = x_{\ell_1} \cdots x_{\ell_c} \in G(I)$ for some $c \leq d$. Since w^σ divides v^σ, we have $(x_b u^\sigma)/x_{i_a + (a-1)} \in I^\sigma$, as desired. $\square$

The operator $I \mapsto I^\sigma$ behaves well with respect to graded Betti numbers.

LEMMA 1.0.28. If I is a strongly stable monomial ideal, then $\beta_{ii+j}(I) = \beta_{ii+j}(I^\sigma)$ for all i and j.

PROOF. The result follows from (1.0.2), and Theorems 1.0.9(a) and 1.0.10(a). $\square$

Next we indicate the proof of condition (S_2) for symmetric algebraic shifting.

THEOREM 1.0.29. Let $I \subset S$ be a squarefree strongly stable ideal of S. Then $I^\sigma = I$.

For the proof we introduce the operation τ which is inverse to σ: For a squarefree monomial $u = x_{i_1} x_{i_2} \cdots x_{i_j} \cdots x_{i_d}$ with $i_1 < i_2 < \cdots < i_j < \cdots < i_d$, we set

$$u^\tau = x_{i_1} x_{i_2 - 1} \cdots x_{i_j - (j-1)} \cdots x_{i_d - (d-1)}.$$

If $I \subset S$ is a squarefree monomial ideal with $G(I) = \{u_1, \dots, u_s\}$, we write I^τ for the monomial ideal generated by $u_1^\tau, \dots, u_s^\tau$ in S.

Similarly to Lemma 1.0.27 we have

LEMMA 1.0.30. Let I be a squarefree strongly stable ideal with $G(I) = \{u_1, \dots, u_s\}$. Then the ideal I^τ is strongly stable with $G(I^\tau) = \{u_1^\tau, \dots, u_s^\tau\}$.

PROOF. Assume that for some $u \in G(I)$ we have $u^\tau \notin G(I^\tau)$. Let $u = x_{i_1} \cdots x_{i_d}$ with $i_1 < \cdots < i_d$. Then for some proper subset $\{j_1, \dots, j_t\}$ of $\{1, 2, \dots, d\}$, where $j_1 < \cdots < j_t$, and for some $1 \le q \le s$, we have $u_q^\tau = \prod_{k=1}^t x_{i_{j_k} - (j_k - 1)}$. Hence $u_q = \prod_{k=1}^t x_{i_{j_k} - (j_k - k)}$. Since $i_k \le i_{j_k} - (j_k - k)$ for $1 \le k \le t$ and I is squarefree strongly stable, we get $x_{i_1} \cdots x_{i_t} \in I$, which contradicts $u \in G(I)$.

Now, we show that I^τ is strongly stable. Let $u = x_{i_1} \cdots x_{i_d} \in G(I)$ with $i_1 < \cdots < i_d$, and consider the monomial $v = (x_b u^\tau)/x_{i_k - (k-1)}$ with $b < i_k - (k-1)$. Let $i_p - (p-1) \le b < i_{p+1} - p$ for some $p < k$. Then

$$v^\sigma = \Big(\prod_{j=1}^{p} x_{i_j}\Big) x_{b+p} \Big(\prod_{j=p+1}^{k-1} x_{i_j+1}\Big)\Big(\prod_{j=k+1}^{d} x_{i_j}\Big).$$

Since $b + p < i_{p+1}$ and $i_j + 1 \le i_{j+1}$ for $p + 1 \le j \le k - 1$, and since I is squarefree strongly stable, we obtain that $v^\sigma \in I$. Say, $v^\sigma = x_{\ell_1} \cdots x_{\ell_d}$ with $\ell_1 < \cdots < \ell_d$. Again, since I is squarefree strongly stable, it follows that $w = x_{\ell_1} \cdots x_{\ell_c} \in G(I)$ for some $c \le d$. Since w^τ divides $(v^\sigma)^\tau = v$, we have $v \in I^\tau$. □

The proof of Theorem 1.0.29 is based on the following lemma. We refer the reader to the original paper **[6]** for the somewhat tedious proof of the lemma.

LEMMA 1.0.31. Let $I \subset S$ be a squarefree strongly stable ideal generated in degree d. Let $G(I) = \{u_1, \dots, u_s\}$ where $u_1 > u_2 > \dots > u_s$ with respect to the reverse lexicographical order. Let $g = (a_{ij})_{1 \le i, j \le n}$ be a generic upper triangular matrix acting on S by $g(x_i) = \sum_{j=1}^i a_{ji} x_j$ for $1 \le i \le n$, and let c_{kj} denote the coefficient of u_j^τ in $g(u_k)$ for $1 \le k, j \le s$. Then the determinant of the matrix $(c_{kj})_{1 \le k, j \le s}$ is different from zero.

PROOF OF 1.0.29. Since the ideal I is squarefree strongly stable, I is componentwise linear **[4]**. Therefore **[7**, Theorem 1.1**]** implies that $\beta_{ii+j}(I) = \beta_{ii+j}(\mathrm{Gin}(I))$ for all i and j. On the other hand, the ideal $\mathrm{Gin}(I)$ being strongly stable, it follows from Lemma 1.0.28 that $\beta_{ii+j}(\mathrm{Gin}(I)) = \beta_{ii+j}((\mathrm{Gin}(I))^\sigma)$. Thus, we obtain the equalities

$$\beta_{i,i+j}(I) = \beta_{i,i+j}((\mathrm{Gin}(I))^\sigma) \quad \text{for all} \quad i, j, \tag{1.0.4}$$

which imply that I and $(\mathrm{Gin}(I))^\sigma$ have the same Hilbert function. Hence it is enough to prove that $I \subseteq (\mathrm{Gin}(I))^\sigma$. By Lemma 1.0.27 and Lemma 1.0.30 this inclusion holds if and only if $I^\tau \subseteq \mathrm{Gin}(I)$.

For any monomial u we denote by $\langle u \rangle$ the smallest squarefree strongly stable ideal containing u. Since $I = \sum_{u \in G(I)} \langle u \rangle$, and $\mathrm{Gin}(\langle u \rangle) \subseteq \mathrm{Gin}(I)$ for every $u \in G(I)$, it is enough to show that the claim is true for squarefree Borel principal ideals. So, we may assume that $I = \langle u \rangle$. Set $d = \deg u$.

Let $G(I) = \{u_1, \dots, u_s\}$ where $u_1 > u_2 > \cdots > u_s$. Then $u_s = u$. We may assume that the claim is true for all u_k, $1 \le k \le s-1$. Then $(u_1^\tau, u_2^\tau, \dots, u_{s-1}^\tau) \subset \mathrm{Gin}(I)$, and since I^τ and $\mathrm{Gin}(I)$ have the same number of minimal monomial generators, one has $G(\mathrm{Gin}(I)) = \{u_1^\tau, u_2^\tau, \dots, u_{s-1}^\tau, v\}$, where v is a monomial of degree d. We have to prove that $v = u^\tau$.

Assume $v > u^\tau$. We will see that this is impossible. First, we show that $m(v) = m(u^\tau)$. It follows from Theorem 1.0.10(a) that

$$\beta_{ii+d}((\mathrm{Gin}(I))^\sigma) = \sum_{j=1}^{s-1} \binom{m(u_j)-d}{i} + \binom{m(v^\sigma)-d}{i};$$

$$\beta_{ii+d}(I) = \sum_{j=1}^{s-1} \binom{m(u_j)-d}{i} + \binom{m(u)-d}{i}.$$

Therefore, according to (1.0.4), we obtain $\binom{m(v^\sigma)-d}{i} = \binom{m(u)-d}{i}$, which implies $m(v^\sigma) = m(u)$, so that $m(v) = m(u^\tau)$.

We fix the following notation: $u = x_{s_1} \cdots x_{s_d}$ where $s_1 < \cdots < s_d$, and $v = x_{j_1} \cdots x_{j_d}$ where $j_1 \leq \cdots \leq j_d$. Since $v > u^\tau$, there exists a k such that $j_i = s_i - (i-1)$ for $k+1 \leq i \leq d$ and $j_k < s_k - (k-1)$. As $j_d = m(v) = m(u^\tau) = s_d - (d-1)$, one has $k < d$. If $j_i + (i-1) \leq s_i$ for $1 \leq i \leq k$, then, in view of the fact that $I = \langle u \rangle$ is squarefree strongly stable, one obtains that $v^\sigma \in I$, which implies $v^\sigma = u_t$ for some $1 \leq t \leq s-1$ and the contradiction $v = u_t^\tau$. Thus, there exists an ℓ, $1 \leq \ell < k$, such that $j_\ell + (\ell - 1) > s_\ell$. Then $j_\ell \leq j_k < s_k - (k-1) \leq s_d - (d-1) = m(v)$; therefore $x_{j_\ell} v / x_{m(v)} \in \mathrm{Gin}(I)$, because $\mathrm{Gin}(I)$ is strongly stable. Since $x_{j_\ell} v / x_{m(v)} > v$, we get $x_{j_\ell} v / x_{m(v)} = u_t^\tau$ for some $1 \leq t \leq s-1$. Say $u_t = x_{t_1} \cdots x_{t_d}$ where $t_1 < \cdots < t_d$. As $I = \langle u \rangle$ is a squarefree Borel principal ideal, we have $t_i \leq s_i$ for $1 \leq i \leq d$, therefore $t_i - (i-1) \leq s_i - (i-1)$ for $1 \leq i \leq d$. This contradicts $j_\ell > s_\ell - (\ell - 1)$.

Hence, $v \leq u^\tau$. Now we apply Lemma 1.0.31 using the same notation. We have $\mathrm{Gin}(I) = \mathrm{in}(g(I))$ and $u_j^\tau \in \mathrm{Gin}(I)$ for $1 \leq j \leq s-1$. Since the rank of the matrix $(c_{kj})_{1 \leq k, j \leq s}$ is maximal, it follows that $v \geq u^\tau$, and so $v = u^\tau$. □

In a recent paper [9] by Babson, Novik and Thomas, the following result was shown:

THEOREM 1.0.32 (Babson-Novik-Thomas). Let S be an operation that associates with every $n \geq 0$ and every simplicial complex Δ on the vertex set $[n]$ a simplicial complex $S(\Delta)$ on the same vertex set satisfying the conditions (S_1), (S_3) and (S_4) and moreover the conditions:

(1) $S(\Delta * \{n+1\}) = S(\Delta) * \{n+1\}$ (here $\Delta * \{n+1\}$ is the *cone* over Δ);
(2) $\sum_{i=0}^{\dim \Delta} \dim_K \tilde{H}_i(\Delta; K) \leq \sum_{i=0}^{\dim S(\Delta)} \dim_K \tilde{H}_i(S(\Delta); K)$.

Then S satisfies also (S_2).

Using this theorem they could give a relatively simple proof of the fact that symmetric algebraic shifting satisfies the property (S_2), avoiding the tedious Lemma 1.0.30.

For symmetric algebraic shifting we can prove the following inequality of graded Betti numbers.

THEOREM 1.0.33. Let Δ be a simplicial complex. Then

$$\beta_{ii+j}(I_\Delta) \leq \beta_{ii+j}(I_{\Delta^s}) \quad \text{for all} \quad i \quad \text{and} \quad j.$$

PROOF. The desired inequalities follow from Theorem 1.0.7 and Lemma 1.0.28. □

Theorem 1.0.33 leads us to conjecture the following inequalities:

CONJECTURE 1.0.34. Let Δ be a simplicial complex. Then for all i and j one has

$$\beta_{ii+j}(I_{\Delta^s}) \leq \beta_{ii+j}(I_{\Delta^e}).$$

By virtue of Theorem 1.0.33 this conjecture implies the yet open inequalities

$$\beta_{ii+j}(I_\Delta) \leq \beta_{ii+j}(I_{\Delta^e})$$

for all i and j, which we conjectured in 1.0.20.

On the other hand, Hibi and Murai [**55**] succeeded recently in proving the following

THEOREM 1.0.35 (Hibi-Murai). *For all i and j one has*

$$\beta_{ii+j}(I_{\Delta^e}) \leq \beta_{ii+j}(I_{\Delta^c}).$$

One should expect that there is a direct proof of Conjecture 1.0.20, avoiding a comparison with the symmmetric shifted ideal. The next result shows that the extremal Betti numbers of the symmetric algebraic shifted ideals behave as expected.

THEOREM 1.0.36. *Let Δ be a simplicial complex. Then for all i and j*

(a) *the following conditions are equivalent:*
   (i) *the (i,j)th Betti number of I_Δ is extremal,*
   (ii) *the (i,j)th Betti number of I_{Δ^s} is extremal;*
(b) *the corresponding extremal Betti numbers of I_Δ and I_{Δ^s} are equal.*

PROOF. By Lemma 1.0.28 we have $\beta_{ij}(\text{Gin}^S(I_\Delta)) = \beta_{ij}(I_{\Delta^s})$. Therefore the assertions follow from Theorem 1.0.11. □

In contrast to exterior and symmetric algebraic shifting, the extremal Betti numbers are not invariant under combinatorial shifting. The following example was communicated to me by Satoshi Murai: Let $I_\Delta = (x_1x_2, x_2x_3, x_3x_4)$. Then either (i) $I_{\Delta^c} = (x_1x_2, x_1x_3, x_2x_3)$, or (ii) $I_{\Delta^c} = (x_1x_2, x_1x_3, x_1x_4, x_2x_3x_4)$. The ideal I_Δ as well as I_{Δ^c} in case (i) have linear resolutions and their extremal Betti numbers are the same, while the extremal Betti numbers of I_{Δ^c} in case (ii) differ from those of I_Δ.

As in the case of exterior algebraic shifting, we deduce from Theorem 1.0.36 the following:

COROLLARY 1.0.37. *Let K be a field of characteristic 0. Then*

$$\tilde{H}_i(\Delta;K) \cong \tilde{H}_i(\Delta^s;K) \quad \text{for all} \quad i.$$

The usefulness of Proposition 1.0.22 and Corollary 1.0.37 is partially explained by the fact that $\tilde{H}_\bullet(\Delta^e;K)$ and $\tilde{H}_\bullet(\Delta^s;K)$ can be computed combinatorially in a simple way. In fact, as noted in [**52**] (in different terminology), one has

LEMMA 1.0.38. *Let Δ be a simplicial complex on the vertex set $[n]$ such that I_Δ is squarefree strongly stable. Then*

$$\begin{aligned}\dim_K \tilde{H}_i(\Delta;K) &= |\{u \in G(I_\Delta)_{i+2} \colon m(u) = n\}| \\ &= |\{\sigma \in \Delta \colon \dim\sigma = i, \quad \sigma \cup \{n\} \notin \Delta\}|.\end{aligned}$$

PROOF. The first equation follows from (1.0.1) and Theorem 1.0.10, while the second equation follows trivially from the definitions. □

CHAPTER 2

Discrete polymatroids

Matroids and polymatroids. Matroid theory is a large area of research in combinatorics. In this lecture we discuss discrete polymatroids and the algebraic objects related to them. Large parts of this lecture are based on my joint paper [**35**] with Takayuki Hibi. There one can also find the proofs that are omitted here.

Matroids are just a special class of polymatroids. We begin our discussion by recalling the definition of matroids: a *matroid* on the ground set $[n]$ is a nonempty subset $\mathscr{M} \subset 2^{[n]}$ satisfying

(M1) if $F_1 \in \mathscr{M}$ and $F_2 \subset F_1$, then $F_2 \in \mathscr{M}$,
(M2) if F_1 and F_2 belong to $\mathscr{M}$ and $|F_1| < |F_2|$, then there is an $x \in F_2 \setminus F_1$ such that $F_1 \cup \{x\} \in \mathscr{M}$.

The members of $\mathscr{M}$ are the *independent sets* of $\mathscr{M}$. A *base* of $\mathscr{M}$ is a maximal independent set of $\mathscr{M}$. It follows from (M2) that if B_1 and B_2 are bases of $\mathscr{M}$, then $|B_1| = |B_2|$. The set of bases of $\mathscr{M}$ possesses the "exchange property" as follows:

(B) If B_1 and B_2 are bases of $\mathscr{M}$ and if $x \in B_1 \setminus B_2$, then there is a $y \in B_2 \setminus B_1$ such that $(B_1 \setminus \{x\}) \cup \{y\}$ is a base of $\mathscr{M}$.

Moreover, the set of bases of $\mathscr{M}$ possesses the "symmetric exchange property" as follows:

(S) If B_1 and B_2 are bases of $\mathscr{M}$ and if $x \in B_1 \setminus B_2$, then there is a $y \in B_2 \setminus B_1$ such that both $(B_1 \setminus \{x\}) \cup \{y\}$ and $(B_2 \setminus \{y\}) \cup \{x\}$ are bases of $\mathscr{M}$.

Given a nonempty set $\mathscr{B} \subset 2^{[n]}$, there exists a matroid $\mathscr{M}$ on $[n]$ with the set of bases $\mathscr{B}$ if and only if $\mathscr{B}$ possesses the exchange property (B).

A typical example of a matroid is the following: let K be a field, V a K-vector space, and $\mathscr{S} = \{v_1, \ldots, v_n\}$ be a finite set of vectors of V. Then we define the matroid $\mathscr{M}$ as the collection of subsets of $\mathscr{S}$ whose elements are linearly independent subsets of vectors of $\mathscr{S}$. The Steinitz exchange property guarantees that $\mathscr{M}$ is indeed a matroid.

Let $\varepsilon_1, \ldots, \varepsilon_n$ denote the canonical basis vectors of $\mathbb{R}^n$. If we associate each $F \subset [n]$ with $\sum_{i \in F} \varepsilon_i$, then a matroid on $[n]$ may be regarded as a set of $(0,1)$-vectors of $\mathbb{R}^n$ satisfying certain conditions. Considering nonnegative integer vectors rather than $(0,1)$-vectors enables us to define the concept of discrete polymatroids. Let $\mathbb{R}^n_+$ denote the set of vectors $u = (u(1), \ldots, u(n)) \in \mathbb{R}^n$ with each $u(i) \geq 0$. If $u = (u(1), \ldots, u(n))$ and $v = (v(1), \ldots, v(n))$ are two vectors belonging to $\mathbb{R}^n_+$, then we write $u \leq v$ if all components $v(i) - u(i)$ of $v - u$ are nonnegative and, moreover, we write $u < v$ if $u \leq v$ and $u \neq v$. We say that u is a *subvector* of v if $u \leq v$. The *modulus* of $u = (u(1), \ldots, u(n)) \in \mathbb{R}^n_+$ is $|u| = u(1) + \cdots + u(n)$. Also, let $\mathbb{Z}^n_+ = \mathbb{R}^n_+ \cap \mathbb{Z}^n$.

DEFINITION 2.0.39. A *discrete polymatroid* on the ground set $[n]$ is a nonempty finite set $P \subset \mathbb{Z}^n_+$ satisfying

(D1) if $u \in P$ and $v \in \mathbb{Z}^n_+$ with $v \leq u$, then $v \in P$,

(D2) if $u = (u(1), \ldots, u(n)) \in P$ and $v = (v(1), \ldots, v(n)) \in P$ with $|u| < |v|$, then there is $i \in [n]$ with $u_i < v_i$ such that $u + \varepsilon_i \in P$.

A *base* of P is a vector $u \in P$ such that $u < v$ for no $v \in P$. It follows from (D2) that if u_1 and u_2 are bases of P, then $|u_1| = |u_2|$. Discrete polymatroids can be also characterized in terms of the exchange property of their bases.

PROPOSITION 2.0.40. A nonempty finite set $B \subset \mathbb{Z}_+^n$ is the set of bases of a discrete polymatroid on $[n]$ if and only if B satisfies:

(i) all $u \in B$ have the same modulus,
(ii) if $u = (u(1), \ldots, u(n)) \in P$ and $v = (v(1), \ldots, v(n)) \in P$ belong to B with $u(i) > v(i)$, then there is $j \in [n]$ with $u(j) < v(j)$ such that $u - \varepsilon_i + \varepsilon_j \in B$.

Surprisingly, one has the following symmetric exchange property for discrete polymatroids.

THEOREM 2.0.41. If $u = (u(1), \ldots, u(n))$ and $v = (v(1), \ldots, v(n))$ are bases of a discrete polymatroid $P \subset \mathbb{Z}_+^n$, then for each $i \in [n]$ with $u(i) > v(i)$ there is $j \in [n]$ with $u(j) < v(j)$ such that both $u - \varepsilon_i + \varepsilon_j$ and $v - \varepsilon_j + \varepsilon_i$ are bases of P.

The following corollary is immediate from Proposition 2.0.40 and the definition of a matroid.

COROLLARY 2.0.42. Let B be a nonempty finite set of integer vectors in $\mathbb{R}_+^n$. The following conditions are equivalent:

(a) B is the set of bases of a matroid,
(b) B is the set of bases of a discrete polymatroid, and for all $u \in B$ one has $u(i) \leq 1$ for $i = 1, \ldots, n$.

The exchange property 2.0.40(c)(ii) suggests the following:

DEFINITION 2.0.43. Let B be a nonempty set of integer vectors that have the same modulus. Then B satisfies:

(W) the *weak exchange property* if for all $u, v \in B$, $u \neq v$, there exist i and j with $u(i) > v(i)$ and $u(j) < v(j)$ such that $u - \varepsilon_i + \varepsilon_j \in B$,
(S) the *strong exchange property* if for all $u, v \in B$, $u \neq v$, and all i and j with $u(i) > v(i)$ and $u(j) < v(j)$, one has $u - \varepsilon_i + \varepsilon_j \in B$.

EXAMPLES 2.0.44. (a) Let B be the set of bases of a polymatroid on the ground set $[n]$ with $n \leq 3$. It is immediate that B satisfies the strong exchange property.

(b) The set $\{(1,1,1,1), (0,2,0,2), (0,1,1,2), (1,2,0,1)\}$ is the set of bases of a discrete polymatroid. It does not satisfy the strong exchange property.

(c) Let $s_1, \ldots, s_n$ and d be nonnegative integers. The set

$$V = \{u \colon u(i) \text{ is an integer with } 0 \leq u(i) \leq s_i \text{ and } |u| = d\}$$

is called of *Veronese type*. It satisfies the strong exchange property.

In fact, let $u, v \in V$ with $u(i) > v(i)$ and $u(j) < v(j)$. Then $u(i) - 1 \geq 0$ and $u(j) + 1 \leq s_j$, so that $u - \varepsilon_i + \varepsilon_j \in V$.

(d) The set $\{(2,1,1), (2,2,0), (3,0,1), (3,1,0), (4,0,0)\}$ satisfies the strong exchange property, but is not of Veronese type.

(e) A set S of integral vectors in $\mathbb{R}_+^n$ of modulus d is called *strongly stable* if for all $u \in S$, all i with $u(i) > 0$ and all $j < i$, one has that $u - \varepsilon_i + \varepsilon_j \in S$. The strongly stable set $\{(3,0,1), (1,3,0), (3,1,0), (2,2,0), (4,0,0)\}$ does not satisfy the weak exchange property, but any strongly stable set S satisfies the following somewhat weaker exchange property: for all $u, v \in S$ there exist integers i and j such that $u(i) > v(i)$ and $u(j) < v(j)$, and either $u - \varepsilon_i + \varepsilon_j \in S$, or $v - \varepsilon_j + \varepsilon_i \in S$.

Indeed, let $u, v \in S$, $u \neq v$. Then there is an integer i such that $u(i) > v(i)$. Let i be the largest such number. If $u(j) = v(j)$ for all $j > i$, then, since u and v have the same modulus, there exists $j < i$ with $u(j) < v(j)$. Since S is strongly stable it follows that $u - \varepsilon_i + \varepsilon_j \in S$. On the other hand, if $u(j) < v(j)$ for some $j > i$, then $v - \varepsilon_j + \varepsilon_i \in S$, since S is strongly stable.

(f) A strongly stable set S is called *principal* if there exists an element $u \in S$ such that the smallest strongly stable set containing u coincides with S. In that case u is called the Borel generator of S. The set in example (d) is a strongly stable principal set with Borel generator $(2,1,1)$.

A strongly stable principal set satisfies the exchange property of Proposition 2.0.40, i.e. it is the set of bases of a polymatroid. On the other hand, the set

$$\{(0,1,0,1),(0,1,1,0),(0,2,0,0),(1,0,0,1),(1,0,1,0),(1,1,0,0),(2,0,0,0)\}$$

is strongly stable principal with Borel generator $(0,1,0,1)$, but does not satisfy the strong exchange property.

In [**69**, Chapter 18] and [**59**, p. 382] we can find the concept of polymatroids, which originated in an article by Edmonds [**24**]. A *polymatroid* on the ground set $[n]$ is a nonempty compact subset $\mathscr{P} \subset \mathbb{R}^n_+$ satisfying

(P1) if $u \in \mathscr{P}$ and $v \in \mathbb{R}^n_+$ with $v \leq u$, then $v \in \mathscr{P}$,

(P2) if $u = (u(1), \ldots, u(n)) \in \mathscr{P}$ and $v = (v(1), \ldots, v(n)) \in \mathscr{P}$ with $|u| < |v|$, then there is $i \in [n]$ with $u_i < v_i$ and $0 < c < v(i) - u(i)$ such that $u + c\varepsilon_i \in \mathscr{P}$.

We say that a polymatroid $\mathscr{P} \subset \mathbb{R}^n_+$ is *integral* if all vertices of $\mathscr{P}$ are integer vectors of $\mathbb{R}^n$.

One has the following fundamental result:

THEOREM 2.0.45. A nonempty finite set $P \subset \mathbb{Z}^n_+$ is a discrete polymatroid if and only if $\operatorname{conv}(P) \subset \mathbb{R}^n_+$ is an integral polymatroid with $\operatorname{conv}(P) \cap \mathbb{Z}^n = P$. Here $\operatorname{conv}(P)$ is the convex hull of P in $\mathbb{R}^n$.

The proof of this theorem uses the following characterization of polymatroids: let $\mathscr{P} \subset \mathbb{R}^n_+$ be a polymatroid on the ground set $[n]$. Let $2^{[n]}$ denote the set of all subsets of $[n]$. The *ground set rank function* of $\mathscr{P}$ is a function $\rho : 2^{[n]} \to \mathbb{R}_+$ defined by setting

$$\rho(A) = \max\{v(A) \colon v \in \mathscr{P}\}$$

for all $\emptyset \neq A \subset [n]$ together with $\rho(\emptyset) = 0$. Here we set $v(A) = \sum_{i \in A} v(i)$ for any vector $v \in \mathbb{R}^N$.

PROPOSITION 2.0.46. *(a)* Let $\mathscr{P} \subset \mathbb{R}^n_+$ be a polymatroid on the ground set $[n]$ and ρ its ground set rank function. Then ρ is nondecreasing; i.e., if $A \subset B \subset [n]$, then $\rho(A) \leq \rho(B)$, and is submodular, i.e.,

$$\rho(A) + \rho(B) \geq \rho(A \cup B) + \rho(A \cap B)$$

for all $A, B \subset [n]$. Moreover, $\mathscr{P}$ coincides with the compact set

$$\{x \in \mathbb{R}^n_+ \colon x(A) \leq \rho(A),\ A \subset [n]\}. \tag{2.0.1}$$

(b) Conversely, given a nondecreasing and submodular function $\rho : 2^{[n]} \to \mathbb{R}_+$ with $\rho(\emptyset) = 0$, the compact set *(2.0.1)* is a polymatroid on the ground set $[n]$ with ρ its ground set rank function.

It follows from Proposition 2.0.46(a) that a polymatroid $\mathscr{P} \subset \mathbb{R}^n_+$ on the ground set $[n]$ is a convex polytope in $\mathbb{R}^n$. In addition, the set of bases of $\mathscr{P}$ is a face of $\mathscr{P}$ with supporting hyperplane

$$\{x = (x_1, \ldots, x_n) \in \mathbb{R}^n \colon \sum_{i=1}^{n} x_i = \text{rank}(\mathscr{P})\}.$$

The rank function of a discrete polymatroid P on the ground set $[n]$ is defined similarly, namely as $\rho(A) = \max\{v(A) \colon v \in P\}$ for $A \subset [n]$. Obviously, it is a submodular integer-valued function on $2^{[n]}$ with $\rho(B) \leq \rho(A)$ for $B \subset A$. Conversely, any such function defines a discrete polymatroid.

We conclude this subsection by quoting the following nice property of polymatroids.

PROPOSITION 2.0.47. Let $\mathscr{P}_1, \ldots, \mathscr{P}_k$ be polymatroids on the ground set $[n]$ and ρ_i the ground set rank function of $\mathscr{P}_i$, $1 \leq i \leq k$. Then the polymatroidal sum

$$\mathscr{P}_1 \vee \cdots \vee \mathscr{P}_k = \{x \in \mathbb{R}^n \colon x = \sum_{i=1}^{k} x_i \quad x_i \mathscr{P}_i\}$$

is a polymatroid on $[n]$ and $\sum_{i=1}^{k} \rho_i$ is its ground set rank function. Moreover, if each $\mathscr{P}_i$ is integral, then $\mathscr{P}_1 \vee \cdots \vee \mathscr{P}_k$ is integral, and for each integer vector $x \in \mathscr{P}_1 \vee \cdots \vee \mathscr{P}_k$ there exist integer vectors $x_i \in \mathscr{P}_i$, $1 \leq i \leq k$, with $x = \sum_{i=1}^{k} x_i$.

Base rings of polymatroids. Let K be a field, and let P be a discrete polymatroid of rank d on the ground set $[n]$ with the set of bases B. The toric ring $K[B]$ generated over K by the monomials $t^u = \prod_i t_i^{u(i)}$, $u = (u(1), \ldots, u(n)) \in B$, is called the *base ring of P*. Let $S = K[x_u \colon u \in B]$ be the polynomial ring in the indeterminates x_u with $u \in B$. Let I_B be the kernel of the K-algebra homomorphism $\xi \colon S \to K[B]$ with $\xi(x_u) = t^u$. There are some obvious elements in I_B, namely, those arising from symmetric exchange. Let $u, v \in B$ with $u(i) > v(i)$ and $u(j) < v(j)$, and such that $u - \varepsilon_i + \varepsilon_j$ and $v - \varepsilon_j + \varepsilon_i$ belong to B. Then clearly $x_u x_v - x_{u-\varepsilon_i+\varepsilon_j} x_{v-\varepsilon_j+\varepsilon_i} \in I_B$. We call such a relation a *symmetric exchange relation*.

White conjectured (cf. [**71**]) that for a matroid, the symmetric exchange relations generate I_B. It is natural to conjecture that this also holds for discrete polymatroids. Only very partial results are known. Here is one of those results. We first recall the notion of sortability, introduced by Sturmfels [**64**]:

Let $u, v \in B$, and write $t^u t^v = t_{i_1} \ldots t_{i_{2d}}$ with $i_1 \leq i_2 \leq \cdots \leq i_{2d}$. Then we set $t^{u'} = \prod_j^d t_{2j-1}$ and $t^{v'} = \prod_j^d t_{2j}$. This defines a map

$$\text{sort} \colon B \times B \to M_d \times M_d, \quad (u, v) \to (u', v'),$$

where M_d denotes the set of all integer vectors of modulus d.

The map 'sort' is called the *sorting operator*, and B is called *sortable* if $\text{Im(sort)} \subseteq B \times B$.

In [**20**], De Negri noticed the following fact that easily follows from a theorem of Sturmfels [**64**].

LEMMA 2.0.48. Suppose $B \subset M_d$ is sortable. Then I_B has a Gröbner basis consisting of the sorting relations $x_u x_v - x_{u'} x_{v'}$ with $u, v \in B$ and $(u', v') = \text{sort}(u, v)$.

One has

THEOREM 2.0.49 (Herzog-Hibi). *(a)* Suppose that each matroid has the property that the toric ideal of its base ring is generated by symmetric exchange relations; then this is also true for each discrete polymatroid.

(b) If P is a discrete polymatroid whose set of bases B satisfies the strong exchange property, then

(i) B is sortable, and hence I_B has a quadratic Gröbner basis and $K[B]$ is Koszul,
(ii) I_B is generated by symmetric exchange relations.

Since P is the set of integer vectors of an integral polymatroid $\mathscr{P}$ (see Theorem 2.0.45), we may study the *Ehrhart ring of* $\mathscr{P}$. To this end one considers the cone $\mathscr{C} \subset \mathbb{R}^{n+1}$ with $\mathscr{C} = \mathbb{R}_+\{(p,1)\colon p \in \mathscr{P}\}$. Then $Q = \mathscr{C} \cap \mathbb{Z}^{n+1}$ is a subsemigroup of $\mathbb{Z}^{n+1}$, and the Ehrhart ring of $\mathscr{P}$ is defined to be the toric ring $K[\mathscr{P}] \subset K[t_1,\dots,t_n,s]$ generated over K by the monomials $t^u s^i$, $(u,i) \in Q$. By Gordan's lemma (cf. [**15**, Proposition 6.1.2]), $K[\mathscr{P}]$ is normal.

Notice that $K[\mathscr{P}]$ is naturally graded if we assign the degree i to $t^u s^i$. We denote by $K[P]$ the K-subalgebra of $K[\mathscr{P}]$ which is generated over K by the elements of degree 1 in $K[\mathscr{P}]$. Since $P = \mathscr{P} \cap \mathbb{Z}^n$ it follows that $K[P] = K[t^u s\colon u \in P]$. Observe that $K[B]$ may be identified with the subalgebra $K[t^u s\colon u \in B]$ of $K[P]$. After this identification, $K[B]$ is an algebra retract of $K[P]$ with retraction map $\pi\colon K[P] \to K[B]$, where π is the residue class map modulo the prime ideal Q generated by the elements $t^u s$, $u \in P$, $|u| < d$.

As an immediate consequence of 2.0.47 we obtain the following important

THEOREM 2.0.50. *$K[P] = K[\mathscr{P}]$. In particular, $K[P]$ is normal.*

COROLLARY 2.0.51. *$K[B]$ is normal.*

PROOF. By [**15**, Theorem 6.1.4], $K[P]$ is normal if and only if the semigroup S generated by $\hat{P} = \{(u,1)\colon u \in P\}$ is normal. This means that if G is the smallest subgroup of $\mathbb{Z}^{n+1}$ containing S and if $ma \in S$ for some $m \in \mathbb{N}$ and $a \in G$, then $a \in S$. We apply the same criterion to show that $K[B]$ is normal. Let H be the subgroup of $\mathbb{Z}^{n+1}$ consisting of all (u,i) with $|u| = id$, and let T be the semigroup generated by $\hat{B} = \{(u,1)\colon u \in B\}$. Then $H \cap G$ is the smallest group containing the semigroup T. Hence if $ma \in T$ for some $m \in \mathbb{N}$ and $a \in H \cap G$, then $a \in S \cap H = T$, as desired. □

Our next aim is to compare algebraic properties of $K[P]$ with those of $K[B]$. We say that a K-algebra A has a quadratic Gröbner basis if the defining ideal of A has this property for some term order.

THEOREM 2.0.52. *Suppose $K[P]$ has quadratic relations, or a quadratic Gröbner basis, or is Koszul; then $K[B]$ has these properties, too.*

In particular it follows from the theorem that the Ehrhart ring of all discrete polymatroids is Koszul if and only if this is the case for all base rings of all discrete polymatroids. Not all properties behave this way. For example, there exist discrete polymatroids for which $K[P]$ is Gorenstein but $K[B]$ is not, and vice versa.

PROOF OF 2.0.52. We have noticed already that $K[B]$ may be viewed as an algebra retract of $K[P]$. Hence the statements concerning quadratic relations and Koszulness follow from [**56**, Proposition 1.4, Corollary 2.5].

Now let $A \subset B$ be an arbitrary retract of standard graded toric K-algebras. Then we may assume that $A = K[x_1,\dots,x_n]/I$, and $B = K[x_1,\dots,x_n,y_1,\dots,y_m]/J$ with toric ideals I and J contained in the square of the corresponding graded maximal ideals, and that the retraction map $B \to A$ is induced by $\pi(y_j) = 0$ for $j = 1,\dots,m$.

We use the following well-known fact: suppose J has a Gröbner basis G with respect to the term order $<$, and suppose that for all $f \in G$ with $\mathrm{in}_<(f) \in K[x_1,\dots,x_n]$ one has that

$f \in k[x_1,\dots,x_n]$. Then $G' = G \cap K[x_1,\dots,x_n]$ is a Gröbner basis of I (with respect to the restricted term order).

Since B is toric, we may assume that the elements of G are binomials. Suppose now that $f \in G$ with $f = m_1 - m_2$ and $\mathrm{in}(f) = m_1 \in K[x_1,\dots,x_n]$, and suppose that $m_2 \notin K[x_1,\dots,x_n]$. Then $m_1 = \pi(f) \in I$, a contradiction since I is a prime ideal. □

Polymatroidal ideals. A monomial ideal generated by monomials corresponding to the base of a discrete polymatroid is called a *polymatroidal ideal*.

It follows from Proposition 2.0.40(ii) that a monomial ideal $I \subset S$ is polymatroidal if all its generators have the same degree and if it satisfies the following exchange property: for all $u, v \in G(I)$, $u = x_1^{a(1)} \cdots x_n^{a(n)}$, $v = x_1^{b(1)} \cdots x_n^{b(n)}$ and all i with $a(i) > b(i)$, there exists an integer j with $b(j) > a(j)$ such that $x_j(u/x_i) \in G(I)$.

EXAMPLES 2.0.53. Let K be a field and let $S = K[x_1,\dots,x_n]$ be the polynomial ring over K in n indeterminates.

(a) An ideal in S is called a *linear ideal* if it is generated by linear forms. A monomial linear ideal is generated by a subset of the variables. Obviously such an ideal is polymatroidal. Hence any finite product of monomial linear ideals is polymatroidal, as follows from Proposition 2.0.56 below. Polymatroidal ideals of this kind are called *transversal polymatroidal ideals*.

(b) Let $I_1,\dots,I_r$ be linear monomial ideals. The squarefree product $I_1 * I_2 * \cdots * I_r$ of these ideals is defined to be the ideal generated by all squarefree monomials $x_{i_1} \cdots x_{i_r}$ with $x_{i_j} \in I_j$ for $j = 1,\dots,r$. It can be checked directly, verifying the exchange property, that $I_1 * I_2 * \cdots * I_r$ is polymatroidal. Since such ideals are squarefree they are called *transversal matroidal ideals*.

(c) Let $s_1,\dots,s_n$ and d be non-negative integers. The monomial ideal generated by all monomials $x_1^{a(1)} \cdots x_n^{a(n)}$ of degree d with $0 \le a(i) \le s_i$ for $i = 1,\dots,n$ is called an *ideal of Veronese type*, and is denoted by $I_{d;s_1,\dots,s_n}$. This ideal is polymatroidal by Example 2.0.44(c).

Let $M = (x_1,\dots,x_n)$. Notice that $I_{d;s_1,\dots,s_n} = M^d$ if $s_i \ge d$ for all i, and that $I_{d;s_1,\dots,s_n} = M * M * \cdots * M$ (d factors) if $s_i = 1$ for all i. So the latter ideal is then also a transversal matroid. In the first case, $I_{d;s_1,\dots,s_n}$ is called the Veronese ideal of degree d, and in the second case, it is called the squarefree Veronese ideal of degree d.

We will show that a polymatroidal ideal has linear quotients. In general we say that a graded S-module M has *linear quotients* if M admits a minimal system of generators $m_1,\dots,m_k$ such that for every $t = 1,\dots,k$ one has that $\langle m_1,\dots,m_{t-1}\rangle :_R m_t$ is an ideal of R generated by linear forms.

Examples of ideals with linear quotients are strongly stable and squarefree strongly stable ideals.

LEMMA 2.0.54. If M has linear quotients, then

$$\mathrm{reg}(M) = \max\{\deg m \colon m \text{ is a minimal generator of } M\}.$$

In particular, if all generators of M have the same degree, then M has a linear resolution over R.

PROOF. Let $m_1,\dots,m_k$ be as in the definition of module with linear quotients. Set $M_t = \langle m_1,\dots,m_t\rangle$. We have an exact sequence

$$0 \to M_{t-1} \to M_t \to M_t/M_{t-1} \to 0$$

and M_t/M_{t-1} is of the form $R/I[-\deg(m_t)]$ with I an ideal of R generated by linear forms. Since $\operatorname{reg}(R/I) = 0$, it follows that $\operatorname{reg}(M_t) \leq \max\{\operatorname{reg}(M_{t-1}), \deg(m_t)\}$ and hence, by induction, the assertion follows. □

The ideal $J = (a^2b, abc, bcd, cd^2)$ has linear quotients, the successive colons being:

$$(0), \quad (a), \quad (a), \quad (b).$$

On the other hand, there are ideals with linear resolution and without linear quotients. The easiest example, taken from [**18**], is the ideal I of 2-minors of the matrix

$$\begin{pmatrix} a & b & c \\ b & c & d \end{pmatrix}.$$

I has a linear resolution but it cannot have linear quotients since it is a prime ideal and hence $(f) : (g) = (f)$ for each $f \in I$ with $\deg(f) = 2$.

A large class of monomial ideals with linear quotients (see [**48**]) is given by

PROPOSITION 2.0.55 (Herzog, Takayama). *A polymatroidal ideal I has linear quotients with respect to the reverse lexicographical order of the generators. In particular, polymatroidal ideals have linear resolutions.*

PROOF. Let $u \in G(I)$, and let J be the ideal generated by all $v \in G(I)$ with $v > u$ (in the reverse lexicographical order). Then

$$J : u = (v/[v,u] : v \in J).$$

Here $[u, v]$ denotes the greatest common divisor of u and v. Thus in order to prove that $J : u$ is generated by monomials of degree 1, we have to show that for each $v > u$ there exists $x_j \in J : u$ such that x_j divides $v/[v,u]$.

In fact, let $u = x_1^{a(1)} \cdots x_n^{a(n)}$ and $v = x_1^{b(1)} \cdots x_n^{b(n)}$. Since $v > u$, there exists an integer i with $a(i) > b(i)$ and $a(k) = b(k)$ for $k = i+1, \ldots, n$, and hence an integer j with $b(j) > a(j)$ such that $u' = x_j(u/x_i) \in I$. Since $j < i$, we see that $u' \in J$, and from the equation $x_i u' = x_j u$ we deduce that $x_j \in J : u$. Finally, since $v_j(v/[u,v]) = b_j - \min\{b(j), a(j)\} = b(j) - a(j) > 0$, we have that x_j divides $v/[v,u]$. □

Note that a monomial ideal I having linear quotients (with respect to the monomial generators) is a purely combinatorial property and hence does not depend on the characteristic of the base field. On the other hand, the minimal free resolution of a monomial ideal, and hence its linearity, depends, in general, on the characteristic of the base field. This also shows that monomial ideals having linear quotients is a stronger property than having linear resolutions. The famous example of the Stanley-Reisner ideal of a triangulation of the real projective plane (see, for example, [**15**, page 236]) gives an example of a squarefree monomial ideal in the situation where the characteristic of K is not 2, it has a linear resolution and does not have linear quotients.

The property of having a linear resolution is not preserved by taking products or powers of ideals. The same thing can happen for the property of having linear quotients: for example, $J = (a^2b, abc, bcd, cd^2)$ has linear quotients, but $(b,c)J$ does not even have a linear resolution.

On the other hand, Proposition 2.0.47 implies

PROPOSITION 2.0.56. *Let I and J be polymatroidal ideals. Then IJ is again polymatroidal.*

Now this result together with Proposition 2.0.55 implies

COROLLARY 2.0.57. Let I be a polymatroidal ideal. Then all powers I^k of I have a linear resolution.

One may ask when a polymatroidal ideal is Cohen-Macaulay. This happens indeed very rarely. We quote the following result [**37**].

THEOREM 2.0.58 (Herzog-Hibi). A polymatroidal ideal I is Cohen–Macaulay if and only if I is

(i) a principal ideal,
(ii) a Veronese ideal, or
(iii) a squarefree Veronese ideal.

CHAPTER 3

Dirac's theorem on chordal graphs

Let Δ be a simplicial complex on the vertex set $[n] = \{1,\dots,n\}$. A *facet* of Δ is a maximal face of Δ. We denote by $\mathscr{F}(\Delta)$ the set of facets of Δ.

Let K be a field, $S = K[x_1,\dots,x_n]$ the polynomial ring over K in n indeterminates. With Δ there are naturally attached two monomial ideals:

- the Stanley-Reisner ideal $I_\Delta = (x_F)_{F \notin \Delta}$, and
- the facet ideal $I(\Delta) = (x_F)_{F \in \mathscr{F}(\Delta)}$.

Here $x_F = x_{i_1} \cdots x_{i_k}$ for $F = \{i_1,\dots,i_k\}$.

If $\mathscr{F}(\Delta) = \{F_1,\dots,F_m\}$ then we write $\Delta = \langle F_1,\dots,F_m\rangle$ and say that $F_1,\dots,F_m$ generate Δ.

EXAMPLE 3.0.59. Let Δ be the simplicial complex shown in Figure 1.

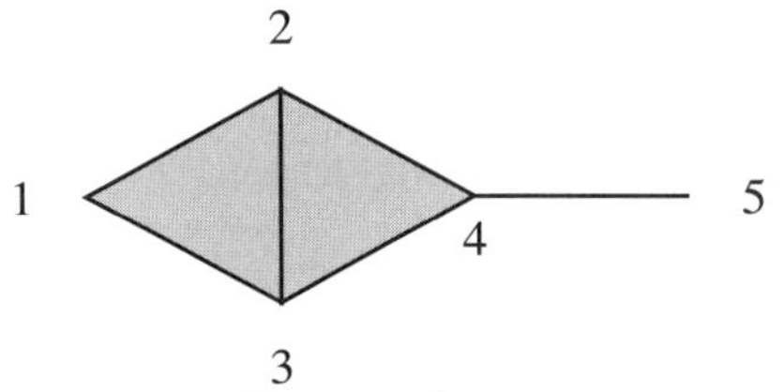

FIGURE 1

Then $I_\Delta = (x_1x_4, x_1x_5, x_2x_5, x_3x_5)$ and $I(\Delta) = (x_1x_2x_3, x_2x_3x_4, x_4x_5)$.

If $\dim \Delta = 1$, then Δ is a graph G, and $I(\Delta) = I(G)$ is the *edge ideal* of the graph.

For the Stanley–Reisner ring $K[\Delta] = S/I_\Delta$, the well-known Reisner criterion tells us when $K[\Delta]$ is Cohen-Macaulay. However, for $S/I(\Delta)$, no such criterion is known.

Alexander Duality. Let Δ be a simplicial complex on the vertex set $[n]$. Recall from Lecture 1 that the collection of subsets

$$\Delta^\vee = \{[n] \setminus F \colon F \notin \Delta\}$$

of $[n]$ is a simplicial complex, called the *Alexander dual* of Δ.

One has

$$(\Delta^\vee)^\vee = \Delta.$$

Set $F^c = [n] \setminus F$, and

$$\Delta^c = \langle F^c \colon F \in \mathscr{F}(\Delta)\rangle.$$

LEMMA 3.0.60. With the notation introduced we have: $I_{\Delta^\vee} = I(\Delta^c)$.

PROOF. By definition, $\Delta^\vee = \langle F^c \colon F$ is a minimal nonface of $\Delta\rangle$. Furthermore, $x_G \in G(I_{\Delta^\vee})$ if and only if G is a minimal subset of $[n]$ such that $G \notin \Delta^\vee$. This means that G^c does not contain any minimal nonface of Δ, and for any proper subset H of G, the complement H^c contains a minimal nonface of Δ. This is equivalent to saying that any subset of G^c is a face of Δ, and for any proper subset H of G, H^c is not a face of Δ. In another words, G^c is a facet of Δ. Hence $I(\Delta^c) = I_{\Delta^\vee}$, as required. □

A *vertex cover* of Δ is a set $G \subset [n]$ such that $G \cap F \neq \emptyset$ for all $F \in \mathcal{F}(\Delta)$. We denote by $\mathcal{C}(\Delta)$ the set of minimal vertex covers of Δ.

For $F = \{i_1, \dots, i_k\} \subset [n]$, set $P_F = (x_{i_1}, \dots, x_{i_k})$. Then we have

LEMMA 3.0.61. Let Γ be the (unique) simplicial complex such that $I_\Delta = I(\Gamma)$. Then

$$I_\Delta = \bigcap_{F \in \mathcal{C}(\Gamma)} P_F \quad \text{and} \quad I_{\Delta^\vee} = (x_F : F \in \mathcal{C}(\Gamma)).$$

PROOF. The monomials x_F with $F \in \Gamma$ generate I_Δ. Note that $x_F \subset P_G$ for all $F \in \Gamma$ if and only if G is a vertex cover of Γ. Thus the first equation follows.

In order to prove the second equation, let $G \in \mathcal{C}(\Gamma)$. Then $G^c \in \mathcal{F}(\Delta)$. Hence the assertion follows from Lemma 3.0.60. □

For example, if $\mathcal{F}(\Delta) = \{\{1,2,3\},\{2,3,4\},\{4,5\}\}$, then

$$I(\Delta) = (x_1x_2x_3, x_2x_3x_4, x_4x_5), \qquad I_\Delta = (x_1x_4, x_1x_5, x_2x_5, x_3x_5),$$

and

$$I(\Delta^\vee) = (x_2x_3x_5, x_2x_3x_4, x_1x_3x_4, x_1x_2x_4), \; I_{\Delta^\vee} = (x_1x_2x_3, x_1x_5, x_4x_5).$$

The following fundamental theorem of Eagon and Reiner (cf. Lecture 1) and its variations reflects the algebraic properties of the Alexander dual of Δ.

THEOREM 3.0.62. Let K be a field, and let Δ be a simplicial complex. Then

(a) *(Eagon–Reiner* [**22**]*)* $K[\Delta]$ is Cohen-Macaulay $\Longleftrightarrow I_{\Delta^\vee}$ has a linear resolution.
(b) Δ is shellable $\Longleftrightarrow I_{\Delta^\vee}$ has linear quotients.
(c) *(Terai* [**67**]*)* $\operatorname{proj\,dim} K[\Delta] = \operatorname{reg}(I_{\Delta^\vee})$.
(d) *(Herzog-Hibi* [**35**]*)* Δ is sequentially Cohen-Macaulay over $K \Longleftrightarrow I_{\Delta^\vee}$ is componentwise linear.

We sketch the proof of (b): recall that Δ is called *shellable* if Δ is pure and there is an order $F_1, \dots, F_m$ of the facets of Δ (called a *shelling order*), such that for all $0 < j < i$ there exists an $x \in F_i \setminus F_j$ and some $k < i$ with $F_i \setminus F_k = \{x\}$, while an ideal I is said to have *linear quotients* if $I = (f_1, \dots, f_m)$ and for all $i > 0$ the colon ideals $(f_1, \dots, f_{i-1}) : f_i$ are generated by linear forms; cf. Lecture 2.

For a monomial ideal I we require that the f_i belong to the unique minimal set of monomial generators $G(I)$ of I. Then I has linear quotients if for all $i > 1$, and any $j < i$, there exists $k < i$ such that $f_k/[f_i, f_k]$ is a monomial of degree 1, say x_ℓ, and $x_\ell | f_j$. Here $[f_i, f_k]$ denotes the greatest common divisor of f_i and f_k. By Lemma 3.0.60 one has $I_{\Delta^\vee} = (x_{F_1^c}, \dots, x_{F_m^c})$; the equivalence of the statements in (b) are obvious.

Quasi-trees and trees. We begin with the following

DEFINITION 3.0.63. Let Δ be a simplicial complex. A facet $F \in \mathcal{F}(\Delta)$ is called a *leaf* if either F is the only facet of Δ, or there exists $G \in \mathcal{F}(\Delta)$, $G \neq F$, such that $H \cap F \subset G \cap F$ for each $H \in \mathcal{F}(\Delta)$ with $H \neq F$. A facet G with this property is called a *branch* of F. A vertex i of Δ is called a *free vertex* if i belongs to precisely one facet.

The simplicial complex in Figure 1 has two leaves, namely $\{1,2,3\}$ and $\{4,5\}$, and has one branch, namely $\{2,3,4\}$. The free vertices are 1 and 5.

DEFINITION 3.0.64 (Faridi). A simplicial complex is called a *forest* if each subcomplex Γ of Δ that is generated by a subset of the facets of Δ has a leaf. A forest is called a *tree* if it is connected.

The simplicial complex in Figure 1 is a tree. However the simplicial complex in Figure 2

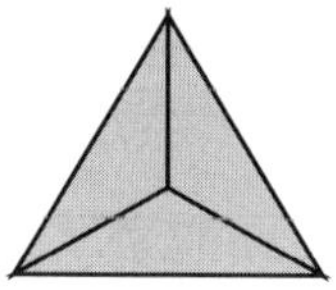

FIGURE 2

is not a tree, because it has no leaves at all.

DEFINITION 3.0.65 (Zheng). A simplicial complex Δ is called a *quasi-forest* if the facets of Δ can be ordered $F_1,\dots,F_m$ such that F_i is a leaf of $\langle F_1,\dots,F_{i-1}\rangle$ for $i=2,\dots,m$. Such an order of the leaves is called a *leaf order.*

If a quasi-forest is connected, then it is called a quasi-tree.

The simplicial complex displayed in Figure 3 is a quasi-tree, but not a tree, because if one removes the inside branch, then the remaining simplicial complex has no leaves anymore.

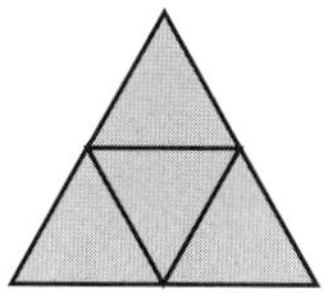

FIGURE 3

One always has: Δ tree $\Rightarrow$ Δ quasi-tree. However, the converse is not true in general.

Of course a graph which is a tree is also a tree in the sense of Faridi and a quasi-tree in the sense of Zheng.

Zheng calls a leaf F of a simplicial complex Δ a *good leaf* if the collection of sets $F\cap G$ with $G\in\mathscr{F}(\Delta)$ is totally ordered with respect to inclusion. In other words, for any two facets G_1,G_2 of Δ, one has

$$F\cap G_1\subset F\cap G_2 \quad\text{or}\quad F\cap G_2\subset F\cap G_1.$$

None of the leaves of the simplicial complex in Figure 3 is a good leaf. It has recently been shown that each tree has a good leaf; see [**39**].

Dirac's theorem. In this subsection we indicate some of the algebraic aspects of Dirac's theorem as it is given in [**41**]. Here Alexander duality and the Hilbert-Burch theorem play essential roles.

Let G be a graph on the vertex set $[n]$ and assume that G has no loops and multiple edges. A subset C of $[n]$ is called a *clique* of G if, for all $i, j \in C$ with $i \neq j$, one has $\{i, j\} \in E(G)$. A *perfect elimination ordering* (or a *simplicial elimination ordering*) is an ordering $v_n, ..., v_2, v_1$ on the vertices of G such that for all i, v_i is a simplicial vertex in the graph induced on vertices $\{v_1, ..., v_i\}$. Here a *simplicial vertex* in a graph is one whose neighbors form a clique.

THEOREM 3.0.66 (Dirac, 1961). Let G be a finite graph without loops and multiple edges. Then the following conditions are equivalent:

(a) G is a chordal graph (i.e. every cycle in the graph of length > 3 has a chord).
(b) G has a perfect elimination ordering.

The collection of all cliques of G form a simplicial complex $\Delta(G)$, called the *clique complex* of G; see Figure 4. It can be shown that G has a perfect elimination ordering if and only if $\Delta(G)$ is a quasi-forest.

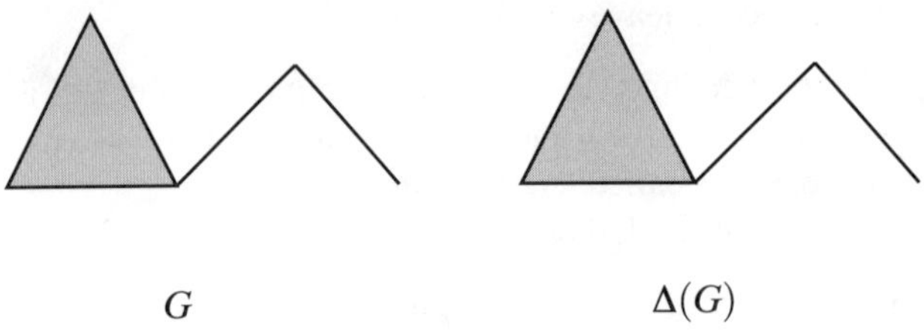

FIGURE 4

We will prove the following equivalent version of Dirac's theorem.

THEOREM 3.0.67. Let G be a finite graph without loops and multiple edges. Then the following conditions are equivalent:

(a) G is a chordal graph.
(b) G is the 1-skeleton of a quasi-forest.

In a first step of the proof we need

THEOREM 3.0.68 (Herzog-Hibi-Zheng). A simplicial complex Δ is a quasi-forest if and only if $\operatorname{proj}\dim I_{\Delta^\vee} = 1$.

The proof uses the *Hilbert–Burch Theorem* as well as the fact that $I_{\Delta^\vee} = I(\Delta^c)$.

Before we continue with the proof of Theorem 3.0.67 we make a digression and study monomial ideals of projective dimension 1. Let I be such a monomial ideal with monomial generators $u_1, \ldots, u_m$. The $\binom{m}{2}$ relations $u_{ji}e_i - u_{ij}e_j$ with $u_{ij} = u_i/\gcd(u_i, u_j)$ for all $i, j \in [m]$ and $i < j$ are called the *Taylor relations*.

A subset of the Taylor relations generates the syzygy module of I. This subset is in general not unique.

For example, let $I = (x_4x_5x_6, x_1x_5x_6, x_1x_2x_6, x_1x_2x_5)$. Then the syzygy module of I can be represented by the matrix

$$\begin{pmatrix} x_1 & -x_4 & 0 & 0 \\ 0 & x_2 & -x_5 & 0 \\ 0 & x_2 & 0 & -x_6 \end{pmatrix},$$

or by

$$\begin{pmatrix} x_1 & -x_4 & 0 & 0 \\ 0 & x_2 & -x_5 & 0 \\ 0 & 0 & x_5 & -x_6 \end{pmatrix} \quad \text{or} \quad \begin{pmatrix} x_1 & -x_4 & 0 & 0 \\ 0 & x_2 & 0 & -x_6 \\ 0 & 0 & x_5 & -x_6 \end{pmatrix}.$$

Nevertheless, for a given choice R of the relation matrix $R = (u_{ij})$, we can define a (1-dimensional) tree Ω with $\{i, j\} \in E(\Omega)$ if $u_{ji}e_i - u_{ij}e_j \in R$ for $i < j$. We call Ω the *relation tree* of R.

This relation tree was first considered by Bruns and Herzog in [**16**]. In our example the relation tree of the first matrix is given in Figure 5, while the relation tree for the other two matrices in given in Figure 6.

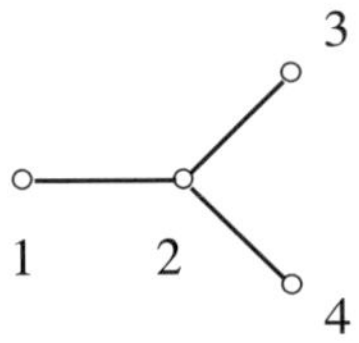

FIGURE 5

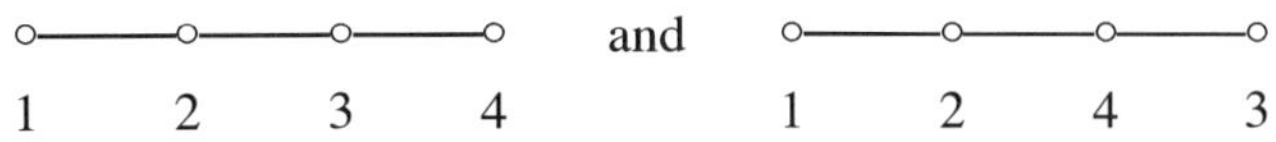

FIGURE 6

If, in our example, we let Δ be the simplicial complex for which $I_{\Delta^\vee} = I$, then by Theorem 3.0.68, Δ is the quasi-tree. In fact, we obtain the quasi-tree shown in Figure 7.

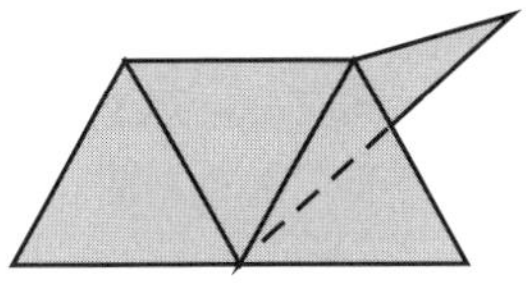

FIGURE 7

The relation trees of the relation matrix can be recovered from the quasi-tree $\Delta = \langle F_1, \ldots, F_m \rangle$ as follows:

- Start with some leaf F_i of Δ, and let F_j be a branch of F_i. Then $\{i, j\}$ will be an edge of Ω. Note that $\langle \mathscr{F}(\Delta) \setminus \{F_i\} \rangle$ is again a quasi-tree.
- Next remove the leaf F_i and continue in the same way with the remaining quasi-tree in order to find the other edges of Ω.

Of course, at each step of the procedure there may be different choices. This gives us the different possible relation trees.

Geometrically a relation tree is obtained from a given quasi-tree by connecting the barycenters of the leaves and branches according to the above rules.

In our example we get the relation trees shown in Figure 8.

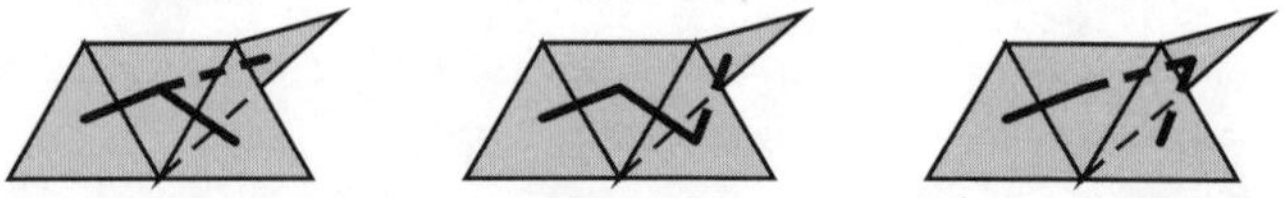

FIGURE 8

We now continue with the next steps in the proof of Theorem 3.0.67. Let G be a graph with no loops and multiple edges. The *complementary graph* $\bar{G}$ is the graph whose edges are the non-edges of G. We shall need the following result of Fröberg for the proof of the next lemma.

THEOREM 3.0.69 (Fröberg). The graph ideal $I(G)$ of G has 2-linear resolution if and only if $\bar{G}$ is chordal.

LEMMA 3.0.70. Let G be a graph. Then

(a) G is the 1-skeleton of $\Delta(G)$,
(b) $I_{\Delta(G)} = I(\bar{G})$,
(c) $\Delta(G)$ is a quasi-forest $\Longleftrightarrow$ G is chordal.

PROOF. (a) is obvious.

(b) Since the 1-skeleton of $\Delta(G) = G$, it follows that $I(\bar{G}) \subset I_{\Delta(G)}$. Conversely, let F be a minimal nonface of $\Delta(G)$. If $|F| > 2$, then each subset $G \subset F$ with $|G| = 2$ is an edge of G. Therefore F is a stable subset of G, and hence $F \in \Delta(G)$, a contradiction. Thus, for every minimal nonface F of $\Delta(G)$, one has $|F| = 2$. This shows that $I_{\Delta(G)} = I(\bar{G})$. Therefore, $\Delta = \Delta(G)$.

(c) The theorem of Fröberg (3.0.69) guarantees that the complementary graph G of $\bar{G}$ is a chordal graph if and only if $I(\bar{G}) = I_\Delta$ has a 2-linear resolution. By Theorem 3.0.62, $\operatorname{reg}(I_\Delta) = \operatorname{proj\,dim} I_{\Delta^\vee} + 1$, and so the ideal $I(\bar{G})$ has a 2-linear resolution if and only if $\operatorname{proj\,dim} I_{\Delta^\vee} = 1$. Since by Lemma 3.0.60, $I_{\Delta^\vee} = I(\Delta^c)$, the assertion follows from Theorem 3.0.68. □

LEMMA 3.0.71. A quasi-forest is a flag complex (i.e. the minimal nonfaces consist of two elements.)

PROOF. Let Δ be a quasi-forest on $[n]$ and fix a leaf ordering of the facets $F_1, \ldots, F_t$ of Δ. We work by induction on t. Let $t > 2$. Since $\Delta' = \langle F_1, \ldots, F_{t-1} \rangle$ is a quasi-forest, by assumption of induction it follows that Δ' is flag. Let F_k with $k < t$ be a branch of F_t. Then Δ' consists of all faces G of Δ with $G \cap (F_t \setminus F_k) = \emptyset$. Suppose H is a minimal nonface of Δ having at least three elements of $[n]$. We then show that H is a minimal nonface of Δ', i.e.,

$H \cap (F_t \setminus F_k) = \emptyset$. Since H is a nonface, there is $p \in H$ with $p \notin F_t$. If $q \in F_t$ belongs to H, then $\{p,q\} \in \Delta$. Thus there is F_j with $j \neq t$ such that $\{p,q\} \subset F_j$. Hence $\{q\} \subset F_t \cap F_j$. Thus $q \in F_k$. Hence $H \cap (F_t \setminus F_k) = \emptyset$, as desired. □

Now we are ready to give the proof of Theorem 3.0.67. The statements (b) and (c) of Lemma 3.0.70 imply that a chordal graph is the 1-skeleton of a quasi-forest.

Conversely, suppose that G is the 1-skeleton of a quasi-forest Γ. Since Γ is flag, the ideal I_Γ is generated by all monomials x_F with $|F| = 2$ and $F \notin \Gamma$. This shows that $I_\Gamma = I(\bar{G})$, and so $\Gamma = \Delta(G)$, by Lemma 3.0.70(a). Hence G is chordal by Lemma 3.0.70(c).

Applications to powers of ideals. In Lecture 2 we have seen that polymatroidal ideals have linear resolutions and that powers of polymatroidal ideals are again polymatroidal. In particular, they again have linear resolutions. In general, however, powers of ideals with linear resolution need not have linear resolutions. The first example of such an ideal was given by Terai. He showed that over a base field of characteristic $\neq 2$ the Stanley Reisner ideal

$$I = (abd, abf, ace, adc, aef, bde, bcf, bce, cdf, def)$$

of the minimal triangulation of the projective plane has a linear resolution, while I^2 has no linear resolution. The example depends on the characteristic of the base field. If the base field has characteristic 2, then I itself has no linear resolution.

Another example, namely $I = (def, cef, cdf, cde, bef, bcd, acf, ade)$ is given by Sturmfels [**65**]. Again I has a linear resolution, while I^2 has no linear resolution. The example of Sturmfels is interesting because of two reasons: 1. it does not depend on the characteristic of the base field, and 2. it is an ideal with linear quotients.

On the other hand, it is known (see [**19**] and [**53**]) that the regularity of powers I^n of a graded ideal I is bounded by a linear function $an+b$, and is a linear function for large n. For ideals I whose generators are all of degree d, one has the bound $\operatorname{reg}(I^n) \leq nd + \operatorname{reg}_x(R(I))$, as shown by Römer [**61**]. Here $R(I)$ is the Rees ring of I that is naturally bigraded, and $\operatorname{reg}_x(R(I))$ is the *x-regularity of* $R(I)$. In order to explain this concept let K be a field, $S = K[x_1, \ldots, x_n]$ the polynomial ring, and $I \subset S$ an equigenerated graded ideal, that is, a graded ideal whose generators $f_1, \ldots, f_m$ are all of the same degree. Then the Rees ring

$$R(I) = \bigoplus_{j \geq 0} I^j t^j = S[f_1 t, \ldots, f_m t] \subset S[t]$$

is naturally bigraded with $\deg(x_i) = (1,0)$ for $i = 1, \ldots, n$ and $\deg(f_i t) = (0,1)$ for $i = 1, \ldots, m$.

Let $T = S[y_1, \ldots, y_m]$ be the polynomial ring over S in the variables $y_1, \ldots, y_m$. We define a bigrading on T by setting $\deg(x_i) = (1,0)$ for $i = 1, \ldots, n$, and $\deg(y_j) = (0,1)$ for $j = 1, \ldots, m$. Then there is a natural surjective homomorphism of bigraded K-algebras $\varphi\colon T \to R(I)$ with $\varphi(x_i) = x_i$ for $i = 1, \ldots, n$ and $\varphi(y_j) = f_j t$ for $j = 1, \ldots, m$.

Let

$$F_\bullet : 0 \to F_p \longrightarrow F_{p-1} \longrightarrow \cdots \longrightarrow F_1 \longrightarrow F_0 \longrightarrow R(I) \to 0$$

be the bigraded minimal free T-resolution of $R(I)$. Here $F_i = \bigoplus_j T(-a_{ij}, -b_{ij})$ for $i = 0, \ldots, p$. The *x-regularity* of $R(I)$ is defined to be the number

$$\operatorname{reg}_x(R(I)) = \max_{i,j}\{a_{ij} - i\}.$$

With the notation introduced, one has the following result [**61**, Theorem 5.3 (i)] of Römer.

THEOREM 3.0.72. $\operatorname{reg}(I^n) \leq nd + \operatorname{reg}_x(R(I))$. In particular, if $\operatorname{reg}_x(R(I)) = 0$, then each power of I admits a linear resolution.

For the reader's convenience we give a simple proof of this theorem: For all n, the exact sequence $F_{\bullet}$ gives the exact sequence of graded S-modules

$$(3.0.1)\ 0 \to (F_p)_{(*,n)} \longrightarrow (F_{p-1})_{(*,n)} \cdots \longrightarrow (F_1)_{(*,n)} \longrightarrow (F_0)_{(*,n)} \longrightarrow R(I)_{(*,n)} \to 0.$$

We note that $R(I)_{(*,n)} = I^n(-dn)$, and that $T(-a,-b)_{(*,n)}$ is isomorphic to the free S-module $\bigoplus_{|u|=n-b} S(-a)y^u$. It follows that (3.0.1) is a (possibly non-minimal) graded free S-resolution of $I^n(-dn)$. This yields at once that $\operatorname{reg}(I^n(-dn)) \leq \operatorname{reg}_x(R(I))$, and thus $\operatorname{reg}(I^n) \leq nd + \operatorname{reg}_x(R(I))$.

We say that I satisfies the x-condition if $\operatorname{reg}_x(R(I)) = 0$.

COROLLARY 3.0.73. Let $I \subset S$ be an equigenerated graded ideal, and let $R(I) = T/P$. Then each power of I has a linear resolution if for some term order $<$ on T the defining ideal P has a Gröbner basis G whose elements are at most linear in the variables $x_1, \ldots, x_n$, that is, $\deg_x(f) \leq 1$ for all $f \in G$.

PROOF. Our assumptions imply that $\operatorname{in}(P)$ is generated by monomials $u_1, \ldots, u_m$ with the property that $\deg_x(u_i) \leq 1$ for $i = 1, \ldots, m$. Let $C_{\bullet}$ be the Taylor resolution of $\operatorname{in}(P)$. The module C_i has the basis e_σ with $\sigma = \{j_1 < i_2 < \ldots < j_i\} \subset [m]$. Each basis element e_σ has the multidegree (a_σ, b_σ) where $x^{a_\sigma} y^{b_\sigma} = \operatorname{lcm}\{u_{j_1}, \ldots, u_{j_m}\}$. It follows that $\deg_x(e_\sigma) \leq i$ for all $e_\sigma \in C_i$. Since the shifts of $C_{\bullet}$ bound the shifts of a minimal multigraded resolution of $\operatorname{in}(P)$, we conclude that $\operatorname{reg}_x(T/\operatorname{in}(P)) = 0$. On the other hand, by semi-continuity one always has $\operatorname{reg}_x(T/P) \leq \operatorname{reg}_x(T/\operatorname{in}(P))$. □

Using Föberg's theorem 3.0.69, which characterizes monomial ideals with 2-linear resolution and the theorem of Dirac 3.0.66 in order to label the variables suitably, one can show

PROPOSITION 3.0.74. Let $I \subset S = K[x_1, \ldots, x_n]$ be a squarefree monomial ideal with 2-linear resolution. Then after suitable renumbering of the variables, one has:

(*) if $x_i x_j \in I$ with $i \neq j$, $k > i$ and $k > j$, then either $x_i x_k$ or $x_j x_k$ belongs to I;
(**) if $x_i^2 \in I$ and $j > i$ for which there is k such that $x_k x_j \in I$, then either $x_i x_j \in I$ or $x_i x_k \in I$.

Ideals generated by quadratic monomials satisfying conditions $(*)$ and $(**)$ are important because one has

THEOREM 3.0.75. Let $I \subset S = K[x_1, \ldots, x_n]$ be an ideal that is generated by quadratic monomials and suppose that I possesses properties $(*)$ and $(**)$. Let $R(I) = T/P$ be the Rees ring of I. Then there exists a lexicographic order $<_{lex}$ on T such that the reduced Gröbner basis G of the defining ideal P with respect to $<_{lex}$ consists of binomials $f \in T$ with $\deg_x(f) \leq 1$.

All this implies

THEOREM 3.0.76 (Herzog-Hibi-Zheng). Let I be a monomial ideal generated in degree 2. The following conditions are equivalent:

(a) I has a linear resolution,
(b) I has linear quotients,
(c) Each power of I has a linear resolution.

PROOF. The implication (c) $\Rightarrow$ (a) is trivial, while (b) $\Rightarrow$ (a) is a general fact. It follows from Proposition 3.0.74 that if I has a linear resolution, then conditions $(*)$ and $(**)$ of Proposition 3.0.74 are satisfied, after a suitable renumbering of the variables. Hence by Corollary 3.0.73 each power of I has a linear resolution.

It remains to prove (a) $\Rightarrow$ (b). Again we may assume that conditions $(*)$ and $(**)$ hold. We denote by $[u,v]$ the greatest common divisor of u and v.

We show that the following condition (q) is satisfied: the elements of $G(I)$ can be ordered such that if $u,v \in G(I)$ with $u > v$, then there exists $w > v$ such that $w/[w,v]$ is of degree 1 and $w/[w,v]$ divides $u/[u,v]$. This condition (q) then implies that I has linear quotients.

The squarefree monomials in $G(I)$ will be ordered by the lexicographical order induced by $x_n > x_{n-1} > \cdots > x_1$, and if $x_i^2 \in G(I)$, then we let $u > x_i^2 > v$, where u is the smallest squarefree monomial of the form $x_k x_i$ with $k < i$, and where v is the largest squarefree monomial less than u.

Now, for any two monomials $u,v \in G(I)$ with $u > v$ corresponding to our order, we need to show that property (q) holds. There are three cases:

Case 1: $u = x_s x_t$ and $v = x_i x_j$ both are squarefree monomials with $s < t$ and $i < j$. Since $u > v$, we have $t \geq j$. If $t = j$, take $w = u$. If $t > j$, then by $(*)$, either $x_i x_t \in G(I)$ or $x_j x_t \in G(I)$. Accordingly, let $w = x_i x_t$ or $w = x_j x_t$.

Case 2: $u = x_t^2$ and $v = x_i x_j$ with $i < j$. Since $u > v$, we have $t > j$. Hence by $(*)$, either $x_i x_t \in G(I)$ or $x_j x_t \in G(I)$. Accordingly, let $w = x_i x_t$ or $w = x_j x_t$.

Case 3: $u = x_s x_t$ with $s \leq t$ and $v = x_i^2$. If $t = i$, then $s \neq t$ and take $w = u$. If $t > i$, then by $(**)$, we have either $x_i x_t \in G(I)$ or $x_i x_s \in G(I)$. Both elements are greater than v in our order. Accordingly, let $w = x_i x_t$ or $w = x_i x_s$. Then again, (q) holds. □

CHAPTER 4

Cohen-Macaulay graphs and finite meet-semilattices

Cohen-Macaulay graphs. Let G be a finite graph on the vertex set $[n]=\{1,2,\cdots,n\}$ without loops and multiple edges and let K be a field, $S=K[x_1,\ldots,x_n]$ being the polynomial ring over K in n variables and $I(G)\subset S$ the edge ideal of G, that is, the ideal generated by all monomials x_ix_j such that $\{i,j\}$ is an edge of G. The graph G is called *Cohen-Macaulay over* K if $S/I(G)$ is a Cohen-Macaulay ring, and it is called Cohen-Macaulay if it is Cohen-Macaulay over any field.

Given a field K, the general problem is to classify the graphs that are Cohen-Macaulay over K. In this generality the problem is as hard as classifying all Cohen-Macaulay simplicial complexes, because given a simplicial complex Δ, one can naturally construct a finite graph G such that G is Cohen–Macaulay if and only if Δ is Cohen–Macaulay. In fact, if P is the face poset of Δ (the poset consisting of all faces of Δ, ordered by inclusion), then Δ is Cohen–Macaulay if and only if the order complex $\Delta(P)$ of P is Cohen–Macaulay. Since the order complex $\Delta(P)$ is flag, i.e., every minimal nonface is a 2-element subset, it follows that there is a finite graph G such that $I(G)$ coincides with the Stanley–Reisner ideal of $\Delta(P)$.

Thus one cannot expect a general classification theorem. On the other hand, the first positive result was given by Villarreal **[68]** who determined all Cohen-Macaulay trees. This result has recently been generalized in **[35]** where all bipartite Cohen-Macaulay graphs have been described. It turned out that the Cohen-Macaulay property of a bipartite graph does not depend on the field K.

In this lecture we will give the classification of the Cohen-Macaulay bipartite graphs.

DEFINITION 4.0.77. A finite graph G on the vertex set $[n]$ is *bipartite* if there exists a partition $[n]=V\cup V'$, $V\cap V'=\emptyset$ such that each edge of G is of the form $\{i,j\}$ with $i\in V$ and $j\in V'$.

A graph G is bipartite if it has no cycles of odd length. Figure 1 gives an example of a bipartite graph.

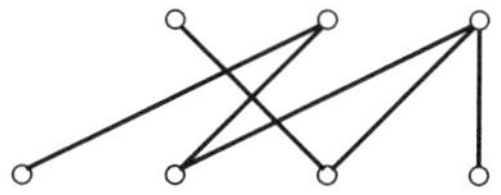

FIGURE 1

We say that a graph G is a *Cohen-Macaulay graph* if $S/I(G)$ is Cohen-Macaulay.

THEOREM 4.0.78 (Herzog-Hibi). Let G be a bipartite graph with vertex partition $V\cup V'$. Then the following conditions are equivalent:

(a) G is a Cohen-Macaulay graph;
(b) $|V| = |V'|$ and the vertices $V = \{x_1,\dots,x_n\}$ and $V' = \{y_1,\dots,y_n\}$ can be labelled such that:
  (i) $\{x_i,y_i\}$ are edges for $i = 1,\dots,n$;
  (ii) if $\{x_i,y_j\}$ is an edge, then $i \le j$;
  (iii) if $\{x_i,y_j\}$ and $\{x_j,y_k\}$ are edges, then $\{x_i,y_k\}$ is an edge.

PROOF. (a) $\Rightarrow$ (b): If G is Cohen-Macaulay, then all minimal prime ideals of $I(G)$ have the same height. If $V = \{x_1,\dots,x_n\}$ and $V' = \{y_1\dots,y_m\}$, then $(x_1,\dots,x_n)$ and $(y_1,\dots,y_m)$ are minimal prime ideals of $I(G)$. Therefore, $n = m$.

(i) Let $U \subset V$, and $N(U) = \{y_j \in V' \colon \{x_i,y_j\} \in E(G)$ for some $x_i \in U\}$. The well-known marriage theorem in graph theory says: if $|U| \le |N(U)|$ for all subsets U of V, then the vertices can be labelled such that $\{x_i,y_i\} \in E(G)$ for all $x_i \in V$.

Now $(V \setminus U) \cup N(U)$ is a vertex cover for all U. Since $I(G)$ is unmixed, all minimal vertex covers have size n. So

$$n - |U| + |N(U)| \ge |(V \setminus U) \cup N(U)| \ge n.$$

Hence $|N(U)| \ge |U|$, and the marriage theorem may be applied.

(ii) Let Δ be the simplicial complex with $I_\Delta = I(G)$. Then V and V' are facets of Δ, while $\{x_i,y_i\}$ (with the labelling as above) are not faces of Δ.

Since Δ is Cohen-Macaulay it is strongly connected. In particular, there exist facets

$$V = F_0,\dots,F_n = V'$$

such that F_i and F_{i-1} intersect in codimension 1 for $i = 1,\dots,n-1$. Then $|F_1 \setminus F_0| = 1$, say $F_1 \setminus F_0 = \{y_1\}$. Therefore $F_1 = \{y_1,x_2,\dots,x_n\}$ because $\{x_1,y_1\}$ is not a face.

Similarly, $|F_2 \setminus F_1| = 1$, say $F_2 \setminus F_1 = \{y_2\}$. Therefore $F_1 = \{y_1,y_2,x_3\dots,x_n\}$ because $\{x_2,y_2\}$ is not a face.

Hence by induction we may assume that $F_i = \{y_1,\dots,y_i,x_{i+1},\dots,x_n\}$ for $i = 0,\dots,n$. In particular, if $j < i$, then $\{x_i,y_j\}$ is a face of Δ, and hence not an edge of G.

(iii) Now we assume that the vertices are labelled such that (i) and (ii) hold.

Let $\{x_i,y_j\}$ and $\{x_j,y_k\}$ be edges of G with $i < j < k$, and suppose that $\{x_i,y_k\} \notin E(G)$.

Since $\{x_i,y_k\}$ is a face of Δ and since Δ is pure, there exists a facet of Δ with $|F| = n$ and $\{x_i,y_k\} \subset F$. Since F is a face of Δ, any 2-element subsets of F are non-edges of G.

We have $y_j \notin F$ since $\{x_i,y_j\} \in E(G)$ and $x_i \in F$, and $x_j \notin F$ since $\{x_j,y_k\} \in E(G)$ and $y_k \in F$.

On the other hand, since $\{x_\ell,y_\ell\} \in E(G)$ for all ℓ, the face F never contains both x_ℓ and y_ℓ, i.e., $F = \{x_{i_1},\dots,x_{i_k},y_{j_1},\dots,y_{j_{n-k}}\}$ with $\{i_1,\dots,i_k,j_1,\dots,j_{n-k}\} = [n]$. So either $y_j \in F$ or $x_j \in F$, a contradiction. □

For the proof of the other direction, we make a detour. A finite poset $\mathscr{L}$ is called a *lattice* if each pair of elements of $\mathscr{L}$ has a least upper bound and a greatest lower bound. The (unique) least upper bound of $p,q \in \mathscr{L}$ is denoted $p \wedge q$ (read 'p meet q'), and the (unique) greatest upper bound is denoted $p \vee q$ (read 'p join q'). A lattice $\mathscr{L}$ is called *distributive* if the operations join $p \vee q$ and meet $p \wedge q$ satisfy the distributive law. An example of a distributive lattice is shown in Figure 2.

Given a finite poset P, a subset $I \subset P$ is a *poset ideal* if with each $p \in I$ all q with $q \le p$ belong to I.

The set $\mathscr{J}(P)$ of poset ideals with union as join and intersection as meet is a distributive lattice. For example, if the elements in P are pairwise incomparable, then $\mathscr{J}(P)$ is a Boolean lattice of rank $|P|$.

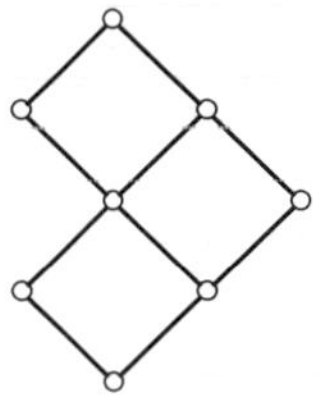

FIGURE 2

The fundamental theorem of Birkhoff [**14**] says: Given a finite distributive lattice $\mathscr{L}$, let $P \subset \mathscr{L}$ be the subposet of join irreducible elements. Then

$$\mathscr{L} \cong \mathscr{J}(P).$$

Figure 3 identifies the lattice in Figure 2 as the ideal lattice of a poset.

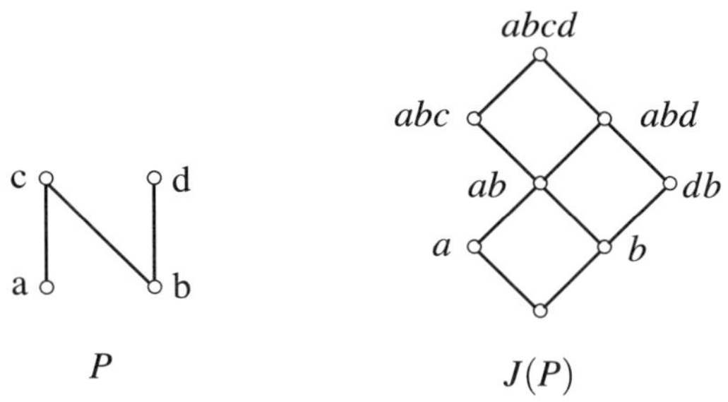

FIGURE 3

Now for any distributive lattice $\mathscr{L} = \mathscr{J}(P)$ we associate an ideal

$$H_{\mathscr{L}} \subset K[\{x_p, y_p\}_{p \in P}]$$

as follows: for each poset ideal $I \subset P$, let

$$u_I = \prod_{p \in I} x_p \prod_{q \notin I} y_q.$$

In our example

$$H_{\mathscr{L}} = (uvwx, avwx, bvwx, abwx, abcx, bdux, abcd).$$

Here we used the variables a, b, c, d, u, v, w, x instead of $x_1, \ldots, x_4, y_1 \ldots, y_4$.

THEOREM 4.0.79. *(a)* $H_{\mathscr{L}}$ has a linear resolution.
(b) Let $b_i =$ number of intervals $[p, q] = \{a \in \mathscr{L} \colon p \leq a \leq q\}$ which are isomorphic to a Boolean lattice of rank i. Then $\beta_i(H_{\mathscr{L}}) = b_i$.

In our example we have $b_0 = 8$, $b_1 = 10$, and $b_2 = 3$.

COROLLARY 4.0.80. $\sum_i (-1)^i b_i = 1$.

PROPOSITION 4.0.81. $H_{\mathscr{L}} = \bigcap_{p \leq q} (x_p, y_q)$.

PROOF. Let (x_p, y_q) with $p \leq q$, and let $u_I \in H_{\mathscr{L}}$. If $p \in I$, then $x_p | u_I$, and hence $u_I \in (x_p, y_q)$. If $p \notin I$, then $q \notin I$, and hence $y_q | u_I$, that is, $u_I \in (x_p, y_q)$. Therefore, $H_{\mathscr{L}} \subset \bigcap_{p \leq q} (x_p, y_q)$.

Conversely, let $\mathfrak{p}$ be a monomial prime ideal with $H_{\mathscr{L}} \subset \mathfrak{p}$. If $\mathfrak{p}$ contains no x_p, then $u_P = \prod_{p \in P} x_p \notin \mathfrak{p}$, a contradiction. If $\mathfrak{p}$ contains no y_q, then $u_\emptyset = \prod_{q \in P} y_q \notin \mathfrak{p}$, a contradiction.

Therefore $x_p \in \mathfrak{p}$ for some $p \in P$, and $y_q \in \mathfrak{p}$ for some $q \in P$. Let p be minimal with this property. We show that $y_q \in \mathfrak{p}$ for some $q \geq p$. (Then $(x_p, y_q) \subset \mathfrak{p}$, as desired.)

Suppose that for all $y_q \in \mathfrak{p}$ we have $q < p$. Let I be the poset ideal $\{r \in P \colon r < p\}$. By the choice of $x_p \in \mathfrak{p}$ we have that x_r does not divide u_I for any $x_r \in \mathfrak{p}$. Also y_q does not divide u_I for any $y_q \in \mathfrak{p}$ since $q \in I$. Hence $u_I \notin \mathfrak{p}$, a contradiction. □

Let Δ be the simplicial complex with $I_\Delta = H_{\mathscr{L}}$, $\mathscr{L} = \mathscr{J}(P)$. Then by Lemma 3.0.61 and Proposition 4.0.81 we have

$$I_{\Delta^\vee} = (x_p y_q)_{p \leq p}.$$

Since $H_{\mathscr{L}}$ has a linear resolution, $S/I_{\Delta^\vee}$ is Cohen-Macaulay, and it is the edge ideal of a bipartite graph, which we denote by $G(P)$.

So for each finite poset P, $G(P)$ is a Cohen-Macaulay bipartite graph (satisfying conditions (b), (i), (ii), and (iii) of Theorem 4.0.78).

Now we can prove Theorem 4.0.78 (b) $\Rightarrow$ (a): Let G be a bipartite graph satisfying the conditions (b) of Theorem 4.0.78 with vertex decomposition

$$V = \{x_p\}_{p \in P} \quad \text{and} \quad V' = \{y_p\}_{p \in P},$$

where $P = [n]$.

On P we define a partial order $<$ by setting $p < q$ if $\{x_p, y_q\} \in E(G)$. Then $G = G(P)$ and hence G is Cohen-Macaulay.

Figure 4 shows a bipartite graph that satisfies the conditions of Theorem 4.0.78, and hence is Cohen-Macaulay.

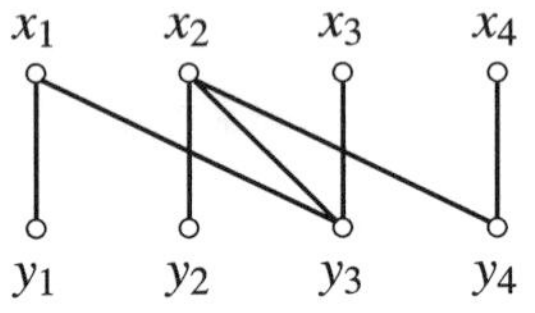

FIGURE 4

An algebraic characterization of meet-distributive meet-semilattices. A finite poset $\mathscr{L}$ is called a *meet-semilattice* if each pair of elements of $\mathscr{L}$ has a meet, and a finite meet-semilattice $\mathscr{L}$ is called *meet-distributive* if each interval $[x, y] = \{p \in \mathscr{L} \colon x \leq p \leq y\}$ of $\mathscr{L}$ such that x is the meet of the lower neighbors of y in this interval is Boolean. Here we call z a lower neighbor of y if y covers z, that is, if $z < y$ and there is no $w \in \mathscr{L}$ such that $z < w < y$. Recall that a lattice is *Boolean* if it is isomorphic to the lattice of all subsets of a finite set.

Let $\mathscr{L}$ be an arbitrary finite meet-semilattice, and $P \subset \mathscr{L}$ the set of join-irreducible elements of $\mathscr{L}$. We denote by $\hat{0}$ and $\hat{1}$ the minimal and maximal element of $\mathscr{L}$. (Since $\mathscr{L}$ is a finite meet-semilattice, it follows [**63**, Proposition 3.3.1] that $\mathscr{L}$ possesses $\hat{1}$ if and only if $\mathscr{L}$ is a lattice.) Recall that $p \in \mathscr{L}$ is *join-irreducible* if $p \neq \hat{0}$ and p is not a join of elements strictly less than p.

To each element $p \in \mathscr{L}$ we associate the subset $\ell(p) = \{q \in P \colon q \leq p\}$ of P. Note that $p \in \ell(p)$ if and only if p is join irreducible. In any case, $\ell(p)$ is a poset ideal of P. The *set of generators* of I is the set of maximal elements in I, denoted by $G(I)$.

We obtain a map

$$\ell \colon \mathscr{L} \longrightarrow \mathscr{B}_P,$$

which we call the *canonical embedding* into the Boolean lattice $\mathscr{B}_P$ consisting of all subsets of P ordered by inclusion.

Let $\hat{\mathscr{L}} \subset \mathscr{B}_P$ be the distributive lattice consisting of all poset ideals of P. Then ℓ induces an injective order preserving map $\ell \colon \mathscr{L} \to \hat{\mathscr{L}}$. We call $\hat{\mathscr{L}}$ the *distributive closure* of $\mathscr{L}$.

The cardinality of $\ell(p)$ is called the *degree* of p, and we denote it by $\deg p$. One always has the inequality $\operatorname{rank} p \leq \deg p$. Recall that the *rank* of p is the maximal length of chains descending from p.

LEMMA 4.0.82. Let $\mathscr{L}$ be a finite meet-semilattice, ℓ the canonical embedding, and $s, t \in \mathscr{L}$ any two elements. We have

(i) $s = t$ if and only if $\ell(s) = \ell(t)$;
(ii) $s \leq t$ if and only if $\ell(s) \subseteq \ell(t)$;
(iii) $\ell(s) \cap \ell(t) = \ell(s \wedge t)$.

PROOF. Note that each element of $\mathscr{L}$ is the join of elements in P. From this observation all assertions follow. □

The lemma implies that ℓ is an injective order preserving map. In general, however, ℓ is not an embedding of lattices. It is not difficult to see that ℓ is an embedding of meet-semilattices if and only if $\mathscr{L}$ is meet-distributive.

The following combinatorial characterization of meet-distributive lattices is discussed in the survey article [**23**]. A finite meet-semilattice is called *graded* if for each element all of its maximal chains have the same length.

LEMMA 4.0.83. For a finite lattice $\mathscr{L}$ the following conditions are equivalent:

(i) $\mathscr{L}$ is meet-distributive;
(ii) $\mathscr{L}$ is graded and $\deg \hat{1} = \operatorname{rank} \hat{1}$;
(iii) $\mathscr{L}$ is graded and $\deg p = \operatorname{rank} p$ for all $p \in \mathscr{L}$;
(iv) each element in $\mathscr{L}$ is a unique minimal join of join-irreducible elements;
(v) $\mathscr{L}$ is lower semimodular, and any upper semimodular sublattice is distributive.

We now introduce the squarefree monomial ideal $H_{\mathscr{L}}$ associated with a finite meet-semilattice $\mathscr{L}$. Let P be the set of join-irreducible elements of $\mathscr{L}$. Let K be a field and $S = K[\{x_p, y_p\}_{p \in P}]$ the polynomial ring in $2|P|$ variables over K. For each element $q \in \mathscr{L}$ write

$$u_q = \prod_{p \in \ell(q)} x_p \prod_{p \in P \setminus \ell(q)} y_p,$$

and set $H_{\mathscr{L}} = (u_q)_{q \in \mathscr{L}}$.

Note that $\operatorname{height}(H_{\mathscr{L}}) = 2$ if $\mathscr{L}$ is a lattice. In fact, $H_{\mathscr{L}} \subset (x_p, y_p)$ for any $p \in P$ while on the other hand $u_{\hat{0}} = \prod_{p \in P} y_p$ and $u_{\hat{1}} = \prod_{p \in P} x_p$ both belong to $H_{\mathscr{L}}$ and have no common factor.

We now come to our algebraic characterization of meet-distributive meet-semilattices as given in [**43**].

THEOREM 4.0.84 (Herzog-Hibi-Zheng). Let $\mathscr{L}$ be an arbitrary finite meet-semilattice. The following conditions are equivalent:

(i) $\mathscr{L}$ is meet-distributive;
(ii) $H_{\mathscr{L}}$ has linear quotients;
(iii) $H_{\mathscr{L}}$ has a linear resolution;
(iv) $H_{\mathscr{L}}$ has linear relations.

PROOF. (i) $\Rightarrow$ (ii): We fix a linear order $\prec$ on $\mathscr{L}$ that extends the partial order given by the degree. We put $u_r < u_q$ if $r \prec q$. For any $u_q \in H_{\mathscr{L}}$ and any $u_r < u_q$, let t be a lower neighbor of q in the interval $[r \wedge q, q]$. Then $u_t/[u_t, u_q] = y_p$, where $\{p\} = \ell(q) \setminus \ell(t)$. We claim that y_p divides u_r. If not, then x_p divides u_r and so $p \in \ell(r) \cap \ell(q) = \ell(r \wedge q)$. Thus $p \in \ell(t)$, since $r \wedge q \leq t$.

(ii) $\Rightarrow$ (iii) and (iii) $\Rightarrow$ (iv) are trivial.

(iv) $\Rightarrow$ (i): Suppose $\mathscr{L}$ is not meet-distributive. Then by Lemma 4.0.83(iii) (which is also valid if $\mathscr{L}$ is a meet-distributive meet-semilattice) there exist $p, q \in \mathscr{L}$ such that q is a lower neighbor of p and $\deg p - \deg q > 1$. The ideal (u_p, u_q) is generated by precisely those monomials in $G(H_{\mathscr{L}})$ that are not divided by x_r for all $r \in P \setminus \ell(p)$, and are not divided by all y_s for all $s \in \ell(q)$. Since we assume that $H_{\mathscr{L}}$ has linear relations, the restriction lemma in [**41**, Lemma 4.4] implies that (u_p, u_q) has linear relations contradicting the fact that $\deg p - \deg q > 1$. □

As an extension of Proposition 4.0.81 we have

PROPOSITION 4.0.85. Let $\mathscr{L}$ be a finite lattice and P its poset of join-irreducible elements. Then

(i) the minimal prime ideals of height 2 of $H_{\mathscr{L}}$ are (x_p, y_q) where $p, q \in P$ and $p \leq q$;
(ii) $H_{\mathscr{L}}$ has only height 2 minimal prime ideals if and only if $\mathscr{L}$ is distributive.

PROOF. Let $\hat{\mathscr{L}}$ be the distributive lattice consisting of all poset ideals of P. Then ℓ induces an injective order preserving map $\ell\colon \mathscr{L} \to \hat{\mathscr{L}}$. Thus $H_{\mathscr{L}} \subset H_{\hat{\mathscr{L}}}$, and equality holds if and only if $\mathscr{L}$ is distributive. This follows from Birkhoff's fundamental structure theorem [**63**].

(i) The minimal prime ideals of $H_{\hat{\mathscr{L}}}$ are precisely the ideals (x_p, y_q) where $p, q \in P$ and $p \leq q$; see Proposition 4.0.81. Of course these are also minimal prime ideals of $H_{\mathscr{L}}$. We claim that there are no other minimal prime ideals of height 2 of $H_{\mathscr{L}}$. Indeed, any such ideal must contain some x_p and some y_q, since $\prod_{p \in P} x_p$ and $\prod_{p \in P} y_p$ belong to $H_{\mathscr{L}}$. Suppose $p \nleq q$; then u_q is not contained in (x_p, y_q).

(ii) It remains to show that if $\mathscr{L}$ is not distributive, then there exists a minimal prime ideal of $H_{\mathscr{L}}$ of height > 2. In fact, the proof of (i) shows that if such a minimal prime ideal does not exist, then $H_{\mathscr{L}} = H_{\hat{\mathscr{L}}}$. Therefore $\mathscr{L} = \hat{\mathscr{L}}$, and hence $\mathscr{L}$ is distributive. □

The Alexander dual of a meet-distributive meet-semilattice. This final section surveys some results of [**38**]. Let $\mathscr{L}$ be an arbitrary finite meet-semilattice with P as its set of join-irreducible ideals. Let $\Delta_{\mathscr{L}}$ be the simplicial complex whose Stanley–Reisner ideal is $H_{\mathscr{L}}$. We denote by $\Gamma_{\mathscr{L}}$ the Alexander dual $(\Delta_{\mathscr{L}})^{\vee}$ of $\Delta_{\mathscr{L}}$, and call it the *Alexander dual* of $\mathscr{L}$.

By Lemma 3.0.60 one has

$$\mathscr{F}(\Gamma_{\mathscr{L}}) = \{F_p : p \in \mathscr{L}\}, \tag{4.0.1}$$

where

$$F_{\alpha} = \{x_q : q \in P \setminus \ell(p)\} \cup \{y_q\colon q \in \ell(p)\}.$$

In particular, $\Gamma_{\mathscr{L}}$ is a pure simplicial complex of dimension $|P|-1$.

As an immediate consequence of Theorem 4.0.84 and Theorem 3.0.62 we obtain

COROLLARY 4.0.86. The following conditions are equivalent:

(i) $\Gamma_{\mathscr{L}}$ is shellable;
(ii) $\Gamma_{\mathscr{L}}$ is Cohen–Macaulay;
(iii) $\mathscr{L}$ is meet-distributive.

The next result describes the h-vector of $\Gamma_{\mathscr{L}}$. For each $p \in \mathscr{L}$ we denote by $N(p)$ the number of lower neighbors in $\mathscr{L}$.

PROPOSITION 4.0.87. Let $\mathscr{L}$ be a finite meet-distributive meet-semilattice and let $h(\Gamma_{\mathscr{L}}) = (h_0, h_1, \dots)$ be the h-vector of $\Gamma_{\mathscr{L}}$. Then, for all i, one has

$$h_i = |\{p \in \mathscr{L} \colon |N(p)| = i\}|.$$

PROOF. Let $p \in \mathscr{L}$ with $|N(p)| = i$ and $\ell(p) = \{q_1, \dots, q_\delta\}$. Let $N(p) = \{r_1, \dots, r_i\}$ with each $\ell(r_j) = \ell(p) \setminus \{q_{\delta-j+1}\}$. Let $r = r_1 \wedge \cdots \wedge r_i$. Thus

$$\ell(r) = \cap_{j=1}^{i} \ell(r_j) = \{q_1, \dots, q_{\delta-i}\}$$

and the interval $[r, p]$ in $\mathscr{L}$ is a Boolean lattice of rank i. Since a subset $A \subset \ell(p)$ is contained in none of the sets $\ell(r_1), \dots, \ell(r_i)$ if and only if A contains $\{q_\delta, q_{\delta-1}, \dots, q_{\delta-i+1}\}$, it follows that the number of subsets $A \subset \ell(p)$ with $|A| = k$ such that $A \subset \ell(q)$ for no $q \in \mathscr{L}$ with $q < p$ is $\binom{\delta-i}{k-i}$. In other words, the number of those faces $F \subset F_p$ of $\Gamma_{\mathscr{L}}$ with $|F| = j+1$ such that $F \subset F_q$ for no $q \in \mathscr{L}$ with $q < p$ is $\sum_{k=i}^{\delta} \binom{|P|-\delta}{j-k+1} \binom{\delta-i}{k-i}$, which is equal to $\binom{|P|-i}{j-i+1} = \binom{|P|-i}{|P|-j-1}$.

Thus the number of faces F of $\Gamma_{\mathscr{L}}$ with $|F| = j+1$ is

$$f_j(\Gamma_{\mathscr{L}}) = \sum_{i=0}^{j+1} \binom{|P|-i}{|P|-j-1} |\{p \in \mathscr{L} \colon |N(p)| = i\}|.$$

On the other hand, in general, one has

$$f_j(\Gamma_{\mathscr{L}}) = \sum_{i=0}^{j+1} \binom{|P|-i}{|P|-j-1} h_i(\Gamma_{\mathscr{L}}).$$

Hence $h_i(\Gamma_{\mathscr{L}}) = |\{p \in \mathscr{L} \colon |N(p)| = i\}|$, as desired. □

COROLLARY 4.0.88. Let $\mathscr{L}$ be a finite meet-distributive meet-semilattice, P the set of join-irreducible elements of $\mathscr{L}$, and $\Gamma_{\mathscr{L}}$ its Alexander dual. Let $n = |P|$ and $(h_0, h_1, \dots, h_n)$ the h-vector of $\Gamma_{\mathscr{L}}$. Then $h_1 = n$, and the a-invariant of $\Gamma_{\mathscr{L}}$ (which is the nonnegative integer $n - \max\{i \colon h_i \neq 0\}$) is equal to $\max\{|N(p)| \colon p \in \mathscr{L}\} - n$.

EXAMPLE 4.0.89. Let $\mathscr{B}_{[n]}$ denote the Boolean lattice of rank n and $\mathscr{L}$ a poset ideal of $\mathscr{B}_{[n]}$ that contains all join-irreducible elements (i.e., $\{1\}, \dots, \{n\}$) of $\mathscr{B}_{[n]}$. Then the meet-distributive meet-semilattice $\mathscr{L}$ is a simplicial complex on $[n]$ and the h-vector of $\Gamma_{\mathscr{L}}$ coincides with the f-vector of $\mathscr{L}$.

(a) By using (4.0.1), the Stanley–Reisner ideal $I_{\Gamma_{\mathscr{L}}}$ of $\Gamma_{\mathscr{L}}$ is generated by those squarefree monomials $\prod_{q \in \ell(\beta)} y_q$ such that $\beta \in \mathscr{B}_{[n]}$ is a minimal nonface of $\mathscr{L}$ and by the quadratic monomials $x_{\{i\}} y_{\{i\}}$ for all $i \in [n]$.

(b) Let $T = K[y_{\{1\}}, \dots, y_{\{n\}}]$ and $J \subset T$ the ideal generated by those squarefree monomials $\prod_{q \in \ell(p)} y_q$ such that $p \in \mathscr{B}_{[n]}$ is a minimal nonface of $\mathscr{L}$ and by $y_{\{i\}}^2$ for all $i \in [n]$. The quotient ring T/J is of 0-dimension and its h-vector is $(f_{-1}, f_0, \dots)$, the f-vector of

$\mathscr{L}$ with $f_{-1} = 1$. It turns out that $I_{\Gamma_{\mathscr{L}}}$ is the polarization [**15**, Lemma 4.3.2] of the ideal J. Since T/J is Cohen–Macaulay, it follows immediately that $\Gamma_{\mathscr{L}}$ is Cohen–Macaulay. This fact is a special case of Corollary 4.0.86.

(c) Since T/J is a level ring [**63**, p. 91] if and only if the simplicial complex $\mathscr{L}$ is pure, it follows that the Alexander dual $\Gamma_{\mathscr{L}}$ of $\mathscr{L}$ is a level complex if and only if the simplicial complex $\mathscr{L}$ is pure.

(d) By using (b) and (c), it follows that every f-vector of a pure simplicial complex is the h-vector of a level complex.

We conclude by quoting a result that characterizes all meet-distributive meet-semilattices $\mathscr{L}$ for which $\Gamma_{\mathscr{L}}$ is level.

For each $p \in \mathscr{L}$ we write $p' \in \mathscr{L}$ for the meet of all $q \in N(p)$. Since $\mathscr{L}$ is a meet-distributive meet-semilattice, it follows that the interval

$$\mathscr{B}_p = [p', p] = \{r \in \mathscr{L} : p' \leq r \leq p\}$$

of $\mathscr{L}$ is a Boolean lattice. Let $S(p) \subset P$ denote the antichain

$$S(p) = \ell(p) \setminus \ell(p').$$

Each element belonging to $S(p)$ is a maximal element of $\ell(p)$, and $t \in \ell(p)$ belongs to $S(p)$ if and only if $\ell(s) = \ell(p) \setminus \{t\}$ for some $s \in N(p)$.

THEOREM 4.0.90 (Herzog-Hibi). (a) The collection of sets $\{S(p) : p \in \mathscr{L}\}$ is a simplicial complex on the vertex set P whose f-vector coincides with the h-vector of $\Gamma_{\mathscr{L}}$.

(b) The simplicial $\Gamma_{\mathscr{L}}$ of a finite meet-distributive meet-semilattice $\mathscr{L}$ is a level complex if and only if the simplicial complex $\{S(\alpha) : \alpha \in \mathscr{L}\}$ is pure. Thus in particular the Alexander dual $\Gamma_{\mathscr{L}}$ of a finite distributive lattice $\mathscr{L} = \mathscr{J}(P)$ is level if and only if the simplicial complex consisting of all antichains of P is pure.

Consider the example of a meet-distributive meet-semilattice $\mathscr{L}$ given in Figure 5 with the poset of join-irreducible elements as shown in Figure 6.

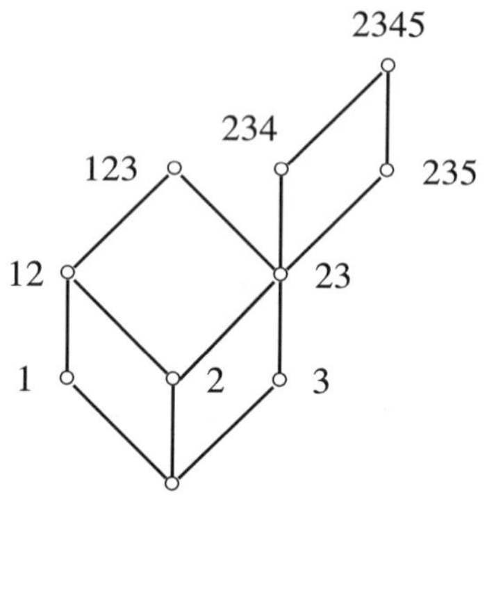

FIGURE 5

The h-vector of $\mathscr{L}$ is $(1, 5, 4)$, and Theorem 4.0.90 implies that $\Gamma_{\mathscr{L}}$ is level.

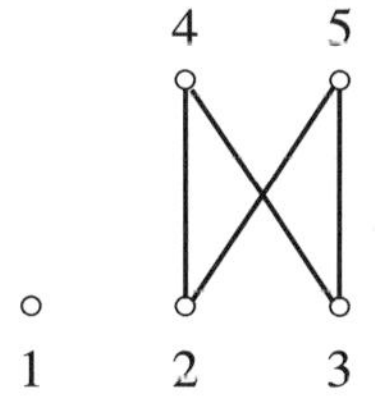

FIGURE 6

By Theorem 4.0.90, the h-vector of the Alexander dual of a meet-distributive meet-semilattice, is just the f-vector of a simplicial complex, and the h-vector of a level simplicial complex coming from a meet-distributive meet-semilattice is just the f-vector of a pure simplicial complex. These facts lead us to the following;

PROBLEMS 4.0.91. (a) Characterize the h-vector of the Alexander dual of a finite distributive lattice.

(b) Characterize the h-vector of the Alexander dual of a meet-distributive lattice.

(c) Find a nice class of level simplicial complexes whose h-vector is not the f-vector of a pure simplicial complex. For example, $(1,3,3)$ is the h-vector of the Alexander dual of the meet-distributive lattice $\mathscr{B}_{[3]} \setminus \{1,3\}$, but is not the h-vector of the Alexander dual of a distributive lattice.

Bibliography

[1] A. Aramova and J. Herzog, Koszul cycles and Eliahou–Kervaire type resolution, J. Alg. **181** (1996), 347 – 370.

[2] A. Aramova and J. Herzog, p-Borel principal ideals, Ill. J. Math. **41**(1) (1997), 103 – 121.

[3] A. Aramova and J. Herzog, Almost regular sequences and Betti numbers, Amer. J. Math. **122** (2000), 689 – 719.

[4] A. Aramova, J. Herzog and T. Hibi, Squarefree lexsegment ideals, Math. Z. **228** (1998), 353 – 378.

[5] A. Aramova, J. Herzog and T. Hibi, Gotzmann Theorems for Exterior Algebras and Combinatorics, J. Alg. **191** (1997), 174 – 211.

[6] A. Aramova, J. Herzog and T. Hibi, Shifting operations and graded Betti numbers, J. Algebraic Combin. **12**(3) (2000), 207 – 222.

[7] A. Aramova, J. Herzog and T. Hibi, Ideals with stable Betti numbers, Adv. Math. **152**(1) (2000), 72 – 77.

[8] A. Aramova, L. L. Avramov and J. Herzog, Resolutions of monomial ideals and cohomology over exterior algebras, Transactions Amer. Math. Soc. **352**(2) (2000), 579–594.

[9] E. Babson, I. Novik and R. Thomas, Reverse lexicographic and lexicographic shifting, J. Alg. Combin. **23**(2) (2006), 107 – 123.

[10] D. Bayer, The division algorithm and the Hilbert scheme, Ph.D. Thesis, Harvard 1982.

[11] D. Bayer and M. Stillman, A theorem on refining division orders by the reverse lexicographical order, Duke Math. J. **55** (1987), 321 – 328.

[12] D. Bayer, H. Charalambous and S. Popescu, Extremal Betti numbers and applications to monomial ideals, J. Alg. **221**(2) (1999), 497 – 512.

[13] D. Bayer and B. Sturmfels, Cellular resolutions of monomial modules, J. reine angew. Math. **502** (1998), 123 – 140.

[14] G. Birkhoff, Lattice Theory, Volume XXV, Third Edition, Second Printing, American Mathematical Society, 1973.

[15] W. Bruns and J. Herzog, Cohen–Macaulay rings, Revised Edition, Cambridge University Press, 1996.

[16] W. Bruns and J. Herzog, On multigraded resolutions, Math. Proc. Camb. Phil. Soc. **118** (1995), 245 – 257.

[17] A. Björner, G. Kalai, An extended Euler-Poincaré theorem, Acta Math. **161** (1988), 279 – 303.

[18] A. Conca and J. Herzog, Castelnuovo-Mumford regularity of products of ideals, Collect. Math. **54** (2003), 137–152.

[19] D. Cutkosky, J. Herzog and N. V. Trung, Asymptotic behaviour of the Castelnuovo-Mumford regularity, Compositio Math. **118** (1999), 243–261.

[20] E. De Negri, Toric rings generated by special stable sets of monomials, Math. Nachr. **203** (1999), 31 – 45.

[21] G. A. Dirac, On rigid circuit graphs, Abh. Math. Sem. Univ. Hamburg, **38** (1961), 71 – 76.

[22] J. A. Eagon and V. Reiner, Resolutions of Stanley-Reisner rings and Alexander duality, J. of Pure and Appl. Alg. **130** (1998), 265 – 275.

[23] P. Edelman, Abstract convexity and meet-distributive lattices, *in* "Combinatorics and Ordered Sets" (Arcata, Calif., 1985), Contemp. Math. **57**, 127–150, Amer. Math. Soc., Providence, RI, 1986.

[24] J. Edmonds, Submodular functions, matroids, and certain polyhedra, *in* "Combinatorial Structures and Their Applications" (R. Guy, H. Hanani, N. Sauer and J. Schonheim, Eds.) Gordon and Breach, New York, 1970, pp. 69 – 87.

[25] D. Eisenbud, Commutative algebra, with a view towards algebraic geometry, Graduate Texts Math., Springer 1995.

[26] S. Eliahou and M. Kervaire, Minimal resolutions of some monomial ideals, J. Alg. **129** (1990), 1 – 25.

[27] P. Erdös, C. Ko and R. Rado, Intersection theorems for systems of finite sets, *Quart. J. Math. Oxford Ser.* **12**(2) (1960), 313–320.

[28] S. Faridi, The facet ideal of a simplicial complex, Manuscripta Mat. **109** (2002), 159 – 174.

[29] S. Faridi, Cohen-Macaulay properties of squarefree monomial ideals, J. Combin. Theory Ser. A **109**(2), (2005), 299–329.

[30] S. Faridi, Simplicial trees are sequentially Cohen-Macaulay, J. Pure and Appl. Alg., **190**(1-3) (2004), 121–136.

[31] R. Fröberg, On Stanley–Reisner rings, in: Topics in Algebra, Banach Center Publications, **26**(2), (1990), 57 – 70.

[32] A. Galligo, Apropos du théorème de préparation de Weierstrass, *in* Fonctions de plusiers variables complexes, Springer Lect. Notes Math. **409** (1974), 543 – 579.

[33] M. Green, Generic initial ideals, *in* Proc. CRM -96, Six Lectures on Commutative Algebra, Barcelona, Spain, **166**, Birkhäuser, 1998, 119 – 186.

[34] J. Herzog, Generic initial ideals and graded Betti-numbers, *in* Computational Commutative Algebra and Combinatorics (T. Hibi, Ed.), Advanced Studies in Pure Math., Volume 33, 2002, 75–120.

[35] J. Herzog and T. Hibi, Distributive Lattices, Bipartite Graphs and Alexander Duality, J. Algebraic Combin. **22** (2005), 289–302.

[36] J. Herzog and T. Hibi, Discrete Polymatroids, J. Algebraic Combin. **16**(3) (2002), 239–268.

[37] J. Herzog and T. Hibi, Cohen–Macaulay polymatroidal ideals, European J. Combin. **27**(4) (2006), 513 – 517.

[38] J. Herzog and T. Hibi, Level rings arising from meet-distributive meet-semilattices, Nagoya Math. J. **181** (2006), 29 – 39.

[39] J. Herzog, T. Hibi, N. V. Trung and X. Zheng, Standard graded vertex cover algebras, cycles and leaves, to appear in Trans. Amer. Math. Soc.

[40] J. Herzog, T. Hibi and M. Vladoiu, Ideals of fiber type and polymatroids, Osaka J. Math. **42** (2005), 1–23.

[41] J. Herzog, T. Hibi and X. Zheng, Dirac's theorem on chordal graphs and Alexander duality, European J. Combin. **25**(7) (2004), 949–960.

[42] J. Herzog, T. Hibi and X. Zheng, Monomial ideals whose powers have a linear resolution, Math. Scand. **95**(1) (2004), 23–32.

[43] J. Herzog, T. Hibi and X. Zheng, The monomial ideal of a finite meet semi-lattice. Trans. Amer. Math. Soc. **358**(9) (2006), 4119–4134.

[44] T. Hibi, Algebraic Combinatorics on Convex Polytopes, Carslaw Publications, Glebe, N.S.W., Australia, 1992.

[45] T. Hibi, Distributive lattices, affine semigroup rings and algebras with straightening laws, *in* Commutative Algebra and Combinatorics (M. Nagata and H. Matsumura, Eds.), Advanced Studies in Pure Math., Volume 11, North–Holland, Amsterdam, 1987, pp. 93 – 109.

[46] J. Herzog and N. Terai, Stable properties of algebraic shifting, Result in Math. **35** (1999), 260 – 265.

[47] J. Herzog and D. Popescu, On the regularity of p-Borel ideals, Proc. AMS. **129** (2001), 2563–2570.

[48] J. Herzog and Y. Takayama, Resolutions by mapping cones, *in* The Roos Festschrift volume Nr.2(2), *Homology, Homotopy and Applications* **4**, (2002), 277 – 294.

[49] M. Hochster, "Cohen–Macaulay rings, combinatorics, and simplicial complexes", *in* Proc. Ring Theory II, Lect. Notes in Pure and Appl Math., **26**, Dekker, New York, 1977, 171 – 223.

[50] G. Kalai, Algebraic shifting, *in* Computational Commutative Algebra and Combinatorics (T. Hibi, Ed.), Advanced Studies in Pure Math., Volume 33, 2002, 75–120.

[51] G. Kalai, Algebraic shifting, Unpublished manuscript, 1993. (http://www.ma.huji.ac.il/~kalai).

[52] G. Kalai, The diameter of graphs of convex polytopes and f-vector theory, *in* Proc. Applied Geometry and Discrete Mathematics, DIMACS Series in Discrete Mathematics and Theoretical Computer Science, Amer. Math. Soc. **4**, (1991), 387 – 411.

[53] V. Kodiyalam, Asymptotic behaviour of Castelnuovo-Mumford regularity, Proc. Amer. Math. Soc. **128** (2000), 407 – 411.

[54] S. Murai and T. Hibi, Combinatorial shifting and graded Betti-numbers, arXiv:math.AC/0509209.

[55] S. Murai and T. Hibi, The behaviour of graded Betti-numbers via algebraic shifting and combinatorial shifting, arXiv:math.AC/0503685, 2005.

[56] H. Ohsugi, J. Herzog and T. Hibi, Combinatorial pure subrings, Osaka J. Math. **37** (2000), 745 – 757.

[57] H. Ohsugi and T. Hibi, Normal polytopes arising from finite graphs, J. Algebra **207** (1998), 409 – 426.

[58] H. Ohsugi and T. Hibi, Toric ideals generated by quadratic binomials, J. Algebra **218** (1999), 509 – 527.

[59] J. G. Oxley, Matroid Theory, Oxford University Press, Oxford, New York, 1992.

[60] K. Pardue, Nonstandard Borel-fixed ideals, Dissertation, Brandeis University, 1994.

[61] T. Römer, Homological properties of bigraded algebras, Ill. J. Math. **45**(2) (2001), 1361–1376.

[62] R. P. Stanley, Combinatorics and Commutative Algebra, Second Edition, Birkhäuser, Boston, MA, 1996.

[63] R. P. Stanley, Enumerative Combinatorics, Volume I, Wadsworth & Brooks/Cole, Monterey, CA, 1986.

[64] B. Sturmfels, Gröbner Bases and Convex Polytopes, Amer. Math. Soc., Providence, RI, 1995.

[65] B. Sturmfels, Four counterexamples in combinatorial algebraic geometry, J. Alg. **230** (2000), 282 – 294.

[66] B. Sturmfels and E. Miller, Combinatorial Commutative Algebra, Graduate Texts in Math. **227**, Springer, 2004.

[67] N. Terai, Generalization of Eagon–Reiner theorem and h-vectors of graded rings, Preprint 2000.

[68] R. H. Villareal, Monomial Algebras, Dekker, New York, NY, 2001.

[69] D. J. A. Welsh, Matroid Theory, Academic Press, London, New York, 1976.

[70] D. West, Introduction to Graph Theory, 2nd Edition, Prentice-Hall, 2001.

[71] N. White, A unique exchange property for bases, Linear Algebra Appl. **31** (1980), 81 – 91.

[72] X. Zheng, Homological properties of monomial ideals associated to quasi-trees and lattices, Ph.D. Thesis, Essen 2004.

[73] X. Zheng, Monomial ideals arising from distributive lattices, Comm. in Alg. **34**(4) (2006), 1361 – 1379.

Notes on Constructive Desingularization

by Orlando Villamayor

Contents

Introduction

Resolution of singularities has been an area of intense research since the late eighties, particularly in the effort to simplify the theory and also in the important task of implementations.

In these notes we prove two important and closely related theorems over fields of characteristic zero: desingularization (or resolution of singularities), and embedded principalization of ideals (or log-resolution of ideals).

Both results, stated in theorems 1.2 and 1.3, due to Hironaka (see [**28**]), are certainly among the most important contributions in algebraic geometry.

We focus here on the proof in [**21**], which is simpler than the original one and avoids the use of Hilbert Samuel functions and normal flatness, and also includes some new contributions like the one of J. Wlodarczyk in [**47**].

Hironaka's proof of both theorems is *existential*; he proves that every singular variety over a field of characteristic zero can be desingularized. The proof of the theorems in these notes is *constructive*, in the sense that we provide an algorithm to achieve such desingularization. We refer to [**10**] and to [**22**] for two computer implementations.

There are several other proofs of these two theorems which also provide an algorithm: [**7**] (E. Bierstone and P. Milman), [**15**] (A. Bravo and O. Villamayor), [**17**] (S. Encinas and H. Hauser), [**34**] (J. Kollár), [**41**] and [**47**] (J. Wlodarczyk).

It is natural to ask why it is interesting to study algorithms of resolution of singularities. Usually we simply need to know the formulation of a theorem in order to apply it. But sometimes a proof of a theorem can be strong enough to be useful as a tool. This is the purpose of developing algorithms to achieve resolution of singularities, a theorem with many applications in algebraic geometry. A very natural application arises, for example, when we want to classify singularities by the way that they can be desingularized. To this end it is not enough to know that singularities can be resolved, it is necessary to have an explicit manner to resolve them (see also [**18**]). This is an advantage of a constructive (or algorithmic) proof over an existential proof.

Resolution of singularities is based on a peculiar form of induction. In the case of resolution of hypersurfaces, this form of induction was stated clearly and explicitly by Abhyankar in what is called a Tschirnhausen transformation.

We will focus on this point in Chapter 1, where we discuss examples of this form of induction, with some indication of how it provides inductive invariants. These invariants are gathered in our *resolution functions*, and we prove the two main theorems 1.2 and 1.3 by extracting natural properties from these functions. This first chapter includes the discussion and definition of basic notions, used in the formulation of main theorems ([**45**]). In Chapter 2 we prove results which were motivated through examples in the first chapter. In Chapter 3 we introduce the *inductive function*, which is the key ingredient for the definition of resolution functions and for the proof of our main theorems. In fact, inductive functions reduce the proof of both theorems to two possible forms of *reduction*.

One is the reduction of the so-called *monomial case*, which is simple to deal with, and the second case is the *simple case*. This last case is called simple because its resolution is of an inductive nature.

Inductive functions are defined in terms of invariants on smooth schemes of a fixed dimension, whereas the resolution function is defined with invariants in several dimensions. The point is to clarify how the resolution function grows from the inductive function. We shall stress this particular fact, namely that the definition of resolution functions, and hence of constructive resolution, relies entirely on the resolution of the so-called simple case and that reduction to the simple case is possible due to properties of the inductive function.

In Chapter 4 we finally present the resolution of the simple case in two different ways. There is a first approach in which we follow the notes of Kollár in [**34**], with some variations. We also sketch a different approach to address resolution of the simple case using a form of *elimination* introduced in [**46**]. Over fields of characteristic zero this form of elimination provides the same information as Abhyankar's notion of Tschirnhausen transformation, and we obtain essentially the same resolution for the simple case. The interest of this second approach is that it is characteristic free, so an important part of the algorithm, namely of the canonical choice of centers discussed at the end of these notes, works in arbitrary characteristic (see also [**9**] and [**6**]).

A very interesting contribution to the problem of resolution of singularities over arbitrary characteristic is that of H. Kawanoue (see [**33**] and also [**38**]). The invariant proposed in his approach is a refinement of Hironaka's τ-invariant. On the contrary, the part of the algorithm that we discuss in the last section, which works over arbitrary fields, is related to the usual τ-invariant. This invariant also appears and is used in Hironaka's announcement [**32**], and the smoothness of the centers described here in Chapter 4 follow from this invariant (codimension τ components of the singular locus are regular) exactly in the same manner as indicated in [**32**].

These notes follow the notation in [**42**] (basic objects). In that paper we prove that the algorithm of desingularization in [**41**] is equivariant and that it also desingularizes schemes in étale topology. However those proofs make use of Hilbert-Samuel functions, whereas in these notes we discuss an algorithm in which such invariants are avoided. Hence the outcome is, in general, a different embedded desingularization. It turns out, however, that both algorithms coincide when it comes to the case of embedded desingularization of hypersurfaces. For this reason we refer to the examples in [**42**], such as the desingularization of the Whitney Umbrella, or to examples that illustrate equivariance of the desingularization of embedded hypersurfaces.

The algorithm in these notes is also equivariant and also extends to étale topology. However we do not study these properties here, and we refer to [**13**] and [**20**] for the study of these and of further properties of this proof.

We finally refer to the works of D. Cutkosky [**16**], H. Hauser [**27**], J. Kollár [**34**], and K. Matsuki [**37**] for other introductions to desingularization theorems.

I thank Ana Bravo, Angélica Benito, Mariluz García, and Santiago Encinas for many suggestions and improvements on the presentation.

CHAPTER 1

First definitions and formulation of the main theorems

1.1. Resolution theorems

We begin with the formulation of the main theorems in desingularization. The set of regular points of a reduced scheme of finite type over a field is a dense open set.

DEFINITION 1.1. We say that a birational morphism of reduced irreducible schemes

$$X \overset{\pi}{\longleftarrow} X' \tag{1.1.1}$$

is a desingularization of X if:

i) π defines an isomorphism over the open set $U = Reg(X)$ of regular points;

ii) π is proper, and X' is regular.

We will prove the existence of desingularizations, over fields of characteristic zero, by proving a theorem of *embedded desingularization* in Theorem 1.2. There we view an irreducible scheme as a closed subscheme in a smooth scheme W.

Let $W_1 \overset{\pi}{\longleftarrow} W_2$ be a proper birational morphism of smooth schemes of dimension n. If a closed point $x_2 \in W_2$ maps to $x_1 \in W_1$, there is a linear transformation of n-dimensional tangent spaces, say $T_{W_2,x_2} \longrightarrow T_{W_1,x_1}$. The set of points $x_2 \in W_2$ for which $T_{W_2,x_2} \longrightarrow T_{W_1,x_1}$ is not an isomorphism defines a hypersurface H in W_2, called the jacobian or exceptional hypersurface. It turns out that there is an open set $U \subset W_1$ such that $U \overset{\pi}{\leftarrow} \pi^{-1}(U)$ is an isomorphism and $\pi^{-1}(U) = W_2 \setminus H$. Examples of proper birational morphisms of this kind are the monoidal transformations, defined by blowing up a closed and smooth subscheme Y in a smooth scheme W_1. In such case a $H = \pi^{-1}(Y)$ is a smooth hypersurface. Let

(1.1.2)
$$\begin{array}{ccccccc} W_0 & \leftarrow & (W_1, E_1 = \{H_1\}) & \leftarrow & (W_2, E_2 = \{H_1, H_2\}) & \leftarrow \cdots \leftarrow & (W_r, E_r = \{H_1, H_2, \dots, H_r\}) \\ Y & & Y_1 & & Y_2 & & Y_r \end{array}$$

be a composition of monoidal transformations, where each $Y_j \subset W_j$ is closed and smooth, $H_j \subset W_j$ is the exceptional hypersurface of $W_{j-1} \longleftarrow W_j$ (the blow up at Y_{j-1}), and $\{H_1, H_2, .., H_r\}$ denote the strict transforms of the hypersurfaces H_i, $1 \leq i \leq r$, in W_r. The composite $W_0 \longleftarrow W_r$ is a proper birational morphism of smooth schemes, and $H = \cup_{i=1}^r H_i$ is the exceptional hypersurface.

THEOREM 1.2 (**Embedded Resolution of Singularities**). *Given W_0 smooth over a field k of characteristic zero, and $X_0 \subset W_0$ closed and reduced, there is a sequence of monoidal transformations, say* (1.1.2), such that

(i) $\cup_{i=1}^r H_i$ have normal crossings in W_r.

(ii) $W_0 \setminus \mathrm{Sing}(X_0) \simeq W_r \setminus \cup_{i=1}^r H_i$, and hence it induces a square diagram

$$\begin{array}{ccc} W_0 & \overset{\Pi_r}{\longleftarrow} & W_r \\ \cup & & \cup \\ X_0 & \overset{\overline{\Pi}_r}{\longleftarrow} & X_r \end{array}$$

of proper birational morphisms, where X_r denotes the strict transform of X_0.

(iii) X_r is regular and has normal crossings with $E_r = \cup_{i=1}^r H_i$.

In particular $\mathrm{Reg}(X_0) \cong \overline{\Pi}_r^{-1}(\mathrm{Reg}(X_0)) \subset X_r$ and $X_0 \xleftarrow{\Pi_r} X_r$ is a desingularization (1.1).

THEOREM 1.3 (**Embedded Principalization of Ideals**). *Given $I \subset \mathscr{O}_{W_0}$, a non-zero sheaf of ideals, there is a sequence of monoidal transformations, say* (1.1.2), such that:

(i) The morphism $W_0 \leftarrow W_r$ defines an isomorphism over $W_0 \setminus V(I)$.

(ii) The sheaf $I\mathscr{O}_{W_r}$ is invertible and supported on a divisor with normal crossings, i.e.,

$$\mathscr{L} = I\mathscr{O}_{W_r} = \mathscr{I}(H_1)^{c_1} \cdot \ldots \cdot \mathscr{I}(H_s)^{c_s}, \tag{1.1.3}$$

where $E' = \{H_1, H_2, \ldots, H_s\}$ are regular hypersurfaces with normal crossings, $c_i \geq 1$ for $i = 1, \ldots, s$, and $E' = E_r$ if $V(I)$ has no components of codimension 1.

Throughout these notes W will denote a smooth scheme of finite type over a field k. Unless it is specifically mentioned, k will be a field of characteristic zero. We first recall here some definitions used in the formulation of the previous theorems.

DEFINITION 1.4. Fix $y \in W$, and let $\{x_1, \ldots, x_d\}$ denote a regular system of parameters in the local regular ring $\mathscr{O}_{W,y}$.

1) A closed subscheme $Y(\subset W)$, defined by $I(Y) \subset \mathscr{O}_W$, is *regular at* $y \in Y$ if there is a regular system of parameters such that $I(Y)_y = \langle x_1, ..., x_s \rangle$ in $\mathscr{O}_{W,y}$.

2) A set $\{H_1, \ldots, H_r\}$ of hypersurfaces in W has *normal crossings at* y if there is a regular system of parameters such that $\cup H_i = V(\langle x_{j_1} \cdot x_{j_2} \cdots x_{j_s} \rangle)$ locally at y, for some $j_i \in \{1, \ldots, r\}$.

3) A closed subscheme Y has normal crossings with E at y if there is a regular system of parameters such that, locally at y:

$$I(Y)_y = \langle x_1, ..., x_s \rangle \text{ and } \cup_{i=1}^r H_i = V(\langle x_{j_1} \cdot x_{j_2} \cdots x_{j_s} \rangle).$$

Y is said to be *regular* if it is regular at any point, and $E = \{H_1, \ldots, H_r\}$ is said to have *normal crossings* if the condition holds at any point.

REMARK 1.5. If

$$\begin{array}{ccc} W_0 & \xleftarrow{\pi} & W_1 \supset H = \pi^{-1}(Y) \\ Y & & \end{array}$$

denotes a monoidal transformation with a closed and regular center $Y(\subset W_0)$, then:

1) π is proper and W_1 smooth.

2) $H = \pi^{-1}(Y)$ is a smooth hypersurface in W_1 (called the exceptional hypersurface).

3) $W_0 \setminus Y \cong W_1 \setminus H$ (i.e. π is birational).

DEFINITION 1.6. The *order* of a non-zero ideal J in a local regular ring (R, M) is the biggest integer $b \geq 0$ such that $J \subset M^b$.

REMARK 1.7. Assume that Y is an irreducible regular center with generic point $y \in W$, and let $h \in W_1$ be the generic point of H in 1.5. Note that $\mathscr{O}_{W,y}$ is a local regular ring and that $\mathscr{O}_{W_1,h}$ is a discrete valuation ring. Let M_y denote the maximal ideal of $\mathscr{O}_{W,y}$.

Set $W_0 \longleftarrow W_1$ and $H \subset W_1$ as above. Then, for an ideal $J \subset \mathscr{O}_W$, the following are equivalent:

a) $J_y \subset M_y^b$ (i.e. the order of J at $\mathscr{O}_{W,y}$ is $\geq b$).

b) $J\mathscr{O}_{W_1} = I(H)^b \cdot J_1$ for some J_1 in $\mathscr{O}_{W_1}$.

c) $J\mathscr{O}_{W_1}$ has order $\geq b$ at $\mathscr{O}_{W_1,h}$.

DEFINITION 1.8. Given a sheaf of ideals $J \subset \mathscr{O}_X$ and a morphism of schemes, $X \leftarrow Y$, the sheaf of ideals $J\mathscr{O}_Y$ is called the *total transform* of J in Y. In the previous remark we considered the total transform by a monoidal transformation, and we do not assume b to be the order of J at the generic point of Y. When such condition holds, b is the highest integer for which an expression $J\mathscr{O}_{W_1} = I(H)^b \cdot J_1$ can be defined, and in such case J_1 is called the *proper transform* of J.

Most morphisms to be considered here arise as a sequence, or composition, of monoidal transformation. In this case we also consider the exceptional components, namely the exceptional hypersurfaces introduced by each monoidal transform. The following result will be used to ensure that E_r has normal crossings in a sequence of monoidal transformations (1.1.2).

PROPOSITION 1.9. *Let W be smooth over k, and let $E = \{H_1, \dots, H_s\}$ be a set of smooth hypersurfaces with normal crossings. Assume that $Y(\subset W)$ is closed, regular, and has normal crossings with $E = \{H_1, \dots, H_s\}$, and set the monoidal transformation*

$$\underset{Y}{(W, E = \{H_1, \dots, H_s\})} \xleftarrow{\pi} (W_1, E_1 = \{H'_1, \dots, H'_s, H_{s+1} = \pi^{-1}(Y)\})$$

where H'_i denotes the strict transform of H_i. Then E_1 has normal crossings in W_1.

1.2. Examples: Tschirnhausen and induction on resolution problems

A variety, or an ideal, is usually presented by equations in a certain number of variables. A key point in resolution problems is to argue by induction on the number of variables involved. In order to illustrate the precise meaning of this form of induction we first consider the polynomial $f = Z^2 + 2XZ + X^2 + XY^2 \in k[Z, X, Y]$ defining a hypersurface $\mathbb{X} \subset \mathbb{A}^3_k$, where k denotes here an algebraically closed field of characteristic zero. We will see that all points in this hypersurface are of multiplicity at most two.

Question: How can we describe the closed set of points of multiplicity 2, say $\mathscr{F}_2 \subset \mathbb{X}$? First recall two definitions:

DEFINITION 1.10. Set $p \in \mathbb{X} = V(\langle f \rangle) \subset \operatorname{Spec}(k[Z, X, Y])$. We say that the hypersurface $\mathbb{X}$ has *multiplicity* b at p, or that p is a *b-fold point* of the hypersurface, if the ideal $\langle f \rangle$ has order b at the local regular ring $k[Z, X, Y]_p$ (see Definition 1.6). We will denote by $\mathscr{F}_b$ the set of points in $\mathbb{X}$ with multiplicity b.

There are now two ways in which we can address our question.

Approach 1): Consider the extension of the ideal $J = \langle f \rangle$, say:

$$J(1) = \langle f, \frac{\partial f}{\partial X}, \frac{\partial f}{\partial Y}, \frac{\partial f}{\partial Z} \rangle.$$

Clearly $V(J(1)) = \mathscr{F}_2$. In fact, by taking Taylor expansions at any closed point q we conclude that $q \in V(J(1))$ if and only if the multiplicity of $\mathbb{X}$ at q is at least 2. Note also that $\mathbb{X}$ has no closed point of multiplicity higher than 2 since $\frac{\partial^2 f}{\partial^2 Z}$ is a unit. So the hypersurface $\mathbb{X}$ has only closed points of multiplicity one and two.

As for the non-closed points of $\mathbb{X}$, recall first that in a polynomial ring any prime ideal is the intersection of all maximal ideals containing it. On the other hand the multiplicity defines an upper-semi-continuous function on the hypersurface. So the multiplicity at a non-closed point, say $y \in \mathbb{X}$, coincides with the multiplicity at closed points in a non-empty open set of the closure $\bar{y}$. This settles our question.

1.11. Approach 2) (linked to the previous): Set $Z_1 = Z + X$. At $k[Z_1, X, Y] = k[Z, X, Y]$:

$$f = Z_1^2 + XY^2. \tag{1.2.1}$$

2i) Note first that $Z_1 \in J(1)$, and hence $\mathscr{F}_2 \subset \overline{W}$, where $\overline{W} = V(Z_1)$ is a smooth hypersurface.

2ii) Set $J^* = \langle XY^2 \rangle \subset \mathscr{O}_{\overline{W}}$. We claim that $\mathscr{F}_2 \subset \overline{W}$ is also defined as the set of points $q \in \overline{W}$ where the order of J^*, at the local regular ring $\mathscr{O}_{\overline{W},q}$, is at least 2.

In fact, if $q \in \operatorname{Spec}(k[Z,X,Y])$ is a point (a prime ideal) where f has order 2, then $J(1) \subset q$, so

$$Z_1 \in q \subset k[Z_1, X, Y].$$

It is clear that among the prime ideals containing Z_1, those where $Z_1^2 + XY^2$ has order 2, are those where XY^2 has order at least 2. So the claim follows by setting $\overline{W} = V(Z_1)$ and $J^* = \langle XY^2 \rangle \subset \mathscr{O}_{\overline{W}}$ as before.

1.12. We will see that the answer to our earlier question, provided in Approach 2, is better adapted to resolution problems, at least over fields of characteristic zero.

We started by asking for those points where the ideal $\langle f \rangle \subset k[Z,X,Y]$ has order at least 2. So we fixed an ideal J ($J = \langle f \rangle$ in this case) and a positive integer b ($b = 2$ in this case), and considered the closed set $\mathscr{F}_2$ of points where this ideal has order 2. We ended up with a new ideal, $J^* = \langle XY^2 \rangle$ in the ring of functions in $\overline{W}$, where

$$\overline{W} = (\operatorname{Spec}(k[X,Y]) =) \operatorname{Spec}(k[Z_1, X, Y]/\langle Z_1 \rangle) \subset \operatorname{Spec}(k[X,Y,Z]),$$

together with an integer $b_1 = 2$, describing the same closed set $\mathscr{F}_2$, but involving one less variable.

DEFINITION 1.13. Fix a scheme W, smooth over a field. A *couple* will be a non-zero sheaf of ideals $J \subset \mathscr{O}_W$ and an integer b and will be denoted by (J, b).

The *set described by the couple* will be the set of points $\{x \in W / \nu_x(J) \geq b\}$, where $\nu_x(J)$ denotes the order of J at the local regular ring $\mathscr{O}_{W,x}$.

1.14. The set described by the couple $(J = \langle Z_1^2 + XY^2 \rangle, 2)$ in $\mathbb{A}^3_k$ is included in a smooth hypersurface $\overline{W} = V(Z_1)$. The dimension of $\overline{W}$ is of course one less than that of W. This inclusion is called the *local inductive principal*. Note that this closed set is also defined by the couple $(J^*, 2)$ $(J^* = \langle XY^2 \rangle \subset \mathscr{O}_{\overline{W}})$.

EXAMPLE 1.15. The fact that $J^* \subset \mathscr{O}_{\overline{W}}$ is principal is just a coincidence of the previous example. Now let $\mathbb{Y} \subset \mathbb{A}^3_k$ be the hypersurface defined by $g = Z^3 + XY^2Z + X^5 \in k[Z,X,Y]$. Define

$$J(2) = \langle g, \frac{\partial g}{\partial x_i}, \frac{\partial^2 g}{\partial x_i \partial x_j} \mid \text{where } x_1 = X, x_2 = Y, x_3 = Z \rangle$$

so $V(J(2)) = \mathscr{F}_3$ is the set of points of multiplicity at least 3. The pattern of this equation is

$$Z^3 + a_2 Z + a_3 \text{ with } a_2, a_3 \text{ in } k[X,Y].$$

One can check that $Z \in J(2)$ and that $\mathbb{Y}$ has at most points of multiplicity 3 since $\frac{\partial^3 g}{\partial^3 Z}$ is a unit.

We can argue as in Approach 2 to show that if $q \in \operatorname{Spec}(k[Z,X,Y])$ is a point (a prime ideal) of multiplicity 3, then $J(2) \subset q$. So

$$Z \in q \subset k[Z,X,Y],$$

and among all prime ideals q containing Z, the polynomial $Z^3 + XY^2Z + X^5$ has order 3 at $k[Z,X,Y]_q$ if and only if XY^2 has order at least 2 and X^5 has order at least 3. In fact Z has order one at $k[Z,X,Y]_q$, and Z, X, and Y are independent variables.

Now set $\overline{W} = V(Z)$, $\overline{a_2} = \overline{XY^2}$, $\overline{a_3} = \overline{X^5}$ (the class of a_2 and a_3 in $\mathscr{O}_{\overline{W}}$), and note that

$$\mathscr{F}_3 = \{x \in \overline{W} / \nu_x(\overline{a_2}) \geq 2; \nu_x(\overline{a_3}) \geq 3\},$$

where $\nu_x(\overline{a_i})$ denotes the order of $\overline{a_i}$ at the local regular ring $\mathscr{O}_{\overline{W},x}$.

Set

$$(J^*, 6), \text{ where } J^* = \langle (\overline{a_2})^3, (\overline{a_3})^2 \rangle \subset \mathscr{O}_{\overline{W}}. \tag{1.2.2}$$

Finally check that $\mathscr{F}_3 \subset \overline{W}$ (local inductive principal (1.14)), and note that we use this fact to show that the closed set $\mathscr{F}_3$ is also defined by the couple $(J^*, 6)$.

REMARK 1.16. (Stability of inductive principal)

Let $\mathbb{Y} \subset \mathbb{A}^3_k$ be the hypersurface defined by $g = Z^3 + XY^2Z + X^5 \in k[Z,X,Y]$, as in Example 1.15. The origin $\overline{0} \in \mathbb{A}^3_k$ is clearly a point of the closed set defined by $(J,3)$. We now define:

$$\mathbb{A}^3_k \longleftarrow W_1 \tag{1.2.3}$$

as the blowup at $\overline{0}$. Let $\overline{W_1}$ be the strict transform of $\overline{W}$, $\mathbb{Y}_1$ the strict transform of $\mathbb{Y}$, and H the exceptional hypersurface. By restriction of the morphism to the subschemes we obtain

$$\overline{W} \longleftarrow \overline{W}_1, \tag{1.2.4}$$

which is also the monoidal transformation at the point $\overline{0} \in \overline{W}$, with exceptional hypersurface $\overline{H} = H \cap \overline{W}_1$. There is a well-defined factorization of the form

$$J\mathscr{O}_{W_1} = I(H)^3 \cdot J_1 \tag{1.2.5}$$

for an ideal $J_1 \subset \mathscr{O}_{W_1}$, defined in terms of (1.2.3), and a factorization

$$J^* \mathscr{O}_{\overline{W}_1} = I(\overline{H})^6 \cdot J_1^* \tag{1.2.6}$$

for $J_1^* \subset \mathscr{O}_{\overline{W}_1}$, defined in terms of (1.2.4). These factorizations hold because $\overline{0}$ is a point of the closed set defined by $(J,3)$, thus of the closed set in $\overline{W}$ defined by $(J^*, 6)$.

Since $\overline{0}$ is a point of order 3 of J (a point of multiplicity 3 of the hypersurface $\mathbb{Y}$), $J_1 \subset \mathscr{O}_{W_1}$ is the ideal defining the strict transform $\mathbb{Y}_1$.

Claim: The set of 3-fold points of the hypersurface $\mathbb{Y}_1$, or say the closed set of points defined by $(J_1, 3)$, is included in $\overline{W}_1$ and coincides with the closed set defined by $(J_1^*, 6)$.

In other words, we claim that the role played by $\overline{W}$ and $(J^*, 6)$ for the hypersurface $\mathbb{Y}$ (the local inductive principal in (1.14)) is now played by $\overline{W}_1$ and $(J_1^*, 6)$ for the hypersurface $\mathbb{Y}_1$. We call this the *stability* of the local inductive principal by transformations.

To check this claim, note first that W can be covered by three charts:

$$U_X = \operatorname{Spec}(k[Z/X, X, Y/X]) = \mathbb{A}^3_k$$

$$U_Y = \operatorname{Spec}(k[Z/Y, X/Y, Y]) = \mathbb{A}^3_k$$

$$U_Z = \mathrm{Spec}(k[Z,X/Z,Y/Z]) = \mathbb{A}^3_k$$

The morphism: $\mathbb{A}^3 \longleftarrow U_Y = \mathrm{Spec}(k[Z/Y,X/Y,Y]) = \mathbb{A}^3_k$, induced by (1.2.3), is defined by the inclusion $k[Z,X,Y] \to k[Z/Y,X/Y,Y]$. At this chart $I(H) = \langle Y \rangle$, the factorization in (1.2.5) is

$$g = Z^3 + XY^2Z + X^5 = Y^3((Z/Y)^3 + (X/Y)Y(Z/Y) + (X/Y)^5Y^2),$$

and $I(\overline{W_1} \cap U_Y) = \langle Z/Y \rangle$. Note that $g_1 = (Z/Y)^3 + (X/Y)Y(Z/Y) + (X/Y)^5Y^2 \in k[Z/Y,X/Y,Y]$ has the same general pattern as g, namely: $(Z/Y)^3 + b_2(Z/Y) + b_3$, with b_2, b_3 in $k[X/Y,Y]$. So the same argument applied to g asserts that:

1) The set of 3-fold points of $\mathbb{Y}_1 \cap U_Y$ is included in $V(\langle Z/Y \rangle)$, or say in

$$\overline{W}_1 \cap U_Y = \mathrm{Spec}(k[Z/Y,X/Y,Y]/\langle Z/Y \rangle) = \mathrm{Spec}(k[X/Y,Y]).$$

2) The set of 3-fold points $\mathbb{Y}_1$ in U_Y is the closed set in $\overline{W}_1 \cap U_Y$ defined by $(\mathscr{A}, 6)$, where

$$\mathscr{A} = \langle (\overline{b_2})^3, (\overline{b_3})^2 \rangle \subset k[X/Y,Y].$$

We are finally ready to address the main property of our form of induction in the number of variables, namely the compatibility of induction with transformations. To this end note that

$$\overline{W} \longleftarrow \overline{W}_1 \cap U_Y$$

is defined by the inclusion $k[X,Y] \to k[X/Y,Y]$, and the transform of the couple $(J^*, 6)$ in (1.2.2), namely (J_1^*, b), with J_1^* defined as in (1.2.6), is such that

$$J_1^* \mathscr{O}_{(\overline{W}_1 \cap U_Y)} = \mathscr{A}.$$

A similar argument applies for $\mathbb{A}^3 \longleftarrow U_X$. To study our claim for $\mathbb{A}^3 \longleftarrow W_1$ it suffices to check the charts U_X, U_Y. In fact, $U_X \cup U_Y$ covers all of W_1 except for one point (the origin at $U_Z = \mathbb{A}^3$), which is not a point of $\mathbb{Y}_1$. So U_Z can be ignored for our purposes.

1.17. **Summarizing: Stability of inductive principal.** Our previous discussion showed that the set of 3-fold points of $\mathbb{Y} \subset \mathbb{A}^3$ (defined by $g = Z^3 + XY^2Z + X^5 \in k[Z,X,Y]$) is included in a smooth hypersurface $\overline{W}$ (defined by $Z \in k[Z,X,Y]$) (see 1.14). From this fact we conclude that the set is also defined by $(J^*, 6)$, where J^* is an ideal in the surface $\overline{W}$. The property that links $\overline{W}$ with 3-fold points of $\mathbb{Y}$ goes beyond this fact. A transformation at a 3-fold point of $\mathbb{Y}$ defines a strict transform $\mathbb{Y}_1$. It also induces a transformation $\overline{W} \longleftarrow \overline{W}_1$, together with a transformation of $(J^*, 6)$, say $(J_1^*, 6)$. $\overline{W}_1$ is the strict transform of $\overline{W}$, and the property is that the set of 3-fold points of $\mathbb{Y}_1$ is included in $\overline{W}_1$. Furthermore, $(J_1^*, 6)$ defines the closed set of 3-fold points of $\mathbb{Y}_1$. In particular, if J_1^* did not have points of order 6 (which is not the case in our example), then $\mathbb{Y}_1$ would not have 3-fold points. Here we have analyzed this stability for one quadratic transformation, but it turns out that the same argument holds for any sequence of monoidal transformations. Defining a sequence of transformations, say

$$\begin{array}{ccccccc} \mathbb{A}^3 & \xleftarrow{\pi_1} & W_1 & \xleftarrow{\pi_2} & \cdots & \xleftarrow{\pi_r} & W_k \\ \mathbb{Y} & & \mathbb{Y}_1 & & & & \mathbb{Y}_k, \end{array} \tag{1.2.7}$$

where each π_{i+1} is a blow up at a closed and smooth center included in the 3-fold points of $\mathbb{Y}_i$ (strict transform of $\mathbb{Y}_{i-1}$), is *equivalent* to defining a sequence of transformations

$$\begin{array}{ccccccc} \overline{W} & \xleftarrow{\pi_1} & \overline{W}_1 & \xleftarrow{\pi_2} & \cdots & \xleftarrow{\pi_r} & \overline{W}_k \\ (J^*, 6) & & (J_1^*, 6) & & & & (J_k^*, 6), \end{array} \tag{1.2.8}$$

where each $J_i^* \subset \mathscr{O}_{\overline{W}_1}$, and $(J_i^*, 6)$ is defined in terms of $(J_{i-1}^*, 6)$ as in (1.2.6). Moreover, each $\overline{W}_i$ is a smooth hypersurface in W_i, and the closed set defined by $(J_i^*, 6)$ in the hypersurface $\overline{W}_i$ is the set of 3-fold points of $\mathbb{Y}_i$. In particular, if the second sequence is defined with the property that J_k^* has no points of order 6 in $\overline{W}_k$, then the hypersurface $\mathbb{Y}_k$ has at most points of multiplicity 2.

This is induction on the dimension of the ambient space, where the lowering of the highest order of an ideal in a smooth scheme of dimension 3 is *equivalent* to a related problem in a smooth scheme of dimension 2. This property of the smooth hypersurface $\overline{W}$ will be discussed in Section 2.4.

1.18. **Tschirnhausen.** Set $f = Z^b + a_1 Z^{b-1} + \cdots + a_b \in k[Z, X_1, \ldots, X_n]$, with $a_i \in k[X_1, .., X_n]$ for $i = 1, \ldots, b$. If the characteristic of k is zero, set $Z_1 = Z + \frac{1}{b} a_1$. Check that $k[Z, X_1, \ldots, X_n] = k[Z_1, X_1, \ldots, X_n]$ and that $f = Z_1^b + c_2 Z_1^{b-2} + \cdots + c_b$, with $c_i \in k[X_1, \ldots, X_n]$ and $c_1 = 0$. One can argue as in Example 1.15 that the b-fold points of $\mathbb{Y}$ are included in the hypersurface $\overline{W} = V(Z_1)(\subset \mathbb{A}^{n+1})$(local inductive principle in (1.14)). Furthermore, $\overline{W}$ will have the stability property discussed above, where the role of $(J^*, 6)$ in Remark 1.16, namely on the previous sequence of transformations (1.2.8), is now played by $(J^*, b!)$, where

$$J^* = \langle c_i^{\frac{b!}{i}}, \, i = 2, 3, \ldots, b \rangle \subset \mathscr{O}_{\overline{W}}.$$

1.3. Resolution functions and the main resolution theorems

Our proofs of the two main theorems 1.2 and 1.3 will be constructive, as opposed to the original existential proofs. We introduce here the notion of resolution algorithm, or resolution functions. Constructive resolutions will be defined in terms of these functions, and the main purpose in this section is to show how both proofs follow easily from natural properties of these functions.

1.19. In 1.15 we study the transform of a hypersurface in $\mathbb{A}^3$ by a monoidal transformation at a 3-fold point. Note that (1.2.5) is an example of a *proper transform* of an ideal as defined in Definition 1.8. However, the ideal J^* has order 9 at the center of the monoidal transformation, so J_1^* in (1.2.6) is not a proper transform. This shows that our form of induction will lead us to transformations of ideals, defined in terms of expressions of the form $J\mathscr{O}_{W_1} = I(H)^b \cdot J_1$, even when b is not the highest possible integer that one could obtain in such an expression (even if J_1 vanishes along the hypersurface H).

We have defined *couples* as pairs (J, b) where $J \subset \mathscr{O}_W$ is a non-zero sheaf of ideals, and $b \in N$ is a positive integer. We introduce now two notions related to couples:

- The **closed set** attached to (J, b):

$$\mathrm{Sing}(J, b) = \{x \in W / \nu_x(J_x) \geq b\},$$

namely the set of points in W where J has order at least b. This is closed in W (see 2.9, ii)).

- **Transformation** of (J, b): Let $Y \subset \mathrm{Sing}(J, b)$ be a closed and smooth subscheme, and let

$$\begin{array}{ccc} W & \xleftarrow{\pi} & W_1 \supset H = \pi^{-1}(Y) \\ Y & & \end{array}$$

be the monoidal transformation at Y. Since $Y \subset \mathrm{Sing}(J, b)$, the total transform $J\mathscr{O}_{W_1}$ can be expressed as a product:

$$J\mathscr{O}_{W_1} = I(H)^b J_1 (\subset \mathscr{O}_{W_1})$$

for a uniquely defined J_1 in $\mathcal{O}_{W_1}$. The new couple (J_1,b) is called the *transform* of (J,b), and the transformation is denoted by

$$\begin{array}{ccc} W & \xleftarrow{\pi} & W_1 \\ (J,b) & & (J_1,b). \end{array} \tag{1.3.1}$$

A sequence of transformations will be denoted as

$$\begin{array}{ccccccc} W & \xleftarrow{\pi_1} & W_1 & \xleftarrow{\pi_2} & \cdots & \xleftarrow{\pi_k} & W_k \\ (J,b) & & (J_1,b) & & & & (J_k,b). \end{array} \tag{1.3.2}$$

Note that in such cases

$$J\mathcal{O}_{W_k} = I(H_1)^{c_1} \cdot I(H_2)^{c_2} \cdots I(H_k)^{c_k} \cdot J_k \tag{1.3.3}$$

for suitable exponents $c_2,\dots,c_k$, and $c_1 = b$. Furthermore, all $c_i = b$ if for any index $i < k$ the center Y_i is not included in $\cup_{j\leq i} H_j \subset W_i$ (the exceptional locus of $W \longleftarrow W_i$).

EXAMPLE 1.20. The ideal $J = \langle x^2 - y^5 \rangle \subset k[x,y]$ has a unique 2-fold point at the origin $(0,0) \in \mathbb{A}^2$. Let $W = \mathbb{A}^2 \longleftarrow W_1$ be the blowup at the origin. The strict transform of the curve has a unique 2-fold point, say $q \in W_1$. Set $W_1 \longleftarrow W_2$ by blowing up q. This defines a sequence,

$$\begin{array}{ccccc} W & \longleftarrow & W_1 & \longleftarrow & W_2 \\ (J,2) & & (J_1,2) & & (J_2,2). \end{array} \tag{1.3.4}$$

Here $J\mathcal{O}_{W_2} = I(H_1)^2 \cdot I(H_2)^4 \cdot J_2$ provides an expression of the total transform of J involving J_2.

REMARK 1.21. The ideal J_1 in the previous example is the proper transform of J, and J_2 is the proper transform of J_1 (Def 1.8). In particular, J_2 does not vanish along H_1 or H_2. However, this is not a general fact as indicated in 1.19. Now set $K = J$, and note the same sequence as before defines $(K,1)$, $(K_1,1)$, and $(K_2,1)$; $K\mathcal{O}_{W_2} = I(H_1)^1 \cdot I(H_2)^2 \cdot K_2$.

In this case the ideal K_2 does vanish along the exceptional hypersurface H_i; in fact, there is a unique and well-defined expression, say

$$K_2 = I(H_1)^a \cdot I(H_2)^b \cdot \overline{K}_2 \tag{1.3.5}$$

in $\mathcal{O}_{W_2}$, so that $\overline{K}_2$ does not vanish along the exceptional hypersurfaces. It follows from 1.20 that $a = 1$, $b = 2$, and $\overline{K}_2 = J_2$.

DEFINITION 1.22. Fix $J \subset \mathcal{O}_W$, W smooth over a field of characteristic zero, and a couple (J,b). A sequence of transformations as in (1.3.2) is said to be a *resolution* of (J,b) if:

i) $\mathrm{Sing}(J_k,b) = \emptyset$.

ii) The exceptional locus of $W \longleftarrow W_k$, namely $\cup_{i=1}^k H_i$, is a union of hypersurfaces with normal crossings.

1.23. We define a *pair*, denoted by $(W, E = \{H_1,\dots,H_r\})$, to be a set of smooth hypersurfaces $H_1,\dots,H_r$ with normal crossings in a smooth scheme W. Let $W \longleftarrow W_1$ be a monoidal transformation at a closed and smooth center Y.

If Y has normal crossings with $\cup_{i=1}^r H_i$, we say that Y is permissible for the pair (W,E), and that

$$(W, E = \{H_1,\dots,H_r\}) \longleftarrow (W_1, E_1 = \{H_1,\dots,H_r,H_{r+1}\})$$

is a *transformation of pairs* (see Prop 1.9).

We define a *basic object* to be a pair $(W, E = \{H_1, \ldots, H_r\})$ together with a couple (J, b), with the condition that $J_x \neq 0 (\subset \mathcal{O}_{W,x})$ at any point $x \in W$. We indicate this basic object by

$$(W, (J, b), E).$$

If a smooth center Y defines a transformation of the pair (W, E), and in addition $Y \subset \mathrm{Sing}(J, b)$, then a transform of the couple (J, b) is defined. In this case we say that

$$(W, (J, b), E) \longleftarrow (W_1, (J_1, b), E_1)$$

is a *transformation* of the basic object. A sequence of transformations

$$(1.3.6) \qquad (W, (J, b), E) \longleftarrow (W_1, (J_1, b), E_1) \longleftarrow \cdots \longleftarrow (W_s, (J_s, b), E_s)$$

is a *resolution* of the basic object if $\mathrm{Sing}(J_s, b) = \emptyset$.

In such case

$$(1.3.7) \qquad J \cdot \mathcal{O}_{W_s} = I(H_{r+1})^{c_1} \cdot I(H_{r+2})^{c_2} \cdots I(H_{r+s})^{c_s} \cdot J_s$$

for some integer c_i, where J_s is a sheaf of ideals of order at most $b-1$, and the union of hypersurfaces $H_j \in E_s$ have normal crossings.

DEFINITION 1.24. Let X be a topological space, and let $(T, \geq)$ be a totally ordered set. A function $g : X \to T$ is said to be *upper semi-continuous* if: **i)** g takes only finitely many values, and **ii)** for any $\alpha \in T$ the set

$$\{x \in X \,/ g(x) \geq \alpha\}$$

is closed in X. The largest value achieved by g will be denoted by:

$$\max g.$$

We also define

$$\underline{\mathrm{Max}}\, g = \{x \in X : g(x) = \max g\},$$

which is a closed subset of X.

1.25. **Resolution functions.** We now define the notion of *constructive* resolutions of basic objects, which will lead us to simple proofs of both main theorems 1.2 and 1.3.

In 1.11 we defined an upper semi-continuous function, say $h_3 : \mathrm{Spec}(k[Z_1, X, Y]) \to \mathbb{Z}$, defined by taking order of the ideal $J = \langle Z_1^2 + XY^2 \rangle$ at the points of this three dimensional scheme. It was shown that $\max h_3 = 2$, and that $\underline{\mathrm{Max}}\, h_3 (= \mathcal{F}_2) \subset \overline{W}$, where $\overline{W} = V(Z_1)$ is a smooth hypersurface isomorphic to $\mathrm{Spec}(k[X, Y])$. Furthermore, an ideal $J^* = \langle XY^2 \rangle \subset \mathcal{O}_{\overline{W}}$ was attached to $\underline{\mathrm{Max}}\, h_3$. We may now take $h_2 : \mathrm{Spec}(k[X, Y]) \to \mathbb{Z}$, defined by taking order of the ideal J^* at this two-dimensional scheme, so that $\underline{\mathrm{Max}}\, h_2$ is included in a smooth hypersurface. In the same way we could also define a function h_1 with values at $\mathbb{Z}$.

With this frame of mind it is conceivable to assign a copy of $\mathbb{Z}$ for each dimension, namely $\mathbb{Z} \times \mathbb{Z} \times \mathbb{Z}$, with lexicographic order, and a function, say $h = (h_3, h_2, h_1)$ with values at this ordered set, so that h is upper semi-continuous. This is not exactly the way we will proceed, but we will define a totally ordered set for each dimension and then take the product of copies of this set, one for each dimension.

We will fix an integer d and define a totally ordered set $(I^d, \geq)$. Moreover, for any basic object

$$B : (W, (J, b), E),$$

with a dimension of $W = d$, an upper semi-continuous function $f_B : \mathrm{Sing}(J, b) \to I^d$ will be defined with the property that $\underline{\mathrm{Max}}\, f_B$ is a smooth subscheme of $\mathrm{Sing}(J, b)$ and a permissible center for the pair (W, E). Thus, a transformation of the basic object can be defined with center $\underline{\mathrm{Max}}\, f_B$.

In this way a unique sequence (1.3.6) is defined inductively by setting centers $\underline{\operatorname{Max}} f_{D_i}$. In addition, this sequence defined by the functions will be a resolution of the basic object. In fact, for some index s, which will depend on the basic object B, we achieve that $\operatorname{Sing}(J_s, b) = \emptyset$.

In other words, the set $(I^d, \geq)$ will be fixed, and the functions on this set defined so as to provide a resolution for any basic object of dimension d. The resolution defined by these functions is what we call a *constructive, or algorithmic, resolution.* We now state the properties that will hold for such a sequence:

Properties:

P1) For each l, $\underline{\operatorname{Max}} f_l$ is closed regular and has normal crossings with $\cup_{H_i \in E_l} H_i$.

P2) For some index k_0 depending on the basic object B, $\operatorname{Sing}(J_{k_0}, b) = \emptyset$.

If $p \in \operatorname{Sing}(J_k, b)$ and $p \notin \underline{\operatorname{Max}} f_k$, then p can be identified with a point in W_{k+1}. Furthermore, $p \in \operatorname{Sing}(J_{k+1}, b)$, and:

P3) $f_k(p) = f_{k+1}(p)$.

Of particular interest will be the case of basic objects with $b = 1$. In such a case, $\operatorname{Sing}(J_0, 1)$ is the underlying topological space of $V(J_0)$ (the subscheme defined by the sheaf of ideals J_0).

P4) There is fixed value $R \in I^d$, and whenever $p \in \operatorname{Sing}(J_0, 1)$ is a point where the subscheme defined by J_0 is smooth, then $f_0(p) = R$ (where $f_0 : \operatorname{Sing}(J_0, 1) \to I^d$).

The definition of $(I^d, \geq)$ and of the functions f will be discussed in Chapter 3, and studied exhaustively in Chapter 4. We now prove our two main theorems 1.2 and 1.3 using the properties of resolution functions.

1.26. Proof of Theorem 1.3.

PROOF. Fix $I \subset \mathcal{O}_W$ as in Theorem 1.3, and consider the basic object

$$(W, (J, 1), E = \emptyset), \tag{1.3.8}$$

with $J = I$ and the resolution defined by the resolution functions. Property **P2)** asserts that $\operatorname{Sing}(J_{k_0}, 1) = \emptyset$ for some index k_0. It follows that $J_{k_0} = \mathcal{O}_{W_{k_0}}$, namely that

$$J\mathcal{O}_{W_{k_0}} = I(H_1)^{c_1} \cdot I(H_2)^{c_2} \cdots I(H_{k_0})^{c_{k_0}}.$$

It is easy to check now that the conditions of the theorem are fulfilled for $W \leftarrow W_{k_0}$. □

1.27. Proof of Theorem 1.2.

PROOF. Let $J \subset \mathcal{O}_{W_0}$ be the sheaf of ideals defining $X \subset W_0$ in Theorem 1.2, and consider, as above, the resolution of the basic object (1.3.8) defined by the resolution functions. So again $J_{k_0} = \mathcal{O}_{W_{k_0}}$, and hence $J\mathcal{O}_{W_{k_0}} = I(H_1)^{c_1} \cdot I(H_2)^{c_2} \cdots I(H_{k_0})^{c_{k_0}}$.

Let $V = W_0 \setminus \operatorname{Sing}(X)$ be the complement of the singular locus of X. Note that V is an open set, dense in W_0, and $f_0(p) = R$ for any $p \in V \cap \operatorname{Sing}(J, 1)$. Here $X = \operatorname{Sing}(J, 1)$, and $V \cap \operatorname{Sing}(J, 1)$ is dense in $\operatorname{Sing}(J, 1)$ since X is reduced. Furthermore, $f_0(p) = R$ for any $p \in V \cap \operatorname{Sing}(J, 1)$ (**P4**). So $\max f_0 \geq R$.

If $\max f_0 = R$, then $\operatorname{Sing}(J, 1) = \underline{\operatorname{Max}} f_0$ and X is smooth in W_0 (**P1**).

If $\max f_0 > R$, then V can be identified with an open set, say V_1, in W_1, and $f_1(p) = R$ for any $p \in V_1 \cap \mathrm{Sing}(J_1, 1)$ **(P3)**.

If $\max f_1 = R$, then the strict transform of X is a union of components of $\underline{\mathrm{Max}}\, f_1$, so the strict transform defines an embedded desingularization **(P1)**.

If $\max f_1 > R$, then V can be identified with an open subset V_2 in W_2.

Note that there must be an index k for some $k < k_0$, so that $\max f_k = R$. In fact, this follows from **P4)**, **P2)**, and the fact that $\mathrm{Sing}(J_{k_0}, 1) = \emptyset$. Note that V can be identified with an open set $V_k \subset W_k$, and that the strict transform of X in W_k fulfills the conditions of the theorem. □

1.28. **Further Properties.** Resolution functions define, for any basic object $(W_0, (J_0, b), E_0 = \emptyset)$, an algorithmic resolution, namely:

$$(W_0, (J_0, b), E_0) \longleftarrow (W_1, (J_1, b), E_1) \longleftarrow \cdots \longleftarrow (W_s, (J_s, b), E_s) \tag{1.3.9}$$

together with functions $f_i : \mathrm{Sing}(J_i, b) \to I^d$, so that each $W_i \leftarrow W_{i+1}$ is a monoidal transformation with center $\underline{\mathrm{Max}}\, f_i$.

Let the algorithmic resolution of $(W'_0, (J', b'), E'_0 = \emptyset)$ be

$$(W'_0, (J'_0, b'), E_0) \longleftarrow (W'_1, (J'_1, b'), E_1) \longleftarrow \cdots \longleftarrow (W'_{s'}, (J'_{s'}, b'), E_{s'}), \tag{1.3.10}$$

defined with functions $f'_i : \mathrm{Sing}(J'_i, b') \to I^d$.

We say that $(W_0, (J, b), E_0)$ and $(W'_0, (J', b'), E'_0)$ *undergo the same resolution* if $s = s'$, $W_i = W'_i$, $\mathrm{Sing}(J_i, b) = \mathrm{Sing}(J'_i, b')$, and $f_i = f'_i$.

If $\Theta : W_0 \to W'_0$ is an isomorphism, we say that (1.3.9) and (1.3.10) are Θ-equivalent, if $s = s'$, there is a natural lifting of Θ to an isomorphism $\Theta_i : W_i = W'_i$, and furthermore, $\Theta_i(\mathrm{Sing}(J_i, b)) = \mathrm{Sing}(J'_i, b')$ and $f_i(x) = f'_i(\Theta_i(x))$.

Let A be a normal domain. Two ideals I and J are said to have the same integral closure if $IS = JS$ for any ring homomorphism $A \to S$, where S is a valuation ring.

The notion extends to sheaves of ideals on a normal scheme. We say that $(W, (J, b), E_0 = \emptyset)$ and $(W, (J', b'), \emptyset)$ are *idealistic equivalent* if $J^{b'}$ and J'^{b} have the same integral closure.

P5) If $(W, (J, b), E_0 = \emptyset)$ and $(W, (J', b'), \emptyset)$ are idealistic equivalent, then they undergo the same resolution.

Let $\Theta : W \to W'$ be an isomorphism, and fix a sheaf of ideals $J \subset \mathscr{O}_W$. Let $\Theta(J) \subset \mathscr{O}_{W'}$ be the ideal defined in terms of the isomorphism.

P6) If $\Theta : W \to W'$ is an isomorphism, then $(W, (J, b), E_0 = \emptyset)$ and $(W', (\Theta(J), b), \emptyset)$ are Θ-equivalent.

Fix a basic object $(W_0, (J_0, b), E_0 = \emptyset)$, d =dim W_0, and a rational closed point $x \in \mathrm{Sing}(J, b)$. The completion of $\mathscr{O}_{W,x}$, say $\hat{\mathscr{O}}_{W,x}$, is a ring of formal power series in d variables, say $\hat{\mathscr{O}}_{W,x} = k[[x_1, \dots, x_d]]$.

The resolution functions will also define resolutions of basic objects of the form $(\hat{W}_0, (J_0, b), E_0 = \emptyset)$, where $\hat{W}_0 = \mathrm{Spec}(k[[x_1, \dots, x_d]])$ is formally smooth over k.

P7) Fix a basic object $(W_0, (J_0, b), E_0 = \emptyset)$, where W_0 is a d-dimensional smooth scheme, and a closed point $x \in \mathrm{Sing}(J_0, b)$. Let (1.3.9) denote the algorithmic resolution of $(W_0, (J_0, b), E_0 = \emptyset)$. Then the algorithmic resolution of $(\hat{W}_0, (J_0 \hat{W}_0, b), \emptyset)$ coincides with that induced by (1.3.9) over $\hat{W}_0$.

CHAPTER 2

Differential operators and monoidal transformations

The objective of this second chapter is to study the local form of induction in the dimension of the ambient space, already suggested through examples in Chapter 1. In 1.17 we discussed a strong link between the set of 3-fold points of the hypersurface $\mathbb{Y} \subset \mathbb{A}^3$, defined by $g = Z^3 + XY^2Z + X^5 \in k[Z,X,Y]$, and the smooth hypersurface $\overline{W}$ defined by $Z \in k[Z,X,Y]$. The link showed that the reduction of 3-fold points of $\mathbb{Y}$, by means of monoidal transformations, was equivalent to a related problem for a suitable ideal in the smooth subscheme $\overline{W}$ (see also 1.18).

This is the key for induction in resolution theorems. In this chapter we justify the discussion in 1.17 (see Example 2.34) and generalize this main form of induction. In 2.3 we study an important preliminary: the behavior of derivations under monoidal transformations.

The main result in this chapter is Proposition 2.31, where induction will be settled locally. The globalization of these local arguments will be addressed in Chapter 4.

2.1. On induction. Where are we heading?

Consider a couple (J,b) on a smooth scheme W and a smooth hypersurface $\overline{W}$ embedded in W. The discussion and examples in the previous chapter motivate the following questions.

Question 1: Given (J,b) and a smooth hypersurface $\overline{W}$, is there a natural way to attach to a couple, say (K,d), $K \subset \mathscr{O}_{\overline{W}}$, so that $\mathrm{Sing}(K,d) = \overline{W} \cap \mathrm{Sing}(J,b)$?

To make the following question precise, recall that in the context of resolutions we consider transformations of couples. So fix a regular subscheme $Y \subset \overline{W} \cap \mathrm{Sing}(J,b)$. The monoidal transform defines, on the one hand, a transform of (J,b), say

$$\begin{array}{ccc} W & \stackrel{\pi}{\longleftarrow} & W_1 \\ (J,b) & & (J_1,b). \end{array}$$

On the other hand, it defines a strict transform of $\overline{W}$ to a subscheme, say $\overline{W}_1 \subset W_1$, so that π induces by restriction a morphism, say

$$\overline{W} \stackrel{\overline{\pi}}{\longleftarrow} \overline{W}_1.$$

By looking at local equations one can check that this restricted morphism is the monoidal transform of $\overline{W}$ at Y. In particular, $\overline{W}_1$ is a smooth subscheme in W_1. We may now consider the intersection $\overline{W}_1 \cap \mathrm{Sing}(J_1,b)$ in W_1. Note also that there is a transform

$$\begin{array}{ccc} \overline{W} & \stackrel{\pi}{\longleftarrow} & \overline{W}_1 \\ (K,d) & & (K_1,d). \end{array}$$

Question 2: Can we define (K,d) so that $\overline{W}_1 \cap \mathrm{Sing}(J_1,b) = \mathrm{Sing}(K_1,d)$ for *any* choice of Y as above?

Given, as before, (J,b) and a smooth hypersurface $\overline{W}$, we could define various sequences, say:

$$\begin{array}{ccccccc} W & \stackrel{\pi_1}{\longleftarrow} & W_1 & \stackrel{\pi_2}{\longleftarrow} & \dots & \stackrel{\pi_k}{\longleftarrow} & W_k \\ (J,b) & & (J_1,b) & & & & (J_k,b) \end{array} \tag{2.1.1}$$

by choosing centers $Y_i \subset \mathrm{Sing}(J_i,b) \cap \overline{W}_i$, where $\overline{W}_i$ is the strict transform of $\overline{W}$. In such a case, each $\overline{W}_i$ is a smooth hypersurface.

Question 3: Can we define (K,d) so that for *any* sequence (2.1.1) (with centers $Y_i \subset \mathrm{Sing}(J_i,b) \cap \overline{W}_i$), the induced sequence, say

$$\begin{array}{ccccccc} \overline{W} & \stackrel{\pi_1}{\longleftarrow} & \overline{W}_1 & \stackrel{\pi_2}{\longleftarrow} & \dots & \stackrel{\pi_k}{\longleftarrow} & \overline{W}_k \\ (K,d) & & (K_1,d) & & & & (K_k,d) \end{array} \tag{2.1.2}$$

is such that $\overline{W}_k \cap \mathrm{Sing}(J_k,b) = \mathrm{Sing}(K_k,d)$?

Suppose now that Question 2 has a positive answer; then:

Main Question: Given only (J,b) $(J \subset \mathcal{O}_W)$, can we find a smooth subscheme $\overline{W}$ so that $\mathrm{Sing}(J,b) = \mathrm{Sing}(K,d)$ (i.e. so that $\mathrm{Sing}(J,b) \subset \overline{W}$) and for *any* sequence (2.1.2) (induced by (2.1.1)) so that $\mathrm{Sing}(J_k,b) = \mathrm{Sing}(K_k,d)$?

Our main question opens the way for induction by reversing the argument. In fact, one can check that any sequence (2.1.2) induces a sequence (2.1.1)! Therefore, if (2.1.2) would be a resolution of (K,d), then (2.1.1) would be a resolution of (J,b). And certainly this would provide a form of induction for resolution of couples.

We will organize these notes around these questions. Note that Question 1 is formulated in terms of the intersections $\mathrm{Sing}(J,b) \cap \overline{W}$, where $\overline{W}$ is an arbitrary smooth hypersurface. Questions 2 and 3 are related to the intersection of the transforms of (J,b) and of $\overline{W}$, namely $\mathrm{Sing}(J_i,b) \cap \overline{W}_i$. The difference of these three questions with the last main question is that there we add a very strong condition on the smooth hypersurface $\overline{W}$, namely that there be an inclusion $\mathrm{Sing}(J,b) \subset \overline{W}$, and that such inclusion be preserved by transformation in the sense of the previous questions. The existence of hypersurfaces with this condition will be guaranteed by the action of differential operators of the sheaf of ideals J.

In 2.2 we discuss our first three questions for an arbitrary smooth hypersurface $\overline{W}$. The clarification of this point is essential for the precise formulation of induction. Roughly speaking, if we fix a regular system of parameters $\{x_1,\dots,x_d\}$ at a closed point $x \in \mathrm{Sing}(J,b)$ so that $\overline{W}$ is locally defined by x_1, then each element $f \in J$ can be expressed as a formal power series:

$$f = \sum x_1^r a_r$$

with coefficients $a_r \in k[[x_2,\dots,x_d]]$. We then show, at least formally, that the pair (K,d) defining the intersection $\mathrm{Sing}(J,b) \cap \overline{W}$ can be defined in terms of the coefficients a_r, more precisely in terms of couples $(\langle a_r \rangle, b-r)$ for $0 \leq r < b$, and b as above. The objective in this section is to define the pair (K,d), and study its transforms so as to settle the first three questions, at least in the context of complete local rings.

In 2.3 we study differential operators, which will play a central role in 2.4, devoted to answer our main question. In fact, in this last section of this chapter we provide a way of choosing smooth hypersurfaces with the containment condition required in that question. We will see that our main question has a positive answer for a particular class of couples (J,b), called here *simple couples*. Unfortunately, even for the case of simple couples, the choice of the smooth subschemes $\overline{W}$ is only local in the sense that such choice is possible in a neighborhood of a point in W. On the other hand, it is not canonical (i.e. there are various ways of choosing smooth hypersurfaces with that nice property). These two apparent draw-backs in our choices of smooth hypersurfaces will be addressed in Chapter 4. However, in Chapter 3 of these notes, we shall show how these simple couples (J,b) provide the invariants required for the definition of the resolution functions in 1.25.

2.2. First steps on the previous questions

The objective is: given (J,b) over a smooth scheme W and a smooth hypersurface $\overline{W}$, define (K,d) with the property discussed in Questions 1, 2, and 3. We shall answer these questions locally.

2.1. Here $\overline{W}$ denotes a smooth hypersurface in W (after open restriction of W), and $y \in \overline{W}$ will be a closed rational point so that the completion $\hat{\mathscr{O}}_{W,y}$ is a ring of formal power series. We can choose a regular system of parameters $\{x_1,\dots,x_d\}$ at $\mathscr{O}_{W,y}$ so that the hypersurface $\overline{W}$ is defined by the ideal spanned by x_1. Note that in such a case, $\hat{\mathscr{O}}_{W,y} = k[[x_1,\dots,x_d]]$, and the regular system of parameters $\{x_1,\dots,x_d\}$ also defines:

a) A natural identification of $\hat{\mathscr{O}}_{\overline{W},y}$ with $k[[x_2,\dots,x_d]]$ ($k[[x_1,\dots,x_d]]/\langle x_1\rangle = k[[x_2,\dots,x_d]]$).

b) An inclusion of rings of formal power series $S = k[[x_2,\dots,x_d]] \subset k[[x_1,x_2,\dots,x_d]]$.

From a geometric point of view, a) is a restriction to a closed hypersurface, and b) is usually called a local retraction. Note here that if we change coordinates in S, say $S = k[[x'_2,\dots,x'_d]]$, then $\hat{\mathscr{O}}_{W,y} = k[[x_1,x'_2\dots,x'_d]]$ and this new system of coordinates fixes both restriction a) and the inclusion b). In particular, there are many choices of regular systems of parameters defining a) and b). In the discussion below we will fix a) and b) and consider coordinates adapted to both. In other words, we consider regular systems of coordinates of the form $\{y_1,\dots,y_d\}$, which fulfill the condition that $\langle y_1\rangle = \langle x_1\rangle$, and that $k[[y_2,\dots,y_d]] = S$.

Note that b) provides an identification, say $\hat{\mathscr{O}}_{W,y} = S[[x_1]]$, and an S-linear partial derivative $\frac{\partial}{\partial x_1}$ that will play a central role.

2.2. Fix a closed point $y \in \mathrm{Sing}(J,b)\cap\overline{W}$. After a suitable restriction of W to an affine neighborhood of y, we assume that:

1) There are global sections $f_i \in \Gamma(\mathscr{O}_W)$, $1 \leq i \leq n$, which are in J.

2) For each index $1 \leq i \leq n$ above, there are auxiliary elements $g_{i,r} \in \Gamma(\mathscr{O}_W)$ for $0 \leq r \leq b-1$.

We also fix at y an inclusion of rings of formal power series, say:

3) $S \subset \hat{\mathscr{O}}_{W,y}$.

We now impose conditions on the previous data, namely that there be a regular system of parameters $\{x_1,\dots,x_d\}$ at $\mathscr{O}_{W,y}$ and that the following (local) conditions (LC) hold:

LCi) $J\mathscr{O}_{W,y}$ is generated by $\{f_i, i = 1,\dots,n\}$.

LCii) $S = k[[x_2,\dots,x_d]]$ and the inclusion of complete rings $S \subset \hat{\mathscr{O}}_{W,y}$ is given by

$$k[[x_1,\dots,x_{d-1}]] \subset k[[x_1,\dots,x_d]].$$

LCiii) x_1 is a local equation of $\overline{W}$ (in particular $\hat{\mathscr{O}}_{\overline{W},y}$ can be identified with $k[[x_2,\dots,x_d]]$).

LCiv) Each f_i in 1) can be expressed in the form

$$f_i = \sum_{r\geq 0} x_1^r a_{i,r}, \qquad a_{i,r} \in S, \tag{2.2.1}$$

where $a_{i,r}$ is the class of $g_{i,r}$ in $S = \hat{\mathscr{O}}_{\overline{W},y}$ for $0 \leq r \leq b-1$ (see 2)).

REMARK 2.3. As indicated in 2.1 the partial derivative $\frac{\partial}{\partial x_1}$ plays a central role. In fact the auxiliary functions in 2) (above) will be defined in terms of partial derivatives, more precisely by $g_{i,r} = \frac{1}{r!}\frac{\partial^r}{\partial^r x_1} f_i$. The inclusion $S \subset \hat{\mathscr{O}}_{W,y}$ in 3) shows that $\frac{\partial}{\partial x_1}$ generates the rank one free module of derivations on $\hat{\mathscr{O}}_{W,y}$ relative to the subring S. We discuss this local condition in 2.5.

2.4. As it stands, we are assuming at any closed point of $\mathrm{Sing}(J,b) \cap \overline{W}$ the existence of a regular system of parameters that fulfill all local conditions (LC). A local analysis shows that:

$$\mathrm{Sing}(J,b) = \cap \mathrm{Sing}(\langle f_i \rangle, b), \tag{2.2.2}$$

at least at an open neighborhood of $\mathrm{Sing}(J,b) \cap \overline{W}$. If we set $\overline{g}_{i,r} \in \Gamma(\mathscr{O}_{\overline{W}})$ as the class of $g_{i,r} \in \Gamma(\mathscr{O}_W)$, then $\overline{g}_{i,r} = a_{i,r}$ in formula (2.2.1), so

$$\mathrm{Sing}(\langle f_i \rangle, b) \cap \overline{W} = \cap_{0 \leq r \leq b-1} \mathrm{Sing}(\langle \overline{g}_{i,r} \rangle, b-r). \tag{2.2.3}$$

It is not hard to produce (K,d) as in Question 1, in terms of the couples $(\langle \overline{g}_{i,r} \rangle, b-r)$ (in terms of $(\langle a_{i,r} \rangle, b-r)$ at the completion), but it now suffices to consider equations (2.2.2) and (2.2.3). We will indicate how these local conditions can be naturally lifted by transformations.

The point is that if these conditions hold, they will also hold after blowing up any smooth subscheme $Y \subset \mathrm{Sing}(J,b) \cap \overline{W}$. To check this, fix a transform

$$\begin{array}{ccc} W & \stackrel{\pi}{\longleftarrow} & W_1 \\ (J,b) & & (J_1,b) \end{array}$$

and a closed point $y' \in \mathrm{Sing}(J_1,b) \cap \overline{W}_1$, mapping to a closed point $y \in Y \subset \mathrm{Sing}(J,b) \cap \overline{W}$, and consider a regular system of parameters $\{x_1, x_2, \ldots, x_d\}$ at $\mathscr{O}_{W,y}$ so that the local conditions (LC) in 2.2 hold. Recall here that there are changes of coordinates that preserve the conditions (LC) (see 2.1). After suitable extension of k, there is a change of coordinates at $S = k[[x_2, \ldots, x_d]]$ so that:

i) $I(Y)_y = \langle x_1, x_2, \ldots, x_r \rangle$ $(r \leq d)$.

ii) $\{\frac{x_1}{x_2}, x_2, \frac{x_1}{x_2}, \ldots, \frac{x_r}{x_2}, x_{r+1}, \ldots, x_d\}$ is a local regular system of parameters at $\mathscr{O}_{W_1,y'}$.

Set $S_1 = k[[x_2, \frac{x_1}{x_2}, \ldots, \frac{x_r}{x_2}, x_{r+1}, \ldots, x_d]]$. Note that there is an inclusion $S \subset S_1$, and:

LCi) $J_1 \mathscr{O}_{W_1,y'}$ is generated by $\{\frac{f_i}{x_2^b}, i = 1, \ldots, n\}$.

LCii) Fix the inclusion of complete rings $S_1 \subset \hat{\mathscr{O}}_{W_1,y'}$.

LCiii) $\frac{x_1}{x_2}$ is a local equation of $\overline{W}_1$ (so $\hat{\mathscr{O}}_{\overline{W}_1,y'}$ can be identified with S_1).

LCiv)

$$\frac{f_i}{x_2^b} = \sum_{r \geq 0} \left(\frac{x_1}{x_2}\right)^r \frac{a_{i,r}}{x_2^{b-r}}, \qquad \frac{a_{i,r}}{x_2^{b-r}} \in S_1,$$

and $a_{i,r} \in S$ is the class of $g_{i,r}$ for $r = 0, \ldots, b-1$.

The previous morphism π induces

$$\begin{array}{ccc} \overline{W} & \stackrel{\pi}{\longleftarrow} & \overline{W}_1 \\ (\langle a_{i,r} \rangle, b-r) & & (\langle a_{i,r} \rangle_1, b-r) \end{array}$$

or, say,

$$\begin{array}{ccc} \overline{W} & \stackrel{\pi}{\longleftarrow} & \overline{W}_1 \\ (\langle \overline{g_{i,r}} \rangle, b-r) & & (\langle \overline{g_{i,r}} \rangle_1, b-r). \end{array}$$

A local analysis at W_1 shows that

(2.2.4) $$\mathrm{Sing}(J_1,b) = \cap\,\mathrm{Sing}(\langle f_i\rangle_1,b)$$

in an open neighborhood of $\mathrm{Sing}(J_1,b)\cap\overline{W}_1$, and that

(2.2.5) $$\mathrm{Sing}(\langle f_i\rangle_1,b)\cap\overline{W}_1 = \cap_{0\leq r\leq b-1}\mathrm{Sing}(\langle\overline{g}_{i,r}\rangle_1,b-r),$$

or, say,

$$\mathrm{Sing}(J_1,b)\cap\overline{W}_1 = \cap_{1\leq i\leq n}\cap_{0\leq r\leq b-1}\mathrm{Sing}(\langle\overline{g}_{i,r}\rangle_1,b-r).$$

Similarly, given

$$\begin{array}{ccccc} W & \stackrel{\pi}{\longleftarrow} & W_1 & \stackrel{\pi'}{\longleftarrow} & W_2 \\ (J,b) & & (J_1,b) & & (J_2,b), \end{array}$$

where π' is defined with center $Y_1\subset\mathrm{Sing}(J_1,b)\cap\overline{W}_1$, then conditions LC can be defined at points of $\mathrm{Sing}(J_2,b)\cap\overline{W}_2$. And these local conditions ensure that the induced transforms

$$\begin{array}{ccccc} \overline{W} & \stackrel{\pi}{\longleftarrow} & \overline{W}_1 & \stackrel{\pi'}{\longleftarrow} & \overline{W}_2 \\ (\langle\overline{g}_{i,r}\rangle,b-r) & & (\langle\overline{g}_{i,r}\rangle_1,b-r) & & (\langle\overline{g}_{i,r}\rangle_2,b-r) \end{array}$$

are such that

$$\mathrm{Sing}(\langle f_i\rangle_2,b)\cap\overline{W}_2 = \cap_{0\leq r\leq b-1}\mathrm{Sing}(\langle\overline{g}_{i,r}\rangle_2,b-r),$$

so

$$\mathrm{Sing}(J_2,b)\cap\overline{W}_2 = \cap_{1\leq i\leq n}\cap_{0\leq r\leq b-1}\mathrm{Sing}(\langle\overline{g}_{i,r}\rangle_2,b-r).$$

For any sequence defined as in (2.1.1), there is a sequence

$$\begin{array}{ccccccc} \overline{W} & \stackrel{\pi_1}{\longleftarrow} & \overline{W}_1 & & \cdots & \stackrel{\pi_k}{\longleftarrow} & \overline{W}_k \\ (\langle a_{i,r}\rangle,b-r) & & (\langle a_{i,r}\rangle_1,b-r) & & & & (\langle a_{i,r}\rangle_k,b-r) \end{array}$$

or, say,

(2.2.6) $$\begin{array}{ccccccc} \overline{W} & \stackrel{\pi_1}{\longleftarrow} & \overline{W}_1 & & \cdots & \stackrel{\pi_k}{\longleftarrow} & \overline{W}_k \\ (\langle\overline{g}_{i,r}\rangle,b-r) & & (\langle\overline{g}_{i,r}\rangle_1,b-r) & & & & (\langle\overline{g}_{i,r}\rangle_k,b-r) \end{array}$$

so that

(2.2.7) $$\mathrm{Sing}(\langle f_i\rangle_k,b)\cap\overline{W}_k = \cap_{0\leq r\leq b-1}\mathrm{Sing}(\langle\overline{g}_{i,r}\rangle,b-r),\text{ and therefore}$$

$$\mathrm{Sing}(J_k,b)\cap\overline{W}_k = \cap_{1\leq i\leq n}\cap_{0\leq r\leq b-1}\mathrm{Sing}(\langle\overline{g}_{i,r}\rangle_k,b-r).$$

2.5. This essentially proves Question 3, up to a precise indication on how to define (K,d) in terms of the couples $(\langle\overline{g}_{i,r}\rangle_k,b-r)$, if we can show the existence of functions f_i and $g_{i,r}$ so that the conditions in 2.2 hold.

Fix a closed point $y\in\mathrm{Sing}(J,b)\cap\overline{W}$ and a regular system of parameters $\{x_1,\dots,x_d\}$ at $\mathcal{O}_{W,y}$, so that $\langle x_1\rangle = I(\overline{W})_x$. The module of differentials is locally free, so there is an open neighborhood, say U, where $\{x_1,\dots,x_d\}$ are global, $I(\overline{W}) = \langle x_1\rangle$ globally, and $\{d(x_1),\dots,d(x_d)\}$ is a basis of this module. In particular, partial derivatives $\{\frac{\partial}{\partial x_1},\dots,\frac{\partial}{\partial x_d}\}$ are defined at U. If $J_x = \langle f_1,\dots f_n\rangle$, we may also assume that $J = \langle f_1,\dots f_n\rangle$ globally at U.

At any closed point $y'\in U$, let α_i denote the class of x_i at the residue field, say k', of $\mathcal{O}_{W,y'}$. Define $x'_i = x_i-\alpha_i$, and note that $\{x'_1,\dots,x'_d\}$ is a regular system of parameters at $\hat{\mathcal{O}}_{W,y'}$. Set

(2.2.8) $$g_{i,r} = \frac{1}{r!}\frac{\partial^r}{\partial^r x_1}f_i$$

for $r = 0, \dots, b-1$, which are also global at U.

Finally, define $S = k'[[x'_2, \dots, x'_d]](\subset \hat{\mathscr{O}}_{W,y'})$, and check that the local conditions in 2.2 hold.

2.3. Derivations and monoidal transformations on smooth schemes

In this section we study the behavior of derivations when applying monoidal transformations. This will be used in Section 2.4, where the inductive properties discussed in 1.17 will be clarified.

Fix W smooth over a field k, and let $y \in W$ be a closed point. Let $\{x_1, \dots, x_n\}$ be a regular system of parameters at $\mathscr{O}_{W,y}$.

We define an operator Δ_y on ideals in $\mathscr{O}_{W,y}$ by setting, for $J_y = \langle f_1, f_2, \dots, f_s \rangle$ in $\mathscr{O}_{W,y}$,

$$\Delta_y(J_y) = \langle f_1, f_2, \dots, f_s, \ \frac{\partial f_j}{\partial x_i} \ /1 \le i \le n; 1 \le j \le s \rangle.$$

Note that $\Delta_y(\Delta_y(J_y)) = \langle f_1, f_2, \dots, f_s, \ \frac{\partial f_j}{\partial x_i}, \ \frac{\partial^2 f_j}{\partial x_i \partial x_j} \ /1 \le i \le n; 1 \le j \le s \rangle$. The whole point of restriction to fields of characteristic zero relies on the following property:

2.6. **(Characteristic zero.)** If k is a field of characteristic zero and $b \ge 1$, then J_y has order b at $\mathscr{O}_{W,y}$ iff $\Delta_y(J_y)$ has order $b-1$.

EXAMPLE 2.7. Let $\mathscr{O}_{W,y} = k[x_1, x_2, x_3]_{\langle x_1, x_2, x_3 \rangle}$.

$$J_y = \langle x_1^3 + x_2^4 + x_3^4 \rangle \subset \Delta_y(J_y) = \langle x_1^2, x_2^3, x_3^3 \rangle \subset \Delta_y^2(J_y) = \langle x_1, x_2^2, x_3^2 \rangle \subset \Delta_y^3(J_y) = \mathscr{O}_{W,y}.$$

Note that if k is of characteristic zero, the orders of these ideals drop by one. In this example the orders are 3, 2, 1, and 0, respectively.

2.8. Other properties of the operator Δ_y are:

i) $J_y \subseteq \Delta_y(J_y) \subseteq \Delta_y(\Delta_y(J_y)) = \Delta_y^2(J_y) \subseteq \Delta_y^3(J_y) \subseteq \dots$.

ii) $J_y \subset \mathscr{O}_{W,y}$ has order $b (\ge 1)$ iff $\Delta_y^{b-1}(J_y)$ has order 1.

iii) The order of $J_y \subset \mathscr{O}_{W,y}$ is $\ge s$ iff $\Delta_y^{s-1}(J_y)$ is a proper ideal in $\mathscr{O}_{W,y}$.

2.9. **On the Δ operator.** The locally defined operators Δ_y can be globalized in the following sense. Fix W smooth over a field k; there is an operator Δ on the class of all $\mathscr{O}_W$-ideals such that:

$$J \subseteq \Delta(J) (\subset \mathscr{O}_W),$$

and at any closed point $y \in W$:

$$\Delta(J)_y = \Delta_y(J_y).$$

Furthermore, the following properties hold

i) $J \subseteq \Delta(J) \subseteq \Delta^2(J) \subseteq \dots$ (hence $V(J) \supseteq V(\Delta(J)) \supseteq V(\Delta^2(J)) \supset \dots$).

ii) $V(\Delta^{s-1}(J)) = \operatorname{Sing}(J, s)$. In fact, $V(\Delta^{s-1}(J))$ is the closed set of points in W where J has order $\ge s$ (i.e. $(\Delta^{s-1}(J))_y = \Delta_y^{s-1}(J_y) \subsetneq \mathscr{O}_{W,y}$) iff the order of $J_y \mathscr{O}_{W,y}$ is $\ge s$).

iii) If b is the biggest order of J, then $V(\Delta^b(J)) = \emptyset$ and $V(\Delta^{b-1}(J))$ is locally included in a smooth hypersurface.

Proof of iii) If b is the biggest order of J, then $\Delta^b(J) = \mathscr{O}_W$ and $\Delta^{b-1}(J)$ has order at most 1. So if $y \in V(\Delta^{b-1}(J))$, then $\Delta^{b-1}(J)\mathscr{O}_{W,y}$ has order 1 at $\mathscr{O}_{W,y}$. If $g \in \Delta^{b-1}(J)$ has order 1 at $\mathscr{O}_{W,y}$, then

$$\overline{W} = V(\langle g \rangle) \supset V(\Delta^{b-1}(J)),$$

and $\overline{W}$ is a smooth hypersurface in a neighborhood of y.

EXAMPLE 2.10. Set $W = A^3_k = \mathrm{Spec}(k[X,Y,Z])$, $g = Z^3 + XY^2Z + X^5$, and $J = \langle g\rangle$, as in 1.17. Then

$$\Delta(J) = \langle 3Z^2 + XY^2, Y^2Z + 5X^4, 2XYZ, F\rangle \subset \Delta^2(J) = \langle Z, XY, Y^2, X^3\rangle \subset \Delta^3(J) = k[X,Y,Z].$$

So, as indicated in 1.17, the 3-fold points of the hypersurface $\mathbb{Y} \subset \mathbb{A}^3$ defined by $V(\langle J\rangle)$ are included in a smooth hypersurface $\overline{W} = V(\langle Z\rangle)$.

2.11. **The Δ operator and monoidal transformations.** We now address the compatibility of Δ operators with monoidal transformations.

Define, at the total quotient field of $\mathbb{Q}[X_1,\dots,X_n]$, $Y_1 = X_1$, $Y_j = \frac{X_j}{X_1}$ for $2 \le j \le r$, and $Y_k = X_k$ for $r+1 \le k \le n$. There is a natural inclusion

$$\mathbb{Q}[X_1,\dots,X_n] \subset \mathbb{Q}[X_1, \frac{X_2}{X_1}, \dots, \frac{X_r}{X_1}, X_{r+1}, \dots, X_n] (= \mathbb{Q}[Y_1,\dots,Y_n]),$$

and $\mathbb{Q}[Y_1,\dots,Y_n]$ is a polynomial ring. A direct computation shows that

$$\frac{\partial}{\partial X_1} = \frac{\partial}{\partial Y_1} - \frac{1}{Y_1}\sum_{2\le j\le r} Y_j \frac{\partial}{\partial Y_j},$$

$$\frac{\partial}{\partial X_j} = \frac{1}{Y_1}\frac{\partial}{\partial Y_j} \qquad 2 \le j \le r,$$

$$\frac{\partial}{\partial X_k} = \frac{\partial}{\partial Y_k} \qquad r+1 \le k \le n.$$

This shows that partial derivatives in $\mathbb{Q}[X_1,\dots,X_n]$ have poles of order at most one along the hypersurface Y_1 when extended to $\mathbb{Q}[Y_1,\dots,Y_n] (\subset \mathbb{Q}(X_1,\dots,X_n))$. Recall that partial derivatives on $\mathbb{Q}[X_1,\dots,X_n]$ extend to derivations on the total quotient field $\mathbb{Q}(X_1,\dots,X_n)$. Our discussion shows that these extended operators do not map the subring $\mathbb{Q}[Y_1,\dots,Y_n]$ into itself.

It is useful, at this point, to view partial derivatives as particular examples of *differential operators* of order one. We refer to **[25]** for the notion and properties of differential operators.

Let $\mathrm{Der}(\mathbb{Q}[X_1,\dots,X_n])$ be the $\mathbb{Q}[X_1,\dots,X_n]$-module of partial derivatives (free module generated by $\frac{\partial}{\partial X_l}$, $l = 1,\dots,n$). Then $\mathrm{Diff}^1(\mathbb{Q}[X_1,\dots,X_n])$ (sheaf of differentials of order one) is the free $\mathbb{Q}[X_1,\dots,X_n]$-module generated by 1 and the partial derivatives $\frac{\partial}{\partial X_l}$, $l = 1,\dots,n$, where elements of $\mathbb{Q}[X_1,\dots,X_n]$ define a differential operator by multiplication. In other words, an element $f(X) \in \mathbb{Q}[X_1,\dots,X_n]$ defines

$$\varphi_f : \mathbb{Q}[X_1,\dots,X_n] \to \mathbb{Q}[X_1,\dots,X_n] \qquad \varphi_f(h(X)) = f(X)\cdot h(X).$$

The composition $\frac{\partial}{\partial X_l}\cdot\varphi_f$ is an element in $\mathrm{Diff}^1(\mathbb{Q}[X_1,\dots,X_n])$, and not a derivation. Differential operators, as is the case for derivations, also extend to the total quotient field $\mathbb{Q}(X_1,\dots,X_n)$. For example, $\mathscr{D}_l^{(1)} = \frac{\partial}{\partial X_l}\cdot X_1$ is a differential operator of order one. Here $\mathscr{D}_l^{(1)} : \mathbb{Q}(X_1,\dots,X_n) \to \mathbb{Q}(X_1,\dots,X_n)$ is defined by $\mathscr{D}_l^{(1)}(h) = \frac{\partial}{\partial X_l} X_1 h$. Recall that $X_1 = Y_1$ and note that for any $l = 1,\dots,n$, the differential operator

$$\mathscr{D}_l^{(1)}(h) = \frac{\partial}{\partial X_l} Y_1 h \tag{2.3.1}$$

does map $\mathbb{Q}[Y_1,\dots,Y_n]$ into itself. Note also that for $l = 1$, $\mathscr{D}_1^{(1)} \in \mathrm{Diff}^1(\mathbb{Q}[Y_1,\dots,Y_n])$ (free $\mathbb{Q}[Y_1,\dots,Y_n]$-module spanned by $\{1, \frac{\partial}{\partial Y_l}, \dots, \frac{\partial}{\partial Y_n}\}$) and $\mathscr{D}_1^{(1)} \notin \mathrm{Der}^1(\mathbb{Q}[Y_1,\dots,Y_n])$.

Define now for any integer $b \geq 1$, $\mathscr{D}_l^{(b)} : \mathbb{Q}(X_1,\ldots,X_n) \to \mathbb{Q}(X_1,\ldots,X_n)$, by setting

$$\mathscr{D}_l^{(b)}(h) = \frac{\partial}{\partial X_l} Y_1^b h.$$

One can check that, for any index $l = 1,\ldots,n$, $\mathscr{D}_l^{(b)}(\mathbb{Q}[Y_1,\ldots,Y_n]) \subset Y_1^{b-1}\mathbb{Q}[Y_1,\ldots,Y_n]$. In fact, it follows that

$$\mathscr{D}_l^{(b)} \in Y_1^{b-1}\mathrm{Diff}^1(\mathbb{Q}[Y_1,\ldots,Y_n]).$$

We express this fact formally as

$$\mathrm{Diff}^1(\mathbb{Q}[X_1,\ldots,X_n]) \cdot Y_1^b \subset Y_1^{b-1}\mathrm{Diff}^1(\mathbb{Q}[Y_1,\ldots,Y_n]). \tag{2.3.2}$$

A similar argument holds when taking completions at the origin, namely for $\mathbb{Q}[[X_1,\ldots,X_n]] \subset \mathbb{Q}[[Y_1,\ldots,Y_n]]$.

REMARK 2.12. Note that for an ideal $J \subset \mathbb{Q}[X_1,\ldots,X_n]$, the extension $\Delta(J)$ can be defined as $\mathrm{Diff}^1(J)$ (the extension defined by the action of all first-order differential operators on J). Similar statements hold for any smooth scheme W over a field k, namely that Diff^1_W is a locally free sheaf, and for an ideal $J \subset \mathscr{O}_W$, $\Delta(J)$ can be naturally expressed as $\mathrm{Diff}^1_W(J)$.

LEMMA 2.13. *Fix a couple* (J,b), *and consider a transformation*

$$\begin{array}{ccc} W & \stackrel{\pi}{\longleftarrow} & W_1 \\ (J,b) & & (J_1,b). \end{array} \tag{2.3.3}$$

Then:

1) If $b \geq 2$, π *induces a transformation of* $(\Delta(J), b-1)$, *say:*

$$\begin{array}{ccc} W & \stackrel{\pi}{\longleftarrow} & W_1 \\ (\Delta(J),b-1) & & ((\Delta(J))_1,b-1). \end{array}$$

2) $(\Delta(J))_1 \subset \Delta_1(J_1)$, *where* Δ_1 *denotes the* Δ*-operator at* W_1.

Proof: Let $Y \subset W$ be the center of the monoidal transformation, and let $H \subset W_1$ be the exceptional locus. By assumption $Y \subset \mathrm{Sing}(J,b)$, so $J \cdot \mathscr{O}_{W_1} = I(H)^b \cdot J_1$. It follows from 2.9 **ii)** that for general b, $\mathrm{Sing}(J,b) \subset \mathrm{Sing}(\Delta(J),b-1)$. Note that $Y \subset \mathrm{Sing}(\Delta(J),b-1)$, which proves 1).

In order to prove 2), we argue locally. Fix a closed point $x' \in H$ mapping to, say $x \in W$, and after a suitable change of base field, there is a regular system of parameters $\{x_1,\ldots,x_n\}$ at $\mathscr{O}_{W,x}$ such that $I(Y)_x = \langle x_1,\ldots,x_h\rangle$, and $\{y_1,\ldots,y_n\}$ is a regular system of parameters at $\mathscr{O}_{W_1,x'}$, where $y_1 = x_1$, $y_j = \frac{x_j}{x_1}$ for $2 \leq j \leq r$, and $y_k = x_k$ for $r+1 \leq k \leq n$.

In this case $I(H) = \langle y_1\rangle$ at $\mathscr{O}_{W_1,x'}$, and for any $h \in J_x \subset \mathscr{O}_{W,x}$, $h\mathscr{O}_{W_1,x'} = y_1^b\langle h_1\rangle$, $h_1 \in J_1$.

Note that $h\mathscr{O}_{W_1,x'} = y_1^b\langle h_1\rangle \subset y_1^{b-1}\Delta_1(J_1)$. It also follows from the discussion in 2.11 that for any index $1 \leq l \leq n$

$$\frac{\partial}{\partial x_l}(y_1^b\langle h_1\rangle) \subset y_1^{b-1}\Delta_1(J_1).$$

This proves that $\Delta(J)\mathscr{O}_{W_1} \subset I(H)^{b-1}\Delta_1(J_1)$, and hence the claim in 2)(see (2.12)).

LEMMA 2.14. *1) A transformation*

$$\begin{array}{ccc} W & \stackrel{\pi}{\longleftarrow} & W_1 \\ (J,b) & & (J_1,b) \end{array}$$

defines, for any $0 < s < b$, *a transformation of* $(\Delta^s(J), b-s)$, *say*

$$\begin{array}{ccc} W & \stackrel{\pi}{\longleftarrow} & W_1 \\ (\Delta^s(J), b-s) & & ((\Delta^s(J))_1, b-s). \end{array}$$

2) $(\Delta^s(J))_1 \subset \Delta_1^s(J_1)$, *where* Δ_1 *denotes the* Δ*-operator at* W_1.

PROOF. The case $s = 1$ is the previous lemma; the general case follows by induction and formula (2.3.2). For example, for $s = 2$ we obtain:

$$\mathrm{Diff}^1(\mathbb{Q}[X_1,\dots,X_n]) \cdot \mathrm{Diff}^1(\mathbb{Q}[X_1,\dots,X_n]) \cdot Y_1^b$$

$$\subset \mathrm{Diff}^1(\mathbb{Q}[X_1,\dots,X_n]) \cdot Y_1^{b-1}\mathrm{Diff}^1(\mathbb{Q}[Y_1,\dots,Y_n])$$

$$\subset Y_1^{b-2}\mathrm{Diff}^1(\mathbb{Q}[Y_1,\dots,Y_n]) \cdot \mathrm{Diff}^1(\mathbb{Q}[Y_1,\dots,Y_n]).$$

This proves that $\Delta(J)^2\mathscr{O}_{W_1} \subset I(H)^{b-2}\Delta_1^2(J_1)$, and hence the claim in 2) for $s = 2$.

REMARK 2.15. Fix $K \subset J$ two ideals in $\mathscr{O}_W$ and couples (J,b) and (K,b). Then clearly

a) $\mathrm{Sing}(J,b) \subset \mathrm{Sing}(K,b)$.

b) Any transformation, as in (2.3.3) of (J,b), induces the transformation

$$\begin{array}{ccc} W & \stackrel{\pi}{\longleftarrow} & W_1 \\ (K,b) & & (K_1,b) \end{array}$$

and $K_1 \subset J_1$.

2.16. We finally extend the previous result to study the behavior of Δ operators with an arbitrary sequence of transformations.

COROLLARY 2.17. *Fix a couple* (J,b) $(J \subset \mathscr{O}_W)$ *and a sequence of transformations*

$$\begin{array}{ccccccc} W & \stackrel{\pi_1}{\longleftarrow} & W_1 & \stackrel{\pi_2}{\longleftarrow} & \dots & \stackrel{\pi_r}{\longleftarrow} & W_r \\ (J,b) & & (J_1,b) & & & & (J_r,b). \end{array} \tag{2.3.4}$$

1) (2.3.4) induces, for any $0 < s < b$, *a sequence of transformations of* $(\Delta^s(J), b-s)$, *say*

$$\begin{array}{ccccccc} W & \stackrel{\pi_1}{\longleftarrow} & W_1 & \stackrel{\pi_2}{\longleftarrow} & \dots & \stackrel{\pi_r}{\longleftarrow} & W_r, \\ (\Delta^s(J), b-s) & & ((\Delta^s(J))_1, b-s) & & & & ((\Delta^s(J))_r, b-s) \end{array}$$

and

2) $(\Delta^s(J))_r \subset \Delta^s(J_r)$.

PROOF. The case $r = 1$ is in 2.14. Consider now the case $r = 2$, namely

$$\begin{array}{ccccc} W & \stackrel{\pi_1}{\longleftarrow} & W_1 & \stackrel{\pi_2}{\longleftarrow} & W_3 \\ (J,b) & & (J_1,b) & & (J_2,b). \end{array}$$

Then Lemma 2.14 asserts that π_1 defines a transform of $(\Delta^s(J), b-s)$, say $((\Delta^s(J))_1, b-s)$, and that $(\Delta^s(J))_1 \subset \Delta^s(J_1)$. The same result says that π_2 defines a transform of $(\Delta^s(J_1), b-s)$, say $((\Delta^s(J_1))_1, b-s)$, and that $(\Delta^s(J_1))_1 \subset \Delta^s(J_2)$. The statement follows in this case from Remark 2.15. The general case $r \geq 2$ follows similarly by induction. □

2.4. Simple couples and a form of induction on resolution problems

2.18. The purpose of this section is the study of *simple couples* (J,b) $(J \subset \mathcal{O}_W)$. Examples of simple couples appear already in Section 1.2. They will play a central role in our inductive arguments (induction on the dimension of the ambient space). The main objective of this section is Theorem 2.22 and the formulation of Proposition 2.31. This proposition will settle the notion of stability of induction discussed in 1.17.

2.19. Fix $J \subset \mathcal{O}_W$, assume that $J_x \neq 0 (\subset \mathcal{O}_{W,x})$ for any $x \in W$, and define a function

$$\mathrm{ord}_J : W \to \mathbb{N}, \tag{2.4.1}$$

where $\mathrm{ord}_J(x)$ denotes the order of J_x in the local ring $\mathcal{O}_{W,x}$.

Note that ord_J is upper-semi-continuous (1.24). In fact, for any positive integer s,

$$\{x \in W / \mathrm{ord}_J(x) \geq s\} = V(\Delta^{s-1}(J)) \text{ (see 2.9).}$$

REMARK 2.20. The following conditions are equivalent:

1) $\max - \mathrm{ord}_J = b$ (where, as in 1.24, $\max - \mathrm{ord}_J$ denotes the maximum value achieved).

2) $V(\Delta^{b-1}(J)) \neq \emptyset$ and $V(\Delta^b(J)) = \emptyset$.

3) $\max - \mathrm{ord}_{\Delta^{b-1}(J)} = 1$.

The equivalence follows from the properties of the Δ operator discussed in 2.9.

DEFINITION 2.21. We say that (J,b) is a *simple couple* if the previous conditions hold for J and b.

The following theorem says that the transform of a simple couple is simple.

THEOREM 2.22. *If (J,b) $(J \subset \mathcal{O}_W)$ is a simple couple, and*

$$\begin{array}{ccc} W & \stackrel{\pi}{\longleftarrow} & W_1 \\ (J,b) & & (J_1,b) \end{array}$$

is a transformation, then either $\mathrm{Sing}(J_1,b) = \emptyset$ *or* (J_1,b) *is a simple couple.*

The case $b = 1$ will be proved in Proposition 2.25, and the case $b \geq 2$ will be proved in Proposition 2.26.

We shall first draw attention to the case of simple couples of the form $(J,1)$.

REMARK 2.23. The following conditions are equivalent:

1) $\max - \mathrm{ord}_J = 1$.

2) $V(J) \neq \emptyset$ and $V(\Delta(J)) = \emptyset$.

3) There is an open covering $\{U_\lambda\}_{\lambda \in \Lambda}$ of W, and for each λ a smooth hypersurface $\overline{W}_\lambda$ in U_λ such that $I(\overline{W}_\lambda) \subset J_\lambda$, where J_λ denotes the restriction of J to U_λ.

For the proof of 3), note that an ideal of order one in a local regular ring $\mathcal{O}_{W,x}$ contains an element of order one, and such an element defines a smooth hypersurface in some open neighborhood of $x \in W$.

REMARK 2.24. Fix, as before, an open covering of W, say $\{U_\lambda\}_{\lambda \in \Lambda}$, and a monoidal transformation with center $Y \subset W$, say $W \longleftarrow W_1$. For each index λ set $U_\lambda^{(1)} \subset W_1$ as the pull-back of U_λ. In this way we get

$$U_\lambda \longleftarrow U_\lambda^{(1)},$$

which is either a monoidal transformation (in case $Y \cap U_\lambda \neq \emptyset$), or the identity map (if $Y \cap U_\lambda = \emptyset$). Note also that $\{U_\lambda^{(1)}\}_{\lambda \in \Lambda}$ is an open cover of W_1.

PROPOSITION 2.25. *Fix $J \subset \mathscr{O}_W$ with maximum order 1 and a sequence of transformations*

$$\begin{array}{ccccccc} W & \xleftarrow{\pi_1} & W_1 & \xleftarrow{\pi_2} & \cdots & \xleftarrow{\pi_r} & W_r \\ (J,1) & & (J_1,1) & & & & (J_r,1). \end{array} \tag{2.4.2}$$

Then the maximum order of J_r is either 1 or 0 (i.e. $J_r = \mathscr{O}_{W_r}$ in the last case).

PROOF. Define an open covering $\{U_\lambda\}_{\lambda\in\Lambda}$ of W and inclusions

$$I(\overline{W}_\lambda) \subset J_\lambda, \tag{2.4.3}$$

where $\overline{W}_\lambda$ is a smooth hypersurface in U_λ, as indicated in Remark 2.23, 3).

The sequence (2.4.2) defines, for each index λ, a sequence of transformations:

$$\begin{array}{ccccccc} U_\lambda & \xleftarrow{\pi_1} & U_\lambda^{(1)} & \xleftarrow{\pi_2} & \cdots & \xleftarrow{\pi_r} & U_\lambda^{(r)} \\ (J_\lambda,1) & & ((J_\lambda)_1,1) & & & & ((J_\lambda)_r,1), \end{array}$$

and also

$$\begin{array}{ccccccc} U_\lambda & \xleftarrow{\pi_1} & U_\lambda^{(1)} & \xleftarrow{\pi_2} & \cdots & \xleftarrow{\pi_r} & U_\lambda^{\lambda\,(r)} \\ (I(\overline{W}_\lambda),1) & & ((I(\overline{W}_\lambda))_1,1) & & & & ((I(\overline{W}_\lambda))_r,1). \end{array}$$

Furthermore, $(I(\overline{W}_\lambda))_r \subset (J_\lambda)_r$ by Remark 2.15. Let $\overline{W}_\lambda^{(r)} \subset U_\lambda^r$ denote the strict transform of $\overline{W}_\lambda$. Since $\overline{W}_\lambda$ is smooth in U_λ, it follows that $\overline{W}_\lambda^{(r)}$ is smooth. Moreover, one can check that this hypersurface is defined by the ideal $(I(\overline{W}_\lambda))_r$. In particular, $(I(\overline{W}_\lambda))_r$ has maximum order at most one, and hence the same holds for $(J_\lambda)_r$. Since the open sets $(U_\lambda)^{(r)}$ cover W_r it follows that J_r has order at most 1. □

PROPOSITION 2.26. *Fix $J \subset \mathscr{O}_W$ with maximum order $b \geq 2$ and consider a sequence of transformations*

$$\begin{array}{ccccccc} W & \xleftarrow{\pi_1} & W_1 & \xleftarrow{\pi_2} & \cdots & \xleftarrow{\pi_r} & W_r \\ (J,b) & & (J_1,b) & & & & (J_r,b). \end{array} \tag{2.4.4}$$

Then the maximum order of $J_r(\subset \mathscr{O}_{W_r})$ is at most b.

PROOF. From the properties in 2.9 we conclude that the maximum order of $\Delta^{b-1}(J)(\subset \mathscr{O}_W)$ is 1. Corollary 2.17 applied for $s = b-1$ says that (2.4.4) defines the sequence of transformations

$$\begin{array}{ccccccc} W & \xleftarrow{\pi_1} & W_1 & \xleftarrow{\pi_2} & \cdots & \xleftarrow{\pi_r} & W_r \\ (\Delta^{b-1}(J),1) & & ((\Delta^{b-1}(J))_1,1) & & & & ((\Delta^{b-1}(J))_r,1), \end{array} \tag{2.4.5}$$

and that $(\Delta^{b-1}(J))_r \subset \Delta^{b-1}(J_r)$. On the other hand, Proposition 2.25 asserts that $(\Delta(J))_r$ has order at most 1, and hence $\Delta^{b-1}(J_r)$ has order at most one. From this and again from 2.9, it follows that J_r has order at most b. □

REMARK 2.27. There is a stronger outcome that follows from the proof of Proposition 2.26 that relates to induction in the dimension of the ambient space. Note that J has highest order b, so $\Delta^{b-1}(J)$ has highest order one. We can argue as in the proof of Proposition 2.25 and define an open cover $\{U_\lambda\}_{\lambda\in\Lambda}$ of W, and for each index λ, a smooth hypersurface $\overline{W}_\lambda \subset U_\lambda$, defined by

$$I(\overline{W}_\lambda) \subset (\Delta^{b-1}(J))_\lambda. \tag{2.4.6}$$

Now use the compatibility of the Δ operator with restriction to open sets and check that

$$(\Delta^{b-1}(J))_\lambda = (\Delta^{b-1}(J_\lambda)).$$

Note also that $\mathrm{Sing}(J,b)\cap U_\lambda \subset \overline{W}_\lambda$. Recall that (2.4.5) defines, for each index λ, a sequence of transformations of $((\Delta^{b-1}(J))_\lambda, 1)$, say

$$\begin{array}{ccccccc} U_\lambda & \xleftarrow{\pi_1} & U_\lambda^{(1)} & \xleftarrow{\pi_2} & \cdots & \xleftarrow{\pi_r} & U_\lambda^{(r)} \\ ((\Delta^{b-1}(J))_\lambda,1) & & (((\Delta^{b-1}(J))_\lambda)_1,1) & & & & (((\Delta^{b-1}(J))_\lambda)_r,1), \end{array}$$

and also

$$\begin{array}{ccccccc} U_\lambda & \xleftarrow{\pi_1} & U_\lambda^{(1)} & \xleftarrow{\pi_2} & \cdots & \xleftarrow{\pi_r} & U_\lambda^{\lambda\,(r)} \\ (I(\overline{W}_\lambda),1) & & ((I(\overline{W}_\lambda))_1,1) & & & & ((I(\overline{W}_\lambda))_r,1). \end{array}$$

Furthermore, $(I(\overline{W}_\lambda))_r \subset ((\Delta^{b-1}(J))_\lambda)_r$, and $(I(\overline{W}_\lambda))_r$ defines a smooth hypersurface $\overline{W}_\lambda^{(r)} \subset U_\lambda^{(r)}$ which is the strict transform of $\overline{W}_\lambda$. We finally note that $\{U_\lambda^{(r)}\}_{\lambda\in\Lambda}$ is a cover of $W^{(r)}$, and taking restriction of the inclusion $(\Delta^{b-1}(J))_r \subset \Delta^{b-1}(J_r)$, we get that

$$((\Delta^{b-1}(J))_\lambda)_r = ((\Delta^{b-1}(J))_r)_\lambda \subset (\Delta^{b-1}(J_r))_\lambda,$$

and hence $(I(\overline{W}_\lambda))_r \subset (\Delta^{b-1}(J_r))_\lambda$. In particular

$$\mathrm{Sing}((J)_r,b)\cap U_\lambda^{(r)} = \mathrm{Sing}((J_\lambda)_r,b) \subset \overline{W}_\lambda^{(r)}.$$

LEMMA 2.28. *Fix $J \subset \mathcal{O}_W$ with maximum order b. There is an open covering, say $\{U_\lambda\}_{\lambda\in\Lambda}$ of W, and for each index λ a smooth hypersurface $\overline{W}_\lambda \subset U_\lambda$ such that the following properties hold:*

P1) $\mathrm{Sing}(J_\lambda,b) \subset \overline{W}_\lambda$.

P2) *For* any *sequence*

$$\begin{array}{ccccccc} W & \xleftarrow{\pi_1} & W_1 & \xleftarrow{\pi_2} & \cdots & \xleftarrow{\pi_r} & W_r \\ (J,b) & & (J_1,b) & & & & (J_r,b) \end{array} \tag{2.4.7}$$

and setting by restriction, for each λ, say

$$\begin{array}{ccccccc} U_\lambda & \xleftarrow{\pi_1} & U_\lambda^{(1)} & \xleftarrow{\pi_2} & \cdots & \xleftarrow{\pi_r} & U_\lambda^{(r)}, \\ (J_\lambda,b) & & ((J_\lambda)_1,b) & & & & ((J_\lambda)_r,b) \end{array} \tag{2.4.8}$$

then $\{U_\lambda^{(r)}\}_{\lambda\in\Lambda}$ is an open covering of W_r, and

$$\mathrm{Sing}(J_r,b)\cap U_\lambda^{(r)} = \mathrm{Sing}((J_\lambda)_r,b) \subset \overline{W}_\lambda^{(r)}, \tag{2.4.9}$$

where $\overline{W}_\lambda^{(r)}$ is the smooth hypersurface defined by the strict transform of $\overline{W}_\lambda$.

PROOF. The case $b=1$ (in which $\mathrm{Sing}(J,1)=V(J)$) is in the proof of Proposition 2.25. The case $b\geq 2$ is in Remark 2.27 and relies entirely on the inclusion (2.4.6). □

2.29. Let $\overline{W}_\lambda^{(i)}$ denote the strict transform of $\overline{W}_\lambda^{(0)}$ in $U_\lambda^{(i)}$(see (2.4.8)). A consequence of (2.4.9) is that all the centers of monoidal transformations involved in (2.4.8) are included in $\overline{W}_\lambda^{(i)}$; hence (2.4.8) defines a sequence of monoidal transformations

$$\overline{W}_\lambda \longleftarrow \overline{W}_\lambda^{(1)} \longleftarrow \cdots \longleftarrow \overline{W}_\lambda^{(r)}. \tag{2.4.10}$$

COROLLARY 2.30. *Let (J,b) be a simple couple in W, and let $R(1)(\mathrm{Sing}(J,b))$ be the union of components of codimension 1 of the closed set* $\mathrm{Sing}(J,b)$. *Then:*

1) $R(1)(\mathrm{Sing}(J,b))$ is open and closed in $\mathrm{Sing}(J,b)$.

2) $R(1)(\mathrm{Sing}(J,b))$ is empty or a smooth hypersurface in W.

3) The blow-up of W at this hypersurface is the identity map, but (J,b) undergoes a transformation

$$\begin{array}{ccc} W & \xleftarrow{\pi} & W_1 \\ (J,b) & & (J_1,b) \end{array}$$

which eliminates $R(1)(\mathrm{Sing}(J,b))$ (eliminates the components of codimension one).

PROOF. Note that, locally, $R(1)(\mathrm{Sing}(J,b)) = \overline{W}_\lambda$ in the previous discussion, and the strict transform becomes empty when blowing up at the same $\overline{W}_\lambda$. □

We formulate the next result in terms of basic objects and of resolution of basic objects (see 1.23). So replace (2.4.7) by

$$(W,(J,b),E=\emptyset) \longleftarrow (W_1,(J_1,b),E_1) \longleftarrow \cdots \longleftarrow (W_r,(J_r,b),E_r), \tag{2.4.11}$$

where we assume again that (J,b) is a simple couple, and consider the open covering and notation as in Lemma 2.28. So for each index λ there is an open set U_λ of W and a smooth hypersurface $\overline{W}_\lambda$, and (2.4.11) defines

$$(\overline{W}_\lambda,\overline{E}_\lambda=\emptyset) \longleftarrow (\overline{W}^{(1)}_\lambda,\overline{E}^{(1)}_\lambda) \longleftarrow \cdots \longleftarrow (\overline{W}^{(r)}_\lambda,\overline{E}^{(r)}_\lambda), \tag{2.4.12}$$

and also, by restriction to U_λ, sequences

(2.4.13)

$$(U_\lambda,(J_\lambda,b),E_\lambda=\emptyset) \longleftarrow (U^1_\lambda,((J_\lambda)_1,b),(E_\lambda)_1) \longleftarrow \cdots \longleftarrow ((W_\lambda)_r,((J_\lambda)_r,b),(E_\lambda)_r).$$

PROPOSITION 2.31. *Fix $J \subset \mathcal{O}_W$ with maximum order b. Assume that* $\mathrm{Sing}(J,b)$ *has no component of codimension one in W. There is an open covering, say $\{U_\lambda\}_{\lambda\in\Lambda}$ of W, and for each index λ a closed and smooth hypersurface $\overline{W}_\lambda \subset U_\lambda$, and a couple $(K^{(0)}_\lambda,d)$ with $K^{(0)}_\lambda \subset \mathcal{O}_{\overline{W}_\lambda}$ such that, in addition to* **P1)** *and* **P2)** *in Lemma 2.28, the following property holds:*

P3) *The sequence (2.4.12) defined by (2.4.11) as above induces a sequence of transformations*

(2.4.14)

$$(\overline{W}_\lambda,((K_\lambda),d),\overline{E}=\emptyset) \longleftarrow (\overline{W}^{(1)}_\lambda,((K_\lambda)_1,d),\overline{E}^{(1)}_\lambda) \longleftarrow \cdots \longleftarrow (\overline{W}^{(r)}_\lambda,((K_\lambda)_r,d),\overline{E}^{(r)}_\lambda)$$

and

$$\mathrm{Sing}((J_\lambda)_r,b) = \mathrm{Sing}((K_\lambda)_r,d)(\subset \overline{W}^{(r)}_\lambda). \tag{2.4.15}$$

REMARK 2.32. *On the converse.* Set $W = U_\lambda$ so that $(J,b) = (J_\lambda,b)$. The equality in (2.4.15) asserts, by induction on r, that any sequence (2.4.14) induces a sequence (2.4.13). Furthermore, if (2.4.14) is a resolution, so is (2.4.13).

We are mainly interested in this converse, since we will argue by increasing induction on the dimension of the ambient space. If we accept, by induction, that there is a resolution, say (2.4.14), for each index λ, then we will have defined a resolution, say (2.4.13), for each λ. We still need to go one step further, namely that these local resolutions patch to a resolution, say (2.4.11), of (J,b).

2.33. **Main Claim.** Resolution of $d-1$-dimensional basic objects can be defined so that the resolutions (2.4.13), induced by $d-1$-dimensional resolutions (2.4.14), patch over the different open set U_λ of W to define a resolution, say (2.4.11), of the simple basic object $(W,(J,b),E=\emptyset)$. Furthermore, this resolution of the d-dimensional simple basic object is independent both of the choice of the open covering and the choices of the smooth hypersurface $\overline{W}_\lambda$.

Full details of the proof of Proposition 2.31 will be given in Chapter 4; however, the following example illustrates the basic idea of the proof.

EXAMPLE 2.34. In Example 2.10 we considered the case $W = A_k^3 = \mathrm{Spec}(k[X,Y,Z])$ and

$$J = \langle Z^3 + XY^2Z + X^5 \rangle,$$

an ideal of maximum order $b=3$. In such an example we noted that $Z \in \Delta^2(J) = \langle Z, XY, Y^2, X^3 \rangle$, and we considered the smooth hypersurface $\overline{W} = V(\langle Z \rangle)$. This is a particular example of Lemma 2.28, where there is no need to consider the open covering $\{U_\lambda\}_{\lambda\in\Lambda}$ of W. In fact here the lemma applies globally in W. In this example $b! = 6$, and Proposition 2.31 applies by setting $K = J^*$ as in (1.2.2).

A similar situation holds, more generally, in 1.18 for $K = J^* = \langle c_i^{\frac{b!}{i}}, i = 2,3,\dots,b \rangle$.

REMARK 2.35. The compatibility of the Δ operator with open restrictions will play a role in our proof of Proposition 2.31, and this will allow us to define the ideals K_λ.

There is yet another context in which there is a natural compatibility of the operator Δ, which are not open restrictions, but will also play a role in the proof of Proposition 2.31. In fact, set $W \longleftarrow W_1 = W \times A_k^1$ where A_k^1 denotes the affine line and the map is the projection on the first coordinate. Note that if J is an ideal in $\mathscr{O}_W$ and if Δ_1 denotes the operator on the smooth scheme W_1, then

$$\Delta_1(J\mathscr{O}_{W_1}) = \Delta(J)\mathscr{O}_{W_1}.$$

Note that a covering $\{U_\lambda\}_{\lambda\in\Lambda}$ of W induces a covering of W_1 by pull-back. The setting of Proposition 2.25 and the inclusions (2.4.3) are compatible with pull-backs, and so are the setting of Proposition 2.26 and the inclusions (2.4.6). This will guarantee the compatibility of all our development for this particular kind of projection.

CHAPTER 3

On inductive and resolution functions

Here we sketch the main ideas and invariants involved in our definition of resolution functions in 1.25, which will lead us to the simple proofs of the main theorems in 1.26 and 1.27. There we claimed that for a fixed integer d, there is a totally ordered set $(I^d, \geq)$, and for any basic object $B : (W, (J, b), E)$ with a dimension of $W = d$, an upper semi-continuous function $f_B : \mathrm{Sing}(J, b) \to I^d$ can be defined with the property that $\underline{\mathrm{Max}}\, f_B$ is a smooth subscheme of $\mathrm{Sing}(J, b)$ and a permissible center for the pair (W, E). So this defines a transformation of the basic object by choosing $\underline{\mathrm{Max}}\, f_B$ as center .

Furthermore, a unique sequence of transformations (1.3.6) was defined inductively, by setting centers $\underline{\mathrm{Max}}\, f_{B_i}$. In addition, this sequence defined by the functions will be a resolution of the basic object. In fact, we require that for some index s (depending on B), $\mathrm{Sing}(J_s, b) = \emptyset$. In other words, the resolution functions define for any basic object $(W_0, (J_0, b), E_0 = \emptyset)$ a resolution

$$(W_0, (J_0, b), E_0) \longleftarrow (W_1, (J_1, b), E_1) \longleftarrow \cdots \longleftarrow (W_s, (J_s, b), E_s) \tag{3.0.1}$$

together with functions $f_i : \mathrm{Sing}(J_i, b) \to I^d$, so that each $W_i \leftarrow W_{i+1}$ is a monoidal transformation with center $\underline{\mathrm{Max}}\, f_i$.

Note first that $f_i : \mathrm{Sing}(J_i, b) \to I^d$ defines a stratification of $\mathrm{Sing}(J_i, b)$ into locally closed sets (the level sets of f_i). The totally ordered set I^d will have coordinates for each dimension $\leq d$. So we will present first the function defined by the d-dimensional components of f_i, or rather, the coordinate corresponding to dimension d. The main ingredient of this d-dimensional component is what we call the inductive function.

We will present the values of f_i by decreasing induction on the dimension d. At dimension d we define an upper-semi-continuous function with the property that the set of points where the function takes the maximum value is either a smooth center or the points where the function takes the maximum value is the closed set attached to a simple basic object. Recall here that, over fields of characteristic zero, a simple basic object can be expressed in terms of $d-1$-dimensional basic objects, as discussed in the previous chapter. This is how induction enters into the definition of the resolution function. Here we will assume the existence of resolution of simple basic objects, which we study in Chapter 4.

The values of the d-dimensional function will be defined in terms of a very natural factorization of J_i, namely

$$J_i = I(H_1)^{b_1} I(H_2)^{b_2} \cdots I(H_s)^{b_s} \cdot \overline{J}_i \tag{3.0.2}$$

so that $\overline{J}_s$ does not vanish along H_i, $1 \leq i \leq s$. We shall consider (J_i, b) together with this factorization of J_i.

The strategy is to define a sequence of transformations, say (3.0.1), so that $\overline{J}_s = \mathscr{O}_{W_s}$ in (3.0.2). In fact, in such a case, called the monomial case in Remark 3.3, it is simple to extend (3.0.1) to a resolution. With this purpose in mind we define the function w-ord, with values in the positive integers, and which takes maximal value zero exactly in the monomial case. In the monomial case we can define the d-dimensional function so that the maximum value is already a smooth center. In fact, in the monomial case the full resolution function is defined entirely by the d-dimensional function, since a resolution of the monomial case is a simple combinatorial choice of centers, which does not require any form of induction on the dimension of the ambient space.

On the other hand, if the maximum value of w-ord is greater then zero (if we are not in the monomial case), we introduce a second function t_d, which we call the inductive function. It has the property that the closed set where t_d takes its maximum value is the closed set defined by a simple basic object, as indicated above. Simple basic objects have been studied in the previous section; recall that the components of codimension one of the singular locus are a smooth hypersurface and a permissible center (see Corollary 2.30). Also in this case we define the full resolution function f_i essentially as the d-dimensional inductive function. In this case the blow-up at this hypersurface induces a transformation, and the maximum value of t_d strictly drops. Finally, if the singular locus has no component of codimension one, which we may also assume by Corollary 2.30, then the simple basic object is locally like a $d-1$-dimensional basic object, and the resolution of this basic object forces the maximum value of t_d to drop (see Proposition 2.31). In this way we will be able to define a unique transformation, which will lead us to the monomial case. In other words, this inductive function t_d, together with induction on resolution problems, will lead us to the monomial case, and hence to resolution. The resolution functions will be defined so as to follow this strategy.

3.1. Back to the main theorems

Recall the notion of a permissible sequence of transformations of pairs, say

$$\begin{array}{ccccccc} W_0 & \longleftarrow & (W_1, E_1 = \{H_1\}) & \longleftarrow & \cdots & \longleftarrow & (W_k, E_k = \{H_1, H_2, \ldots, H_k\}), \\ Y & & Y_1 & & & & \end{array}$$

in which we require that each monoidal transformation $W_i \leftarrow W_{i+1}$ be defined so that all exceptional hypersurfaces introduced have normal crossings (Prop. 1.9).

Given $J \subset \mathcal{O}_W$, there is an expression of the total transform (Def. 1.8), say

$$J\mathcal{O}_{W_k} = I(H_1)^{a_1} I(H_2)^{a_2} \cdots I(H_k)^{a_k} \cdot \mathcal{A}_k.$$

This factorization is unique if we require the a_i to be the highest possible exponents in any such expression. In 1.26 we want to achieve $\mathcal{A}_k = \mathcal{O}_{W_k}$ with the conditions stated in Theorem 1.3. This requires a suitable sequence of monoidal transformations, and each transformation will introduce an exceptional hypersurface. This will lead us to consider a pair $(W, E = \{H_1, \ldots, H_r\})$, rather than simply W, and also permissible transformations of pairs

$$(W, E) \leftarrow (W_1, E_1) \leftarrow \cdots \leftarrow (W_k, E_k) \tag{3.1.1}$$

always in the conditions of Prop. 1.9.

In 1.23 we have defined a basic object as a couple (J, b), $J \subset \mathcal{O}_W$, together with a pair (W, E). Recall that a sequence of transformations, say

$$(W, (J, b), E) \leftarrow (W_1, (J_1, b), E_1) \leftarrow \cdots \leftarrow (W_k, (J_k, b), E_k), \tag{3.1.2}$$

is a sequence of transformations of couples, say

$$\begin{array}{ccccccc} W & \leftarrow & W_1 & \leftarrow & \cdots & \leftarrow & W_k \\ (J, b) & & (J_1, b) & & & & (J_k, b) \end{array}$$

(see (1.3.2)), which also defines a sequence of transformations of pairs, as in (3.1.1), and that (3.1.2) is a resolution of $(W,(J,b),E)$ if $V(\Delta^{b-1}(J_k)) = \emptyset$. Note that $V(\Delta^{b-1}(J_k)) = \emptyset$ is equivalent to $\mathrm{Sing}(J_k, b) = \emptyset$, and also to the condition $\max - \mathrm{ord}_{J_k} < b$ (maximum value achieved by the function ord_{J_k}) (see (2.4.1)).

So the resolution would provide an expression of the form

$$J\mathscr{O}_{W_k} = I(H_1)^{a_1} I(H_2)^{a_2} \cdots I(H_k)^{a_k} \cdot J_k, \text{ and } \max - \mathrm{ord}_{J_k} = b' < b.$$

If $b' = 0$ we have achieved embedded principalization (Theorem 1.3). If not we repeat the argument and try to produce a resolution of (J_k, b') and (W_k, E_k). It is clear that ultimately we come to the case $b' = 0$.

Our task is to produce a resolution of (J,b) and (W,E) in some explicit manner, in which centers of monoidal transformations are defined by an upper-semi-continuous function. In some particular cases this will be clear from the data involved (see Remark 3.3). But, in general, the strategy will be to reduce to the case in which $b = \max - \mathrm{ord}_J$, namely to the case of simple couples (2.21).

In case $b = \max - \mathrm{ord}_J$, Proposition 2.31 says that there is $\overline{W} \subset W$, at least locally, and that a resolution (3.1.2) induces

(3.1.3)

$$(\overline{W},(K,d),\overline{E} = \emptyset) \quad \leftarrow \quad (\overline{W}_1,(K_1,d),\overline{E}_1) \quad \leftarrow \quad \cdots \quad \leftarrow \quad (\overline{W}_k,(K_k,d),\overline{E}_k)$$

such that $V(\Delta^{d-1}(K_k)) = \emptyset$. It is important to point out here that we will argue by induction, and hence we would like to reverse the argument, namely, to define (3.1.2) in terms of (3.1.3). We now indicate the difficulties to overcome.

The three difficulties for an inductive argument:

D1) (K,d) encodes information of (J,b), but not of the set of hypersurfaces E in W. Proposition 2.31 asserts that, after restriction to an open subset of W, (3.1.3) will define a sequence of transformations of couples, say

$$\begin{array}{ccccccc} W & \leftarrow & W_1 & \leftarrow & \cdots & \leftarrow & W_k \\ (J,b) & & (J_1,b) & & & & (J_k,b), \end{array}$$

such that $V(\Delta^{b-1}(J_k)) = \emptyset$. However this sequence might not define a sequence (3.1.2). In fact, it might not be permissible over (W,E) because of the presence of hypersurfaces of E.

This is an important difficulty to overcome. As indicated above, since we will argue in steps, we introduce hypersurfaces with normal crossings (those in E) and we want to preserve this condition of normal crossings in all exceptional hypersurfaces to be introduced in forthcoming steps.

D2) The couple (K,d) might not be a simple couple (might not be such that $d = \max - \mathrm{ord}_K$). Take for example the case $J = \langle z^3 - x^2y^2 \rangle$ and the couple $(J,3)$ in the affine 3-space. Clearly $3 = \max - \mathrm{ord}_J$, so the couple is simple. Since $z \in \Delta^2(J)$, we may take $\overline{W}$ as the affine plane, and $(K,d) = (\langle x^2y^2 \rangle, 3)$. Note that $\max - \mathrm{ord}_K = 4$, so that (K,d) is not a simple couple (2.21).

D3) If (J,b) is a simple couple (i.e. if $\max - \mathrm{ord}_j = b$), then $\overline{W}$ is defined by choosing, locally at a point $x \in V(\Delta^{b-1}(J))$, an element of order one in $\Delta^{b-1}(J)_x$ (see (2.4.6)). In general this choice is not unique, and the definition of (K,d) ($K \subset \mathscr{O}_{\overline{W}}$) is only local at x. Our form of induction should provide a resolution 3.1.2, with independence of open restrictions and of choices of $\overline{W}$.

3.1. Set $J \subset \mathcal{O}_W$ and (W,E) as before. Assume, in accordance with D2), that $b \geq \max-\mathrm{ord}_J$. So here (J,b) might not be simple. Consider a sequence of transformations, say

$$(3.1.4) \qquad (W,(J,b),E) \;\leftarrow\; (W_1,(J_1,b),E_1) \;\leftarrow\; \cdots \;\leftarrow\; (W_s,(J_s,b),E_s).$$

This provides a factorization of J_s, say

$$(3.1.5) \qquad J_s = I(H_1)^{b_1} I(H_2)^{b_2} \cdots I(H_s)^{b_s} \cdot \overline{J}_s$$

so that $\overline{J}_s$ does not vanish along H_i, $1 \leq i \leq s$. As indicated before, we consider (J_s,b) together with this factorization of J_s. This extra structure on (J_s,b) will allow us to overcome D2), namely to reduce the general case to the case of simple couples.

EXAMPLE 3.2. Set $J = \langle x_1, x_2^2\rangle^4$, $W = \mathbb{A}_k^2$

$$\begin{array}{ccccc} (W,(J,3),E=\emptyset) & \leftarrow & (W_1,(J_1,3),E_1=\{H_1\}) & \leftarrow & (W_2,(J_2,3),E_2=\{H_1,H_2\}) \\ & & J\mathcal{O}_{W_1} = I(H_1)^4 \cdot \mathcal{M}_p^4 & & J_1\mathcal{O}_{W_2} = I(H_1)^2 \cdot I(H_2)^6 \\ & & J_1 = I(H_1)^2 \mathcal{M}_p^4 & & J_2 = I(H_1)^2 \cdot I(H_2)^3. \end{array}$$

Here $W \leftarrow W_1$ is the blow-up at $0 \in \mathbb{A}_k^2$, $p \in W_1$ is a point in the exceptional line H_1, $\mathcal{M}_p$ is the sheaf of functions that vanish at p, and finally $W_1 \leftarrow W_2$ is the blow-up at p.

REMARK 3.3. If $\overline{J}_s = \mathcal{O}_{W_s}$ in (3.1.5), we say that (J_s,b) is within the *monomial case*. In this case it is easy to extend (3.1.4) to a resolution, namely, to define for some $k \geq s$:

(3.1.6)

$$(W,(J,b),E) \;\leftarrow \cdots \;\leftarrow\; (W_s,(J_s,b),E_s) \;\leftarrow\; \cdots \;\leftarrow\; (W_k,(J_k,b),E_k)$$

so that $V(\Delta^{b-1}(J_k)) = \emptyset$. The following example illustrates this fact. Note that in the previous example $\overline{J}_2 = \mathcal{O}_{W_r}$.

EXAMPLE 3.4. Consider transformations with centers Y_j:

$$\begin{array}{ccc} (W_2,(J_2,3),E_2=\{H_1,H_2\}) & \overset{id}{\leftarrow} & (W_3,(J_3,3),E_3=\{H_1,H_2\}) \\ J_2 = I(H_1)^2 \cdot I(H_2)^3 & & J_3 = I(H_1)^2 \cdot I(H_2) \\ Y_2 = H_2 & & Y_3 = H_1 \cap H_2 \end{array}$$

$$\begin{array}{cc} \leftarrow & (W_4,(J_4,3),E_4=\{H_1,H_2,H_4\}) \\ & J_3 = I(H_1)^2 \cdot I(H_4)^0 \cdot I(H_2) \\ & V(\Delta^2(J_4)) = \emptyset. \end{array}$$

The first transformation is defined with center at the hypersurface H_2. So the morphism is the identity map, but here we take $H_2 \in E_2$ to be the exceptional locus. Note that J_3 is not J_2.

3.2. On the function w-ord

Given a sequence of transformations of basic objects, say (3.1.4), we have defined an expression

$$J_s = I(H_1)^{b_1} I(H_2)^{b_2} \cdots I(H_s)^{b_s} \cdot \overline{J}_s$$

so that $\overline{J}_s$ does not vanish along H_i, $1 \leq i \leq s$. Now define

$$\text{w-ord}_s^d : \quad \mathrm{Sing}(J_s,b) = V(\Delta^{b-1}(J_s)) \;\to\; \mathbb{Q}$$

$$\text{w-ord}_s^d(x) = \frac{\nu_{\overline{J}_s}(x)}{b},$$

where $\nu_{\overline{J}_s}(x)$ denotes the order of $\overline{J}_s$ at $\mathcal{O}_{W_s,x}$. Compare with (2.4.1) and note that:
1) The function is upper-semi-continuous. In particular, $\underline{\mathrm{Max}}$ w-ord is closed.

2) For any index $i \leq s$, there is an expression

$$J_i = I(H_1)^{b_1} I(H_2)^{b_2} \cdots I(H_i)^{b_i} \cdot \overline{J}_i,$$

and hence a function w-ord$_i : V(\Delta^{b-1}(J_i)) \to \mathbb{Q}$ can also be defined.

Another property of these functions is:

3) If each step $(W_i, E_i) \leftarrow (W_{i+1}, E_{i+1})$ in (3.1.4) is defined with center $Y_i \subset \underline{\mathrm{Max}}$ w-ord$_i$, then

$$\max \text{w-ord} \geq \max \text{w-ord}_1 \geq \cdots \geq \max \text{w-ord}_s .$$

If max w-ord$_s = \frac{d}{b}$, and $d \geq b$, set

$$(J'_s, b') = (\overline{J}_s, d). \tag{3.2.1}$$

If max w-ord$_s = \frac{d}{b}$, and $d < b$, set

$$(J'_s, b') = (J^d + (\overline{J}_s)^b, bd). \tag{3.2.2}$$

One can check that if max w-ord$_s = \frac{d}{b}$, then $\underline{\mathrm{Max}}$ w-ord$_s = \mathrm{Sing}(J'_s, b')$.

Property 3) follows from the fact that $\mathrm{Sing}(J'_s, b') \subset \mathrm{Sing}(\overline{J_s}, d)$, and $(\overline{J_s}, d)$ is, by definition, a simple couple.

EXAMPLE 3.5. Set $(J,1)$, $J = \langle x^2 - y^5 \rangle$, and $W = \mathbb{A}^2_k$. Let C denote the curve defined by J.

$$\begin{array}{ccccc} (W,(J,1),E=\emptyset) & \leftarrow & (W_1,(J_1,1),E_1=\{H_1\}) & \leftarrow & (W_2,(J_2,1),E_2=\{H_1,H_2\}) \\ & & J_1 = I(H_1)\cdot I(C') & & J_2 = I(H_1)^1 \cdot I(H_2)^1 \cdot I(C'') \\ Y = 0 \in \mathbb{A}^2_k & & Y_1 = C' \cap H_1 & & Y_2 = H_1 \cap H_2 \end{array}$$

$$\begin{array}{cc} \leftarrow & (W_3,(J_3,1),E_3) \\ & J_3 = I(H_1)^1 \cdot I(H_2)^1 \cdot I(H_3)^2 \cdot I(C'''). \end{array}$$

Here the Y_i are the centers of the monoidal transformations, and C', C'', and C''' are strict transforms of C. In this example,

$$\max \text{w-ord}_J = 2\ ,\ \max \text{w-ord}_{J_1} = 1\ , \max \text{w-ord}_{J_2} = 1, \max \text{w-ord}_{J_3} = 1,$$

and the sequence is defined by setting

$$Y = \underline{\mathrm{Max}}\,\text{w-ord}_J, Y_1 = \underline{\mathrm{Max}}\,\text{w-ord}_{J_1}\ , \text{and } Y_2 = \underline{\mathrm{Max}}\,\text{w-ord}_{J_2}\ .$$

REMARK 3.6. Closely related to the function w-ord^d_s defined above in terms of the expression $J_s = I(H_1)^{b_1} I(H_2)^{b_2} \cdots I(H_s)^{b_s} \cdot \overline{J}_s$ is the function

$$ord^d_{(J_s,b)} : \quad \mathrm{Sing}(J_s, b) = V(\Delta^{b-1}(J_s)) \quad \to \quad \mathbb{Q} \geq 1$$

$$ord^d_{(J_s,b)}(x) = \frac{\nu_{J_s}(x)}{b},$$

where $\nu_{J_s}(x)$ denotes the order of J_s at $\mathcal{O}_{W_s,x}$.

3.3. On the inductive function t

Consider, as before, a sequence

$$(W,(J,b),E) \quad\leftarrow\quad (W_1,(J_1,b),E_1) \quad\leftarrow\quad \cdots \quad\leftarrow\quad (W_s,(J_s,b),E_s), \tag{3.3.1}$$

where each $W_i \leftarrow W_{i+1}$ is defined with center $Y_i \subset \underline{\mathrm{Max}}\,\text{w-ord}_i$, so that

$$\max \text{w-ord} \geq \max \text{w-ord}_1 \geq \cdots \geq \max \text{w-ord}_s .$$

We now define a function t, only under the assumption that $\max \text{w-ord}_s > 0$. In fact, $\max \text{w-ord}_s = 0$ in the monomial case, which is easy to resolve.

Set $s_0 \leq s$ such that

$$\max \text{w-ord} \geq \cdots \geq \max \text{w-ord}_{s_{0-1}} > \max \text{w-ord}_{s_0} = \max \text{w-ord}_{s_{0+1}} = \cdots = \max \text{w-ord}_s,$$

and set

$$E_s = E_s^+ \sqcup E_s^- \text{ (disjoint union)},$$

where E_s^- are the strict transforms of hypersurfaces in E_{s_0}. Define

$$t_s : V(\Delta^{b-1}(J_s)) \to \mathbb{Q} \times \mathbb{N} \text{ (ordered lexicographically)},$$
$$t_s(x) = (\text{w-ord}_s(x), n_s(x)),$$
$$n_s(x) = \sharp\{H_i \in E_s^-, / x \in H_i\},$$

where $\sharp S$ denotes the total number of elements of a set S.

One can check that:

1) the function is upper-semi-continuous. In particular, $\underline{\mathrm{Max}}\,t_s$ is closed.

2) There is a function t_i for any index $i \leq s$.

Example 3.7 illustrates the following properties which also hold for this function:

• If each $(W_i,E_i) \leftarrow (W_{i+1},E_{i+1})$ in 3.3.1 is defined with center $Y_i \subset \underline{\mathrm{Max}}\,t_i$, then

$$\max t \geq \max t_1 \geq \cdots \geq \max t_s.$$

• If $\max t_s = (\frac{d}{b}, r)$ (here $\max \text{w-ord}_s = \frac{d}{b}$), then $\underline{\mathrm{Max}}\,t_s \subset \underline{\mathrm{Max}}\,\text{w-ord}_s$.

• If $\underline{\mathrm{Max}}\,t_s$ is of pure codimension 1 in W_s, then it is smooth. Moreover, in such a case $Y_s = \underline{\mathrm{Max}}\,t_s$ is a permissible center, defining

$$(W_s,(J_s,b),E_s) \quad\leftarrow\quad (W_{s+1},(J_{s+1},b),E_{s+1}),$$

and $\max t_s > \max t_{s+1}$ (hence $\max \text{w-ord}_s \geq \max \text{w-ord}_{s+1}$).

EXAMPLE 3.7.

0) Consider $(J,1)$, $J = \langle x^2 - y^3 \rangle$ defining a curve $C \subset W = \mathbb{A}_k^2$. Here $t(x) = (1,0)$ at any $x \in C$ except at $\overline{0} \in \mathbb{A}_k^2$, $t(\overline{0}) = (2,0)$. So

$$\max t = (2,0) \text{ and } \underline{\mathrm{Max}}\,t = \overline{0} \in \mathbb{A}_k^2. \text{ Now let}$$

$$(W,(J,1),E=\emptyset) \quad\leftarrow\quad (W_1,(J_1,1),E_1 = \{H_1\})$$

be the quadratic transformation at $\overline{0} \in \mathbb{A}_k^2$.

1) Let $C' \subset W_1$ denote the strict transform of C. Here

$$J_1 = I(H_1) \cdot \overline{J}_1,$$

where $\overline{J}_1 = I(C')$ and $t_1(x) = (1,0)$ at any $x \in C'$ except for $p = C' \cap H_1$, where $t_1(p) = (1,1)$. So

$$\max t_1 = (1,1) \text{ and } \underline{\mathrm{Max}}\,t_1 = p.$$

Set

$$(W_1,(J_1,1),E_1) \quad \leftarrow \quad (W_2,(J_2,1),E_2=\{H_1,H_2\})$$

with center at $p \in W_1$.

2) If $C'' \subset W_2$ denotes the strict transform of C, then

$$J_2 = I(H_1)\cdot I(H_2)\cdot \overline{J}_2$$

where $\overline{J}_2 = I(C'')$.

Now $t_2(x) = (1,0)$ at any $x \in C''$ except for $q = C'' \cap H_1 \cap H_2$, where $t_2(q) = (1,1)$. So

$$\max t_2 = (1,1) \text{ and } \underline{\mathrm{Max}}\, t_2 = q.$$

Set

$$(W_2,(J_2,1),E_2=\{H_1,H_2\}) \quad \leftarrow \quad (W_3,(J_3,1),E_3=\{H_1,H_2,H_3\})$$

with center at $q \in W_2$.

3) Now

$$J_3 = I(H_1)\cdot I(H_2)\cdot I(H_3)^2 \cdot \overline{J}_3,$$

where $\overline{J}_3 = I(C''')$ (ideal of the strict transform). Finally, check that $t_3(x) = (1,0)$ at any $x \in C'''$. So

$$\max t_3 = (1,0) \text{ and } \underline{\mathrm{Max}}\, t_3 = C'''.$$

This is a case in which $\underline{\mathrm{Max}}\, t$ has codimension 1. Note that $\underline{\mathrm{Max}}\, t_3$ is a smooth hypersurface, and the blow-up at $\underline{\mathrm{Max}}\, t_3$ defines a permissible transformation (the identity map)

$$(W_3,(J_3,1),E_3=\{H_1,H_2,H_3\}) \quad \leftarrow \quad (W_3,(J_4,1),E_3=\{H_1,H_2,H_3\})$$

with $J_4 = I(H_1)\cdot I(H_2)\cdot I(H_3)^2$.

3.8. **Overcoming difficulties D1) and D2)**

We finally indicate a further property of the function t_s, which leads to constructive desingularization by induction. To this end, set

(3.3.2) $$(W,(J,b),E) \quad \leftarrow \quad (W_1,(J_1,b),E_1) \quad \leftarrow \quad \cdots \quad \leftarrow \quad (W_s,(J_s,b),E_s)$$

so that

$$\max \text{w-ord} \geq \max \text{w-ord}_1 \geq \cdots \geq \max \text{w-ord}_s\,.$$

And define, as before, the function

$$t_s : V(\Delta^{b-1}(J_s)) \to \mathbb{Q}\times\mathbb{N}.$$

This last property can be stated as follows:

There is a couple (J''_s, b'') (see 3.9) with the following properties:

• $V(\Delta^{b''-1}(J'')) = \underline{\mathrm{Max}}\, t_s$, and $\max \mathrm{ord}_{J''_s} = b''$ (i.e. (J''_s, b'') is a simple couple; see 2.20).

• Any resolution

(3.3.3)
$$(W_s,(J''_s,b''),E''_s=\emptyset) \quad \leftarrow \quad (W_{s+1},(J''_{s+1},b''),E''_{s+1}) \quad \leftarrow \cdots \leftarrow \quad (W_{s+k},(J''_{s+k},b''),E''_{s+k})$$

(i.e. $\mathrm{Sing}(J''_{s+k}, b'') = \emptyset$) induces an extension of (3.3.2), say

(3.3.4)
$$(W_s,(J_s,b),E_s) \quad \leftarrow \quad (W_{s+1},(J_{s+1},b),E_{s+1}) \quad \leftarrow \quad \cdots \quad \leftarrow \quad (W_{s+k},(J_{s+k},b),E_{s+k}),$$

such that

$$\max t_s = \max t_{s+1} = \cdots = \max t_{s+k-1} > \max t_{s+k}.$$

Furthermore

$$\underline{\mathrm{Max}}\, t_{s+i} = \mathrm{Sing}(J''_{s+i}, b'')$$

for $i = 0, \cdots, k-1$.

The major simplification here is due to the fact that in (3.3.3) $E''_s = \emptyset$, whereas E_s might be non-empty, and also to the fact that (J''_{s+k}, b'') is a simple couple. In fact, at least locally, we can find a smooth hypersurface $\overline{W}_s$ containing $V(\Delta^{b''-1}(J''))$ and (K,d) $(K \subset \mathscr{O}_{\overline{W}_s})$ as in Proposition 2.31, so that (3.3.3) is equivalent to a resolution, say

(3.3.5)

$$(\overline{W}, (K,d), \overline{E}_s = \emptyset) \leftarrow (\overline{W}_1, (K_1,d), \overline{E}_{s+1}) \leftarrow \cdots \leftarrow (\overline{W}_k, (K_k,d), \overline{E}_{s+k})$$

(i.e. $V(\Delta^{d-1}(K_k)) = \emptyset$). So, at least locally, the sequence (3.3.4), with the previous properties, is induced by (3.3.5). And we may assume the existence of a resolution (3.3.5) by induction.

3.9. Fix the sequence (3.3.1) and E_s^- as in 3.3. Given a positive integer r, let G_r be the set of all subsets $\mathscr{F} \subset E_s^-$, $\mathscr{F} = \{H_{j_1}, \ldots, H_{j_r}\}$ (with r hypersurfaces). For each positive integer e, define

$$\mathscr{H}_r(e) = \sum_{\mathscr{F} \in G_r} \sum_{H_{i_j} \in \mathscr{F}} I(H_{i_j})^e.$$

Set $\max t_s = (\frac{d}{b}, r)$.

If $d \geq b$, set $(J'', b'') = (J'_s + \mathscr{H}_r(d), d)$, with J'_s as in (3.2.1).

If $d < b$, set $(J'', b'') = (J'_s + \mathscr{H}_r(bd), bd)$, with J'_s as in (3.2.2).

Note that b'' is the same as b' in (3.2.1) and (3.2.2). Note also the (J'', b'') is simple, and $\mathrm{Sing}(J'', b'') = \underline{\mathrm{Max}}\, t_s \subset \mathrm{Sing}(J', b') = \underline{\mathrm{Max}}\, \text{w-ord}_s$.

3.10. **Example of constructive resolution.**

EXAMPLE 3.11. The curve C defined by $J = \langle x^2 - y^5 \rangle$ in $W = \mathbb{A}^2_k$ is irreducible, in particular reduced. We attach to it the basic object

$$(W, (J,1), E = \emptyset)$$

and the function

$$t : V(\Delta^0(J)) = V(J) \to \mathbb{Q} \times \mathbb{N}.$$

Here $t(x) = (1,0)$ except at the origin $\overline{0} \in \mathbb{A}^2_k$, $t(\overline{0}) = (2,0)$.

Note that in Example 3.5:

- $\max t = (2,0)$ and $Y = \underline{\mathrm{Max}}\, t = \overline{0}$;
- $\max t_1 = (1,1)$ and $Y_1 = \underline{\mathrm{Max}}\, t_1$;
- $\max t_2 = (1,1)$ and $Y_2 = \underline{\mathrm{Max}}\, t_2$
- $\max t_3 = (1,0)$ and $\underline{\mathrm{Max}}\, t_3 = C'''$ is a smooth hypersurface (see 3.8). Thus, this defines an embedded desingularization.

Compare with the proof of Theorem 1.2.

EXAMPLE 3.12. The hypersurface $Z^2 + X^2 + Y^3 = 0$ is irreducible with an isolated singularity at $\overline{0} \in \mathbb{A}^3_k$. Set $W = \mathbb{A}^3_k$, $J = \langle Z^2 + X^2 + Y^3 \rangle$. According to the proof of Theorem 1.2 in 1.27, desingularization is achieved at some intermediate step of the resolution of the basic object

$$(W, (J,1), E = \emptyset).$$

The function $t : V(J) \to \mathbb{Q} \times \mathbb{N}$ takes value $t(x) = (1,0)$ except at the singular point, $t(\overline{0}) = (2,0)$. In this case, and following the notation in 3.8:

- $\max t = (2,0)$.
- (J'', b'') can be defined as $(J, 2)$.
- $\overline{W} = V(\langle Z \rangle)$ (in fact $Z \in \Delta^1(J)$).
- (K,d) can be defined by $(\langle X^2 + Y^3 \rangle, 2)$.

Here $\overline{W} = \mathbb{A}^2_k$, and the blow-up at $\overline{0} \in \mathbb{A}^2_k$ defines a resolution, namely

$$(\overline{W}, (K,2), E = \emptyset) \quad \longleftarrow \quad (\overline{W}_1, (K_1, 2), E_1 = \{\overline{H}_1\})$$

and $V(\Delta(K_1)) = \emptyset$. According to 3.8, this defines

$$(W, (J,1), E = \emptyset) \quad \longleftarrow \quad (W_1, (J_1, 1), E_1 = \{H_1\}) \qquad \text{and } \max t > \max t_1.$$

In fact, $\max t_1 = (1,1)$. So again, we argue as in 3.8 and attach a couple (J'', b'') to the value $\max t_1 = (1,1)$. Moreover, a smooth hypersurface $\overline{W}$ and a couple (K,d) can be defined so that a resolution, say:

(3.3.6)
$$(\overline{W}, (K,d), \overline{E}_s = \emptyset) \leftarrow (\overline{W}_1, (K_1, d) \overline{E}_{s+1}) \leftarrow \cdots \leftarrow (\overline{W}_k, (K_k, d), \overline{E}_{s+k})$$

(such that $V(\Delta^{d-1}(K_k)) = \emptyset$), induces

(3.3.7) $$(W_1, (J_1, 1), E_1) \leftarrow (W_2, (J_2, 1), E_2) \leftarrow \cdots \leftarrow (W_s, (J_s, 1), E_s)$$

such that

$$(1,1) = \max t_1 = \max t_2 = \cdots = \max t_{s-1} > \max t_s = (1,0).$$

Note that J_s is the sheaf of ideals of the strict transform of the hypersurface and that $\underline{\mathrm{Max}}\, t_s = V(J_s)$. So $\underline{\mathrm{Max}}\, t_s$ is a hypersurface, and the last property in 3.3 says that this is an embedded desingularization.

3.4. From inductive to resolution functions

3.13. The strong property of the inductive function t in 3.3 is that it reduces resolution of basic objects to resolution of simple basic objects, namely those of the form $(W, (J,b), E)$, where (J,b) is a simple couple, and $E = \emptyset$ (3.8). Resolution of simple basic objects will follow by induction, essentially from Proposition 2.31 and Corollary 2.30, as we shall prove in this last part.

The argument in 3.8 shows that if we assume resolution of simple basic objects, then we can force $\max t$ to drop. As the second coordinate of $\max t$ can drop only finitely many times, at some point the first coordinate must drop, and ultimately we come to the monomial case. Finally, the monomial case is simple to resolve (3.3).

Set $J \subset \mathcal{O}_W$ and (W,E) as before, with W smooth of dimension d, and consider a sequence of transformations, say:

(3.4.1) $$(W, (J,b), E) \leftarrow (W_1, (J_1, b), E_1) \leftarrow \ldots \leftarrow (W_s, (J_s, b), E_s)$$

which we consider together with the natural factorization

(3.4.2) $$J_s = I(H_1)^{b_1} I(H_2)^{b_2} \cdots I(H_s)^{b_s} \cdot \overline{J}_s$$

so that $\overline{J}_s$ does not vanish along H_i, $1 \leq i \leq s$.

Recall that $\max \text{w-ord} = 0$ if and only if $\overline{J}_s = \mathcal{O}_{W_s}$. In such case, it is not hard to define a totally ordered set Γ and an upper-semi-continuous function $h_s : \mathrm{Sing}(J_s, b) \to \Gamma$, so that $\underline{\mathrm{Max}}\, h_s$ is smooth and that a resolution be achieved by blowing up successively at these centers. In this case no form of induction on the dimension of the ambient space is required (see Remark 3.3 or Chapter 5 in **[20]** for a precise formulation of h).

DEFINITION 3.14. Set $T^d = \{\infty\} \sqcup (\mathbb{Q}\times\mathbb{Z}) \sqcup \Gamma$, where this disjoint union is totally ordered by setting ∞ as the biggest element, and that $\alpha < \beta$ if $\beta \in (\mathbb{Q}\times\mathbb{Z})$ and $\alpha \in \Gamma$. We now set $I_d = T^d \times I_{d-1}$ ordered lexicographically, and define $f_r^d : \mathrm{Sing}(J_r, b) \to I_d$.

(i) If $\max \text{w-ord}_s = 0$: $f_s^d(x) = (h(x), \infty_{d-1})$.

(ii) If $\max \text{w-ord}_s > 0$, then it will be enough to define the function for $x \in \underline{\mathrm{Max}}\, t_s \subset \mathrm{Sing}(J_s, b)$; in fact, we will define the resolution function, and hence the resolution sequence, so that centers be included in $\underline{\mathrm{Max}}\, t_s$. Let (J'', b'') be the simple basic object attached to $\max t_s$ (defining the closed set $\underline{\mathrm{Max}}\, t_s$), and finally set:

A) $f_r^d(x) = (\max t_r, \infty_{d-1})$ if $x \in R(1)(\mathrm{Sing}(J'', b''))$ (see 2.30).

B) $f_r^d(x) = (\max t_r, \overline{f}_r^{d-1}(x))$, if $x \notin R(1)(\mathrm{Sing}(J'', b''))$, where $\overline{f}_r^{d-1}(x)$ is defined in accordance with the $d-1$-dimensional basic object attached to (J'', b''). The precise definition in this case B) will be discussed below.

3.15. **The notion of birth and uniqueness of the sequence.**

The previous definition leads to a unique sequence 3.4.1, for some index s, so that $\max \text{w-ord} \geq \max \text{w-ord}_1 \geq \cdots \geq \max \text{w-ord}_s$.

If $\max \text{w-ord}_s = 0$, define the *birth* to be the smallest index $s' \leq s$ for which $\max \text{w-ord}_{s'} = 0$. In such a case, we assume that

$$(W_{s'}, (J_{s'}, b), E_{s'}) \leftarrow (W_{s'+1}, (J_{s'+1}, b), E_{s'+1}) \leftarrow \cdots \leftarrow (W_s, (J_s, b), E_s)$$

is defined by blowing up at $\underline{\mathrm{Max}}\, h_j$, $s' \leq j \leq s$, and we will extend 3.4.1 to a constructive resolution by continuing to blow up at $\underline{\mathrm{Max}}\, h_s$.

If $\max \text{w-ord}_s > 0$, then we may assume that $\max t \geq \max t_1 \geq \cdots \geq \max t_s$. Define the *birth* to be the smallest index $r' \leq s$ for which

$$\max t_{r'} = \max t_{r'+1} = \cdots = \max t_s.$$

At level r' let (J'', b'') be as in 3.9, and define a basic object by setting $(W_{r'}, (J'', b''), E = \emptyset)$.

Recall that (J'', b'') is a simple couple. Corollary 2.30 indicates a canonical choice of center, given by $R(1)(\mathrm{Sing}(J'', b''))$: the components of codimension one of $\mathrm{Sing}(J'', b'')$. Furthermore, Corollary 2.30 together with Proposition 2.31 show that after blowing up at $R(1)(\mathrm{Sing}(J'', b''))$ (if not empty), then (J'', b'') can be replaced (at least locally) by couples in a smaller-dimensional smooth scheme.

We assume that

$$(W_{r'}, (J_{r'}, b), E_{r'}) \leftarrow (W_{r'+1}, (J_{r'+1}, b), E_{r'+1}) \leftarrow \cdots \leftarrow (W_s, (J_s, b), E_s)$$

is defined by blowing first at $R(1)(\mathrm{Sing}(J'', b''))$, if not empty, and then successively at $\underline{\mathrm{Max}}\, \overline{f}_r^{d-1}$, where $\overline{f}_r^{d-1}(x)$ is defined by the $d-1$-dimensional basic object attached to (J'', b''). In such a case, 3.4.1 will be extended to a resolution of (J'', b'') defined by continuing to blow up at $\underline{\mathrm{Max}}\, \overline{f}_j^{d-1}$, say:

$$(W_s, (J_s, b), E_s) \leftarrow (W_{s+1}, (J_{s+1}, b), E_{s+1}) \leftarrow \cdots \leftarrow (W_v, (J_v, b), E_v).$$

In this way we get that $\max t_s = \cdots = \max t_{v-1} > \max t_v$.

Recall that $\max t$ can drop only finitely many times, after which we must come to the case of $\max \text{w-ord} = 0$ (or to a resolution).

3.16. The definition of the set I_d and of the resolution functions $f_s^d : \mathrm{Sing}(J_s, b) \to I_d$ will follow from the previous discussion. As the set I_d is totally ordered and lexicographic on the dimension, one can check that resolution of d-dimensional basic objects is in fact achieved by blowing up at the closed sets $\underline{\mathrm{Max}}\, \overline{f}_j^d$.

For any index s, and $x \in \operatorname{Sing}(J_s, b)$, the value $f_s^d(x)$ is either of the form

$$((\text{w-ord}_s^d(x), n_s^d(x)), (\text{w-ord}_s^{d-1}(x), n_s^{d-1}(x)), \dots, (\text{w-ord}_s^j(x), n_s^j(x)), \infty^{j-1}, \infty^{j-2}, \dots, \infty^1)$$

(all $(\text{w-ord}_s^i(x), n_s^i(x)) \in (\mathbb{Q} \times \mathbb{Z})$,
or of the form

$$((\text{w-ord}_s^d(x), n_s^d(x)), (\text{w-ord}_s^{d-1}(x), n_s^{d-1}(x)), \dots,$$
$$(\text{w-ord}_s^{j+1}(x), n_s^{j+1}(x)), h^j(x), \infty^{j-1}, \infty^{j-2}, \dots, \infty^1),$$

with only one value $h^j(x) \in \Gamma$.

3.5. Main features of the resolution functions

The key for constructive resolution is the definition of a totally ordered set, and an assignment of an upper-semi-continuous function from the singular locus on this ordered set. This is the strategy of the previously defined functions on the singular locus of basic objects.

Such functions also stratify the singular locus by the level sets; this stratification also deserves attention.

3.17. **The two steps of the procedure**

The outcome of our algorithm has two different and distinguishable steps.

Step 1): Given a basic object $(W, (J, b), E)$, the algorithm defines first a sequence of transformation

$$(W, (J, b), E) \leftarrow (W_1, (J_1, b), E_1) \leftarrow \dots \leftarrow (W_s, (J_s, b), E_s) \tag{3.5.1}$$

so that

$$J_s = I(H_1)^{b_1} I(H_2)^{b_2} \cdots I(H_s)^{b_s}. \tag{3.5.2}$$

In this first step, the ideal J is replaced by J_s, where locally at any point of W_s this ideal is generated by a monomial on a suitable regular system of coordinates at such point. So we call this first step *monomialization* of the basic object.

This first step is achieved because of the inductive function t. Recall that if the dimension of W is d, the sequence is defined so as to come to the case of $\max \text{w-ord}_s^d = 0$, which is the case of local monomialization.

Step 2): (*Resolution of locally monomial basic objects*). It is reasonable that in this case there be no form of induction required in the definition of the functions $h : \operatorname{Sing}(J_s, b) \to \Gamma$.

We have not drawn special attention to the definition of this function; however, it plays an important role, particularly because it also appears in Step 1. Recall that in that step the invariants involved in the resolution function are defined by decreasing induction on the dimension. The next main feature will clarify to what extent the function h (i.e. resolution of the locally monomial case) affects the procedure of Step 1).

3.18. We finally address the full description of the resolution function. Assume as before that W is smooth of dimension d, and $(W, (J, b), E)$ is a basic object. It follows from the discussion in 3.16 that for each index r in the resolution sequence, the value $\max f_r^d$ has the first coordinates in $\mathbb{Q} \times \mathbb{Z}$.

Let e be a positive integer such that the first e coordinates of $\max f_r^d$, say $(\max f_r^d)_e$, are all in $\mathbb{Q} \times \mathbb{Z}$ (i.e. that $(\max f_r^d)_e \in (\mathbb{Q} \times \mathbb{Z})^e$).

We define the e-birth of $\max f_r^d$ as the smallest index $r'(\leq r)$ for which

$$(\max f_{r'}^d)_e = (\max f_r^d)_e.$$

Let $F(r,e)$ be the set of points where the first e coordinates of f_r^d take this highest value $(\max f_r^d)_e$. As the function is upper-semi-continuous and I^d is ordered lexicographically, one can check that $F(r,e)$ is closed in $\mathrm{Sing}(J_r,b)$ and that $(\max f_{r'}^d)_e = (\max f_{r'+1}^d)_e = \cdots = (\max f_r^d)_e$. It follows from our construction that:

1) Locally at a point $x \in F(r',e)$ there are smooth schemes $W_{r'} \supset (\overline{W}^1)_{r'} \supset (\overline{W}^2)_{r'} \supset \cdots (\overline{W}^{e-1})_{r'}$, where $(\overline{W}^1)_{r'}$ is a hypersurface in $W_{r'}$ (after open restriction), and each $(\overline{W}^{i+1})_{r'}$ is a smooth hypersurface in $(\overline{W}^i)_{r'}$. It is convenient to set $W_{r'} = (\overline{W}^0)_{r'}$; note that for $e=1$ the only member of the flag is $W_{r'}$.

2) All centers of monoidal transformations between the levels r' and r are included in the strict transform of this flag, inducing, in particular, a sequence

$$(\overline{W}^{e-1})_{r'} \leftarrow (\overline{W}^{e-1})_{r'+1} \cdots \leftarrow (\overline{W}^{e-1})_r.$$

3) There is a simple couple $(J'_{r'},b')$, $J'_{r'} \subset \mathscr{O}_{(\overline{W}^{e-1})_{r'}}$ attached to the value $(\max f_{r'}^d)_e$, so that the previous sequence induces permissible transformations

(3.5.3)

$$((\overline{W}^{e-1})_{r'},(J'_{r'},b'),\overline{E}_{r'}=\emptyset) \leftarrow ((\overline{W}^{e-1})_{r'+1},(J'_{r'+1},b'),\overline{E}_{r'+1}) \leftarrow \cdots \leftarrow ((\overline{W}^{e-1})_r(J'_r,b'),\overline{E}_r)$$

and

$$F(j,e) = \mathrm{Sing}(J'_j,b')$$

for $r' \leq j \leq r$.

4) Sequence (3.5.3) is defined so that the first transformation has center at the set of points where $\mathrm{Sing}(J'_{r'},b')$ has codimension one at $(\overline{W}^{e-1})_{r'}$, if such components exist, and then in accordance with a longer flag, say $W_{r'} \supset (\overline{W}^1)_{r'} \supset (\overline{W}^2)_{r'} \supset \cdots (\overline{W}^{e-1})_{r'} \supset (\overline{W}^e)_{r'}$ and a couple $(K'_{r'},d')$, where both $(\overline{W}^e)_{r'}$ and $(K'_{r'},d')$ are defined in terms of $(J'_{r'},b')$ as in Proposition 2.31. In fact, the proposition applies because $((\overline{W}^{e-1})_{r'},(J'_{r'},b'),\overline{E}_{r'}=\emptyset)$ is a simple basic object on the smooth scheme $(\overline{W}^{e-1})_{r'}$.

Note that the previous properties were discussed in 3.8 (and in 3.15) for the case $e=1$, and in fact these properties follow from the case $e=1$ together with induction.

3.19. **Canonical choice of centers.** Another main feature of the resolution function is the choice of centers at each level r. The discussion in 3.15 ensures that $\max f_r^d$ has the first coordinates in $(\mathbb{Q}\times\mathbb{Z})$ followed by coordinates which are ∞, or $\max f_r^d$ has one (and only one) coordinate in Γ. So there are two possibilities:

A) $\max f_r^d = (\alpha_1,\alpha_2,\dots,\alpha_e,\infty,\infty,\dots,\infty)$, $\alpha_i \in (\mathbb{Q}\times\mathbb{Z})$.

B) $\max f_r^d = (\alpha_1,\alpha_2,\dots,\alpha_e,\gamma,\infty,\infty,\dots,\infty)$, $\alpha_i \in (\mathbb{Q}\times\mathbb{Z})$, $\gamma\in\Gamma$.

We consider in both cases the first e-coordinates, say $(\max f_r^d)_e = (\alpha_1,\alpha_2,\dots,\alpha_e)$, and let r' denote the e-birth of $(\max f_r^d)_e$.

In case A), $r'=r$; namely, the e-birth is at the same level r. This case occurs when the simple basic object $((\overline{W}^{e-1})_r,(J'_r,b'),\overline{E}_{r'}=\emptyset)$, which was defined above in 3), is such that $\mathrm{Sing}(J'_j,b') = R(1)(\mathrm{Sing}(J'_j,b'))$ (locally at a point).

In this case $F(r,e) = \mathrm{Sing}(J'_j,b') = R(1)(\mathrm{Sing}(J'_j,b'))$, and this is precisely the smooth center defined by the resolution function (see Corollary 2.30).

In case B), $r'<r$, and there is a simple basic object, namely $((\overline{W}^{e-1})_{r'},(J'_{r'},b')\overline{E}_{r'} = \emptyset)$, and $R(1)(\mathrm{Sing}(J'_{r'},b'))) = \emptyset$ at an open neighborhood.

There is a smooth hypersurface $(\overline{W}^e)_{r'}$ in the smooth scheme $(\overline{W}^{e-1})_{r'}$, and a couple $(K^{(r')}, d)$ with $K^{(r')} \subset \mathcal{O}_{(\overline{W}^e)_{r'}}$, such that (3.5.3) induces a permissible sequence

(3.5.4)
$$((\overline{W}^e)_{r'}, (K_{r'}, d), \overline{E}_{r'} = \emptyset) \leftarrow ((\overline{W}^e)_{r'+1}, (K_{r'+1}, d), \overline{E}_{r'+1}) \leftarrow \cdots \leftarrow ((\overline{W}^e)_r (K_r, d), \overline{E}_r)$$

and
$$\mathrm{Sing}(K_j, d) = \mathrm{Sing}(J'_j, b')$$
for $r' \leq j \leq r$ (see Proposition 2.31).

In this case $((\overline{W}^e)_r (K_r, d), \overline{E}_r)$ is locally monomial. The maximal value of h along $\mathrm{Sing}(K_r, d)$ is γ, and $\underline{\mathrm{Max}}\, h$ is a smooth subscheme of $(\overline{W}^e)_r$. Note that this center is also smooth in W_r, and this will be the center of the resolution function in case B).

CHAPTER 4

Globalization of local reductions

In the previous chapter we defined resolution functions, which already lead to the proof of the main theorems, as indicated at the end of Chapter 1. However, in defining resolution functions we assumed that simple basic objects could be resolved by induction on the dimension of the ambient space. This form of induction is only local, so up to induction we may assume there is a constructive resolution of the restriction of the basic object to open sets (in an open cover). It remains to prove that these locally defined resolutions globalize to a resolution of the basic object. The objective in this last chapter is to present the proof of this globalization formulated in Proposition 2.31, at least in the formal context. We also outline, in the final sections, an alternative proof of resolution of simple basic objects, different from that in Proposition 2.31, using the notions of elimination algebras.

The key ingredients for both approaches are the notion of Rees algebras and differential algebras. They turn out to be natural reformulations of the notions of couples and basic objects.

In Hironaka's definition of idealistic exponents, two ideals are equivalent when they have the same integral closure. In the context of Rees algebras the natural analog will be that two algebras will be equivalent if they have the same integral closure, now in the sense of algebras. "Equivalent" means that they will be treated similarly in the procedure of transformations and resolutions to be defined.

4.1. Rees algebras over smooth schemes

4.1. We now define a *Rees algebra* over a smooth scheme W to be a graded noetherian subring of the polynomial algebra $\mathscr{O}_W[T]$, say

$$\mathscr{G} = \bigoplus_{k\geq 0} I_k T^k,$$

where $I_0 = \mathscr{O}_W$ and each I_k is a sheaf of ideals. We assume that at any affine open set $U \subset W$, there is a finite set

$$\mathscr{F} = \{f_1 T^{n_1}, \dots, f_s T^{n_s}\},$$

$n_i \geq 1$ and $f_i \in \mathscr{O}_W(U)$, so that the restriction of $\mathscr{G}$ to U is

$$\mathscr{O}_W(U)[f_1 T^{n_1}, \dots, f_s T^{n_s}] (\subset \mathscr{O}_W(U)[T]).$$

To a Rees algebra $\mathscr{G}$ we attach a closed set

$$\mathrm{Sing}(\mathscr{G}) := \{x \in W / \nu_x(I_k) \geq k, \text{ for any } k \geq 1\},$$

where $\nu_x(I_k)$ denotes the order of the ideal I_k at the local regular ring $\mathscr{O}_{W,x}$.

REMARK 4.2. Rees algebras are related to Rees rings. A Rees algebra is a Rees ring if, for any affine open set $U \subset W$ and $\mathscr{F} = \{f_1 T^{n_1}, \dots, f_s T^{n_s}\}$ as above, all degrees n_i are one.

Rees algebras are integral closures of Rees rings defined in a suitable way. In fact, if N is a positive integer divisible by all n_i, it is easy to check that

$$\mathcal{O}_W(U)[f_1T^{n_1},\dots,f_sT^{n_s}] = \oplus_{r\geq 0} I_r(U)T^r (\subset \mathcal{O}_W(U)[T])$$

is integral over the Rees sub-ring $\mathcal{O}_W(U)[I_NT^N](\subset \mathcal{O}_W(U)[T^N])$.

PROPOSITION 4.3. *Given an affine open $U \subset W$, and $\mathcal{F} = \{f_1T^{n_1},\dots,f_sT^{n_s}\}$ as above,*

$$\mathrm{Sing}(\mathcal{G}) \cap U = \cap_{1\leq i\leq s} \mathrm{Sing}(f_i, n_i).$$

PROOF. It is clear that $\nu_x(f_i) \geq n_i$ for $x \in \mathrm{Sing}(\mathcal{G})$, $0 \leq i \leq s$. So

$$\mathrm{Sing}(\mathcal{G}) \cap U \subset \cap_{1\leq i\leq s} \mathrm{Sing}(f_i, n_i).$$

On the other hand, for any index $N \geq 1$, we have that $I_N(U)T^N$ is generated by elements of the form $G_N(f_1T^{n_1},\dots,f_sT^{n_s})$, where $G_N(Y_1,\dots,Y_s) \in \mathcal{O}_U[Y_1,\dots,Y_s]$ is weighted homogeneous of degree N, provided each Y_j has weight n_j. The reverse inclusion is now clear. □

4.4. A monoidal transformation is said to be *permissible* for $\mathcal{G}$ if $Y \subset \mathrm{Sing}(\mathcal{G})$. In such a case, for each index $k \geq 1$, there is a sheaf of ideals, say $I_k^{(1)} \subset \mathcal{O}_{W_1}$, so that

$$I_k\mathcal{O}_{W_1} = I(H)^k I_k^{(1)}.$$

One can easily check that

$$\mathcal{G}_1 = \bigoplus_{k\geq 0} I_k^{(1)}T^k$$

is a Rees algebra over W_1, which we call the *transform* of $\mathcal{G}$.

Let $\mathcal{G} = \bigoplus_{k\geq 0} I_kT^k$ be a Rees algebra on W, $U \subset W$ an affine open set, and let

$$\mathcal{F} = \{f_1T^{n_1},\dots,f_sT^{n_s}\}$$

be such that the restriction of $\mathcal{G}$ to U is

$$\mathcal{O}_W(U)[f_1T^{n_1},\dots,f_sT^{n_s}](\subset \mathcal{O}_W(U)[T]).$$

PROPOSITION 4.5. *Let $W \leftarrow W_1$ be a permissible transformation of $\mathcal{G}$. There is an open covering of $\pi^{-1}(U)$ by affine sets $U^{(l)}$ so that:*
1) $\langle f_i\rangle = I(H\cap U^{(l)})^{n_i}\langle f_i'\rangle$ for suitable $f_i' \in \mathcal{O}_{W_1}(U^{(l)})$.
2) The restriction of $\mathcal{G}_1$ to $U^{(l)}$ is

$$\mathcal{O}_{W_1}(U^{(l)})[f_1'T^{n_1},\dots,f_s'T^{n_s}](\subset \mathcal{O}_{W_1}(U^{(l)})[T]).$$

PROOF. 1) follows from Prop 4.3. For 2), argue as in the proof of Prop 4.3 by using the fact that each ideal I_N is generated by weighted homogeneous polynomials on the element of $\mathcal{F}$.

4.2. Resolution functions on Rees algebras and integral closures

4.6. In the previous chapter we studied resolution functions on basic objects $(W,(J,b),E)$. Here we will extend the inductive function in 3.3, now as a function on the class of Rees algebras. The point is that the two invariants involved in the definition of the inductive function t will be well defined in the class of Rees algebras and will allow us to consider Rees algebras up to integral closure. In other words, we will see that two Rees algebras with the same integral closure will have the same singular locus and the same function t.

Define the function ord for Rees algebras by setting, for $\mathscr{G} \subset \mathscr{O}_W[T]$,

$$\mathrm{ord}_{\mathscr{G}} : \mathrm{Sing}(\mathscr{G}) \to \mathbb{Z} \qquad \mathrm{ord}_{\mathscr{G}}(x) = \inf_{n\geq 1}\{\frac{\nu(I_n)}{n}\},$$

where ν denotes the order at the local regular ring $\mathscr{O}_{W,x}$.

If $U \subset W$ is an affine open neighborhood of x, and if $\mathscr{F} = \{f_1T^{n_1},\dots,f_sT^{n_s}\}$ generate $\mathscr{G}$ (i.e. if $\mathscr{O}_W(U)[f_1T^{n_1},\dots,f_sT^{n_s}] = \oplus_{r\geq 0} I_rT^r$), then $\mathrm{ord}(x) = min\{\frac{\nu(f_i)}{n_i}, 1 \leq i \leq s\}$ (argue as in the proof of Prop 4.3). In particular, if N is any positive integer divisible by all n_i, then $\mathrm{ord}_{\mathscr{G}}(x) = \frac{\nu(I_N)}{N}$. As W can be covered by a finite number of affine open sets, one can always choose an integer N so that $\mathrm{Sing}(\mathscr{G}) = \mathrm{Sing}(J_N,N)$, and $\mathrm{ord}_{\mathscr{G}}$ is independent of the previous choice of N.

There is a natural way to embed couples in the class of Rees algebras. Given $J \subset \mathscr{O}_W$ and a couple (J,b), define

$$\mathscr{G}_{(J,b)} = \mathscr{O}_W[JT^b], \tag{4.2.1}$$

the Rees algebra generated by JT^b, as a subalgebra of $\mathscr{O}_W[T]$. In this way

$$\mathrm{Sing}(J,b) = \mathrm{Sing}(\mathscr{G}_{(J,b)}) \tag{4.2.2}$$

and

$$\mathrm{ord}_{(J,b)} = \mathrm{ord}_{\mathscr{G}_{(J,b)}}, \tag{4.2.3}$$

where the left-hand side is as in 3.6. Furthermore, such equalities are preserved by transformations. In other words, a sequence of transformations of couples over (J,b), say

$$\begin{array}{ccccccc} W & \xleftarrow{\pi_1} & W_1 & \xleftarrow{\pi_2} & \dots & \xleftarrow{\pi_k} & W_k \\ (J,b) & & (J_1,b) & & & & (J_k,b), \end{array} \tag{4.2.4}$$

induces a sequence of transformations of the Rees algebra $\mathscr{G} = \mathscr{G}_{(J,b)}$:

$$\begin{array}{ccccccc} W & \xleftarrow{\pi_1} & W_1 & \xleftarrow{\pi_2} & \dots & \xleftarrow{\pi_k} & W_s, \\ \mathscr{G} & & (\mathscr{G})_1 & & & & (\mathscr{G})_s \end{array} \tag{4.2.5}$$

and it turns out that $\mathscr{G}_{(J_i,b)} = (\mathscr{G})_i$ for each positive index $i \leq s$. In particular

$$\mathrm{Sing}((J_i,b)) = \mathrm{Sing}(\mathscr{G}_i).$$

In the setting of basic objects we also take into account a pair (W,E), with W smooth of dimension d, and instead of a sequence (4.2.4), we take permissible sequences of the form

$$(W,(J,b),E) \leftarrow (W_1,(J_1,b),E_1) \leftarrow \dots \leftarrow (W_s,(J_s,b),E_s), \tag{4.2.6}$$

which we consider together with the natural factorization

$$J_s = I(H_1)^{b_1} I(H_2)^{b_2} \cdots I(H_s)^{b_s} \cdot \overline{J}_s \tag{4.2.7}$$

so that $\overline{J}_s$ does not vanish along H_i, $1 \leq i \leq s$. In fact, the function w-ord$_s$ is defined in terms of this expression (see 3.2).

It is important to check that if we replace $(W,(J,b),E)$ by $(W,(J^2,2b),E)$, or more generally by any basic object of the form $(W,(J^r,rb),E)$ for some positive integer r, then (4.2.6) is permissible for the new basic object, and the singular loci at all levels $i \leq s$ are the same. Furthermore, we also claim that the functions w-ord$_s$ are the same. To check this last assertion, note that for any index $0 \leq i \leq s$ in (4.2.6), $\mathrm{ord}_{(J_i,b)} = \mathrm{ord}_{(J_i^r,rb)}$ as functions on $\mathrm{Sing}(J_i,b) = \mathrm{Sing}(J_i^r,rb)$ (see 3.6). And both functions w-ord$_s$: $\mathrm{Sing}(J_s,b) \to \mathbb{Z}$ and w-ord$_s$: $\mathrm{Sing}(J_s^r,b) \to \mathbb{Z}$ can be defined in terms of the previous functions (see [**20**], p. 187).

Consider now the Rees algebra $\mathscr{G} = \mathscr{G}_{(J,b)}$; then (4.2.6) induces:

$$(4.2.8) \qquad \begin{array}{ccccccc} (W,E) & \stackrel{\pi_1}{\longleftarrow} & (W_1,E_1) & \stackrel{\pi_2}{\longleftarrow} & \cdots & \stackrel{\pi_k}{\longleftarrow} & (W_s,E_s) \\ \mathscr{G} & & (\mathscr{G})_1 & & & & (\mathscr{G})_s \end{array}$$

and define at $\mathrm{Sing}((J_i,b)) = \mathrm{Sing}(\mathscr{G}_i)$ the function

$$(4.2.9) \qquad \text{w-ord}_{\mathscr{G}_i} = \text{w-ord}_{(J_i,b)}$$

for $0 \leq i \leq s$.

It is clear that the second coordinate of the inductive function in 3.3, and hence the inductive function itself, also extends naturally to the case of Rees algebras.

4.7. We now claim that the resolution function in 1.25 extends to the class of Rees algebras, namely to the case in which couples (J,b) are replaced by Rees algebras $\mathscr{G}$. The argument in proving so is similar to that of the previous discussion. Fix an affine open set $U \subset W$ as above, so that $\mathscr{F} = \{f_1T^{n_1},\dots,f_sT^{n_s}\}$ generate $\mathscr{G}$ (i.e. $\mathscr{O}_W(U)[f_1T^{n_1},\dots,f_sT^{n_s}] = \oplus_{r\geq 0} I_rT^r$), and let N be divisible by all n_i. In such a case, $(f_iT^{n_i})^{\frac{N}{n_i}} \in I_NT^N$. So the restriction of $\mathscr{G}$ to U is a finite extension of the Rees *ring* of the ideal I (i.e. of $\mathscr{O}_W(U)[I_NT^N]$). As W can be covered by finitely many affine open sets we may assume that, for a suitable integer b, $\mathscr{G} = \oplus I_rT^r$ is finite over the subalgebra $\mathscr{O}_W[I_bT^b]$.

One can check, from the local description in terms of generators, that setting $I_b = J$, then $\mathrm{Sing}(J,b) = \mathrm{Sing}(\mathscr{G})$, and any sequence (4.2.6) induces a sequence (4.2.5), and $\mathrm{Sing}(J_i,b) = \mathrm{Sing}(\mathscr{G}_i)$. In particular, if (4.2.6) is a resolution, then so is (4.2.5).

In order to extend the resolution functions in 1.25 to Rees algebras, we now prove that the resolution of $(W,(J,b),E=\emptyset)$ is independent of the choice of b with the previous conditions. Let b and b' be two positive integers so that $\mathscr{G} = \oplus I_kT^k$ is finite over the subalgebras $\mathscr{O}_W[I_bT^b]$ and $\mathscr{O}_W[I_{b'}T^{b'}]$. We claim that the algorithmic resolution of $(W,(J,b),E=\emptyset)$ is the same as that of $(W,(J',b'),E=\emptyset)$, where $J = I_b$ and $J' = I_{b'}$.

A particular feature of Rees rings of ideals, say $\oplus_{k\geq 0} I^kT^k$, is that the homogeneous component of degree k of the integral closure is the integral closure of the ideal I^k.

In our case the Rees algebra $\mathscr{G}$ is a finite extension of both Rees rings; in particular, both Rees rings have the same integral closure in $\mathscr{O}_W[T]$. The components of these two Rees rings at degree bb' are $(I_b)^{b'}T^{bb'}$ and $(I_{b'})^bT^{bb'}$, respectively. Therefore, $(I_b)^{b'}$ and $(I_{b'})^b$ have the same integral closure, and the claim follows finally from property **P5)** in 1.28.

The previous discussion shows that two Rees algebras with the same integral closure undergo the same constructive resolution, as the singular locus and resolution functions are the same.

4.3. Differential algebras and Kollár's tuned ideals

4.8. Here W is smooth over a field k, so for each non-negative integer r there is a locally free sheaf of differential operators of order r, say Diff_k^r.

DEFINITION 4.9. We say that a Rees algebra $\bigoplus I_nT^n$ is a Diff-algebra relative to the field k if:

i) $I_n \supset I_{n+1}$ for $n \geq 0$.

ii) There is open covering of W by affine open sets $\{U_i\}$, and for any $D \in \mathrm{Diff}^{(r)}(U_i)$ and any $h \in I_n(U_i)$, $D(h) \in I_{n-r}(U_i)$, provided $n \geq r$.

Let $\bigoplus I_nT^n$ be any Rees algebra, let $U \subset W$ be an affine open set, and let $\mathscr{F} = \{f_1T^{n_1}, \dots, f_sT^{n_s}\}$ generate $\mathscr{G}$ over U (i.e. $\mathscr{O}_W(U)[f_1T^{n_1}, \dots, f_sT^{n_s}] = \oplus_{r\geq 0} I_r(U)T^r$). In principal, condition i) will not hold; however, it will be achieved if we enlarge the Rees algebra by considering the algebra generated over U by $\mathscr{F}' = \{f_1T^{n'_1}, \dots, f_sT^{n'_s}\} | 0 \leq n'_i \leq n_i\}$. One can also check that this extension of Rees algebras is finite. Recall that we consider Rees algebras up to integral closure.

The formulation of (ii) shows that, given a sheaf of ideals $I \in \mathscr{O}_W$, there is an extension, say $\mathrm{Diff}^{(r)}(I)$. With this notation, (ii) can be reformulated by

ii') $\mathrm{Diff}^{(r)}(I_n) \subset I_{n-r}$ for each n, and $0 \leq r \leq n$.

Fix a closed point $x \in W$ and a regular system of parameters $\{x_1, \dots, x_n\}$ at $\mathscr{O}_{W,x}$. The residue field, say k', is a finite extension of k, and the completion $\hat{\mathscr{O}}_{W,x} = k'[[x_1, \dots, x_n]]$.

The Taylor development is the continuous k'-linear ring homomorphism

$$Tay : k'[[x_1, \dots, x_n]] \to k'[[x_1, \dots, x_n, T_1, \dots, T_n]]$$

that maps x_i to $x_i + T_i$, $1 \leq i \leq n$. So for $f \in k'[[x_1, \dots, x_n]]$, $Tay(f(x)) = \sum_{\alpha \in \mathbb{N}^n} g_\alpha T^\alpha$, with $g_\alpha \in k'[[x_1, \dots, x_n]]$.

Define, for each $\alpha \in \mathbb{N}^n$, $\Gamma^\alpha(f) = g_\alpha$. It turns out that each Γ^α maps $\mathscr{O}_{W,x}$ into itself (i.e. $\Gamma^\alpha(\mathscr{O}_{W,x}) \subset \mathscr{O}_{W,x}$), and that $\{\Gamma^\alpha, \alpha \in (\mathbb{N})^n, 0 \leq |\alpha| \leq c\}$ generate the $\mathscr{O}_{W,x}$-module $\mathrm{Diff}_k^c(\mathscr{O}_{W,x})$ (i.e. generate the $\mathscr{O}_W$-module $\mathrm{Diff}_k^c(W)$ locally at x).

In the particular case in which k is a field of characteristic zero, it is well known that

$$\Gamma^\alpha = (\frac{1}{\alpha_1!} \cdots \frac{1}{\alpha_n!}) \frac{\partial^{\alpha_1}}{\partial^{\alpha_1} X_1} \cdots \frac{\partial^{\alpha_n}}{\partial^{\alpha_n} X_n}. \tag{4.3.1}$$

So over fields of characteristic zero, $\mathrm{Diff}^{(r)}(I) = \Delta^r(I)$ (see 2.12).

REMARK 4.10. Recall that, up to integral closure, any Rees algebra is of the form $\mathscr{G} = \mathscr{G}_{(J,b)}$. We are to consider $\mathscr{G}$ at a smooth scheme W together with a set of closed and smooth hypersurfaces E having only normal crossings. We have defined permissible transformations of such data:

$$\begin{array}{ccccccc} (W,E) & \xleftarrow{\pi_1} & (W_1,E_1) & \xleftarrow{\pi_2} & \dots & \xleftarrow{\pi_k} & (W_s,E_s). \\ \mathscr{G} & & (\mathscr{G})_1 & & & & (\mathscr{G})_s \end{array} \tag{4.3.2}$$

This sequence of transformations induces transformations of basic objects:

$$(W,(J,b),E) \leftarrow (W_1,(J_1,b),E_1) \leftarrow \dots \leftarrow (W_s,(J_s,b),E_s), \tag{4.3.3}$$

which we consider together with the natural factorization

$$J_s = I(H_1)^{b_1} I(H_2)^{b_2} \cdots I(H_s)^{b_s} \cdot \overline{J}_s \tag{4.3.4}$$

so that $\overline{J}_s$ does not vanish along H_i, $1 \leq i \leq s$. In fact the function w-ord$_s$ is defined in terms of this expression.

It is sometimes convenient to consider differential operators on W_s, that are, in some natural sense, compatible with such factorization, namely, the so-called differential operators with logarithmic poles along the hypersurfaces of E_s, say Diff_{E_s} . These differential operators arise naturally in the study of the toroidal structure on W_s defined by E_s. To obtain a local description, fix a closed point $x \in \mathrm{Sing}(\mathscr{G}_s) = \mathrm{Sing}(J_s, b)$, and assume that $H_1, \dots, H_r$ are the hypersurfaces in E_s that contain x.

Take a regular system of parameters $\{x_1, \dots, x_n\}$ at $\mathscr{O}_{W,x}$ so that $\langle x_i \rangle = I(H_i)$ for $1 \leq i \leq r$. If k' denotes the residue field, the completion is $\hat{\mathscr{O}}_{W,x} = k'[[x_1, \dots, x_n]]$. We now consider the continuous k'-linear ring homomorphism, say

$$Tay_{E_s} : k'[[x_1, \dots, x_n]] \to k'[[x_1, \dots, x_n, T_1, \dots, T_n]]$$

that map x_i to $x_i + x_i T_i$, $1 \leq i \leq r$, and x_j to $x_j + T_j$, $r+1 \leq j \leq n$. So for $f \in k'[[x_1, \dots, x_n]]$, $Tay_{E_s}(f(x)) = \sum_{\alpha \in \mathbb{N}^n} g_\alpha T^\alpha$, with $g_\alpha \in k'[[x_1, \dots, x_n]]$.

Define, for each $\alpha \in \mathbb{N}^n$, $\Gamma^\alpha_{E_s}(f) = g_\alpha$. It turns out that each $\Gamma^\alpha_{E_s}$ maps $\mathscr{O}_{W,x}$ into itself (i.e. $\Gamma^\alpha_{E_s}(\mathscr{O}_{W,x}) \subset \mathscr{O}_{W,x}$), and that $\{\Gamma^\alpha_{E_s}, \alpha \in (\mathbb{N})^n, 0 \leq |\alpha| \leq c\}$ generate the locally free sheaf of $\mathscr{O}_{W_s,x}$-module $\mathrm{Diff}^c_{E_s}$.

One can also define, for any ideal K in $\mathscr{O}_{W_s}$ and any index c, an extension, say $K \subset \mathrm{Diff}^c_{E_s}(K)$, by taking the ideal spanned by differential operators of order c at elements of K. The main property of these differential operators is that $K = \mathrm{Diff}^c_{E_s}(K)$ whenever K is an invertible sheaf of ideals supported on E_s. So for instance, in our case in which $J_s = I(H_1)^{b_1} I(H_2)^{b_2} \cdots I(H_s)^{b_s} \cdot \overline{J}_s$, we get $\mathrm{Diff}^c_{E_s}(J_s) = I(H_1)^{b_1} I(H_2)^{b_2} \cdots I(H_s)^{b_s} \cdot \overline{J}'_s$ and $\overline{J}_s \subset \overline{J}'_s$.

THEOREM 4.11. *For any Rees algebra $\mathscr{G}$ over a smooth scheme W, there is a Diff-algebra, say $G(\mathscr{G})$, such that:*

i) $\mathscr{G} \subset G(\mathscr{G})$.

ii) If $\mathscr{G} \subset \mathscr{G}'$ and $\mathscr{G}'$ is a Diff-algebra, then $G(\mathscr{G}) \subset \mathscr{G}'$.

Furthermore, if $x \in W$ is a closed point, $\{x_1, \dots, x_n\}$ is a regular system of parameters at $\mathscr{O}_{W,x}$, and $\mathscr{G}$ is locally generated by

$$\mathscr{F} = \{f_{n_i} T^{n_i}, n_i > 0,\ 1 \leq i \leq m\},$$

then

(4.3.5)

$$\mathscr{F}' = \{\Gamma^\alpha(f_{n_i}) T^{n'_i - \alpha} / f_{n_i} T^{n_i} \in \mathscr{F}, \alpha = (\alpha_1, \alpha_2, \dots, \alpha_n) \in (\mathbb{N})^n, \textit{ and } 0 \leq |\alpha| < n'_i \leq n_i\}$$

generates $G(\mathscr{G})$ locally at x.

PROOF. To simplify notation, set $\mathscr{F}' = \{h_{n_j} T^{n_j} / 1 \leq j \leq M\}$. Note that for $1 \leq n' \leq n_j$, $h_{n_j} T^{n'} \in \mathscr{F}'$, and if $|\alpha| < n_j$, $\Gamma^\alpha(h_{n_j}) T^{n'} \in \mathscr{F}'$. It is clear that

$$\mathscr{O}_{W,x}[\mathscr{F}'] = \mathscr{O}_{W,x}[\{h_{n_j} T^{n_j} / 1 \leq j \leq M\}]$$

is included in any Diff-algebra containing $\mathscr{G}$. Our claim is that it already is a Diff-algebra. Set $\mathscr{O}_{W,x}[\mathscr{F}'] = \bigoplus I_k T^k$.

We prove that for any $h \in I_N$ and any multi-index α, such that $|\alpha| < N$, $\Delta^\alpha(h) \in I_{N-|\alpha|}$.

The ideal I_N is generated by all elements of the form

(4.3.6) $$H_N = h_{n_{i_1}} \cdot h_{n_{i_2}} \cdots h_{n_{i_p}} \qquad n_{i_1} + n_{i_2} + \cdots n_{i_p} = N,$$

with the $h_{n_{i_j}} T^{n_{i_j}} \in \mathscr{F}'$ not necessarily different.

Since the operators Γ^α are linear, it suffices to prove that $\Gamma^\alpha(a \cdot H_N) \in I_{N-|\alpha|}$, for $a \in \mathscr{O}_{W,x}$, H_N as in 4.3.6, and $0 \leq |\alpha| \leq N$. We proceed in two steps, by proving:

1) $\Gamma^{\alpha}(H_N) \in I_{N-|\alpha|}$.
2) $\Gamma^{\alpha}(a \cdot H_N) \in I_{N-|\alpha|}$.
We first prove 1). Consider, for any element $h_{n_{i_l}} T^{n_{i_l}} \in \mathscr{F}'$,

$$Tay(h_{n_{i_l}}) = \sum_{|\beta| \geq 0} \Gamma^{\beta}(h_{n_{i_l}}) T^{\beta} \in k'[[x_1, \ldots, x_n, T_1, \ldots, T_n]].$$

The definition of $\mathscr{F}'$ is such that, if β is a multi-index, and if $0 \leq |\beta| < n_{i_l}$, then $\Gamma^{\beta}(h_{n_{i_l}}) T^{n_{i_l} - \beta} \in \mathscr{F}'$.

On the one hand,

$$Tay(H_N) = \sum_{|\alpha| \geq 0} \Gamma^{\alpha}(H_N) T^{\alpha},$$

and on the other hand,

$$Tay(H_N) = Tay(h_{n_{i_1}}) \cdot Tay(h_{n_{i_2}}) \cdots Tay(h_{n_{i_p}})$$

in $k'[[x_1, \ldots, x_n, T_1, \ldots, T_n]]$. This shows that for a fixed α ($0 \leq |\alpha| \leq N$), $\Gamma^{\alpha}(H_N)$ is a sum of elements of the form

$$\Gamma^{\beta_1}(h_{n_{i_1}}) \cdot \Gamma^{\beta_2}(h_{n_{i_2}}) \cdots \Gamma^{\beta_p}(h_{n_{i_p}}), \qquad \sum_{1 \leq s \leq p} \beta_s = \alpha.$$

So it suffices to show that each of these summands is in $I_{N-|\alpha|}$.

Note here that

$$\sum_{1 \leq s \leq p} (n_{i_s} - |\beta_s|) = N - |\alpha|,$$

and that some of the integers $n_{i_s} - |\beta_s|$ might be zero or negative. Set

$$G = \{r, 1 \leq r \leq p, \text{ and } n_{i_r} - |\beta_r| > 0\}.$$

So

$$N - |\alpha| = \sum_{1 \leq s \leq p} (n_{i_s} - |\beta_s|) \leq \sum_{r \in G} (n_{i_r} - |\beta_r|) = N'.$$

Hypothesis on $\mathscr{F}'$ ensures that $\Gamma^{\beta_r}(h_{n_{i_r}}) \in I_{n_{i_r} - |\beta_r|}$ for every index $r \in G$, in particular

$$\Gamma^{\beta_1}(h_{n_{i_1}}) \cdot \Gamma^{\beta_2}(h_{n_{i_2}}) \cdots \Gamma^{\beta_p}(h_{n_{i_p}}) \in I_{N'}.$$

Finally, since $N' \geq N - |\alpha|$, $I_{N'} \subset I_{N-|\alpha|}$, and this proves Case 1).

For Case 2), fix α such that $0 \leq |\alpha| \leq N$. We claim that $\Gamma^{\alpha}(a \cdot H_N) \in I_{N-|\alpha|}$, for $a \in \mathscr{O}_{W,x}$ and H_N as in 4.3.6. On the one hand,

$$Tay(a \cdot H_N) = \sum_{\alpha} \Gamma^{\alpha}(a \cdot H_N) T^{\alpha},$$

and on the other hand,

$$Tay(a \cdot H_N) = Tay(a) \cdot Tay(H_N).$$

This shows that $\Gamma^{\alpha}(a \cdot H_N)$ is a sum of terms of the form $\Gamma^{\alpha_1}(a) \cdot \Gamma^{\alpha_2}(H_N)$, $\alpha_i \geq 0$, and $\alpha_1 + \alpha_2 = \alpha$. In particular, $|\alpha_2| \leq |\alpha|$, and by Case 1), $\Gamma^{\alpha_2}(H_N) \in I_{N-|\alpha_2|}$. On the other hand, $N - |\alpha_2| \geq N - |\alpha|$, so $\Gamma^{\alpha_2}(H_N) \in I_{N-|\alpha|}$, and hence $\Gamma^{\alpha}(a \cdot H_N) \in I_{N-|\alpha|}$. □

4.12. In general $\mathscr{G} \subset G(\mathscr{G})$, and equality holds if $\mathscr{G}$ is already a Diff-algebra.

Let $\mathscr{G} = \bigoplus_{k \geq 0} I_r W^r$ be a Diff-algebra; in particular, it is integral over a Rees subring, say $\mathscr{O}_W[I_N W^N]$ for suitable N (see 4.2). These ideals I_N are called *tuned ideals* in [**34**], page 45.

4.13. The local description in the theorem shows that

$$\operatorname{Sing}(\mathscr{G}) = \operatorname{Sing}(G(\mathscr{G})). \tag{4.3.7}$$

In fact, as $\mathscr{G} \subset G(\mathscr{G})$, it is clear that $\operatorname{Sing}(\mathscr{G}) \supset \operatorname{Sing}(G(\mathscr{G}))$. For the converse, note that if $\nu_x(f_{n_i}) \geq n_i$, then $\Gamma^{\alpha}(f_{n_i})$ has order at least $n_i - |\alpha|$ at the local ring $\mathscr{O}_{W,x}$.

Let Y be a smooth subscheme in $\operatorname{Sing}(\mathscr{G}) = \operatorname{Sing}(G(\mathscr{G}))$. The monoidal transformation defines transforms of both, say

$$\begin{array}{ccc} W & \xleftarrow{\pi} & W_1 \\ \mathscr{G} \subset G(\mathscr{G}) & & \mathscr{G}_1 \subset (G(\mathscr{G}))_1, \end{array} \tag{4.3.8}$$

where $\mathscr{G}_1$ and $(G(\mathscr{G}))_1$ are the transforms, and the inclusion $\mathscr{G}_1 \subset (G(\mathscr{G}))_1$ is clear from the definition of transforms.

We should keep in mind that the object of interest is $\mathscr{G}$, as well as the monoidal transforms of this Rees algebra. In fact, we are searching for a constructive resolution of $\mathscr{G}$. In what follows we will replace $\mathscr{G}$ by $G(\mathscr{G})$, so we need to understand the relation of the inclusion $\mathscr{G} \subset G(\mathscr{G})$ after one or more transformations.

Note that Lemma 2.14, together with the local description of generators in (4.3.5), asserts that

$$(G(\mathscr{G}))_1 \subset G(\mathscr{G}_1). \tag{4.3.9}$$

From this formula and (4.3.8), it follows that $\operatorname{Sing}((\mathscr{G})_1) = \operatorname{Sing}((G(\mathscr{G}))_1)$. And furthermore, for any permissible sequence for $\mathscr{G}$, we obtain inclusions

(4.3.10)

$$\begin{array}{ccccccc} W & \xleftarrow{\pi_1} & W_1 & \xleftarrow{\pi_2} & \cdots & \xleftarrow{\pi_r} & W_r \\ \mathscr{G} \subset G(\mathscr{G}) & & \mathscr{G}_1 \subset (G(\mathscr{G}))_1 \subset G(\mathscr{G}_1) & & & & \mathscr{G}_r \subset (G(\mathscr{G}))_r \subset G(\mathscr{G}_r) \end{array}$$

and

$$\operatorname{Sing}((\mathscr{G})_r) = \operatorname{Sing}((G(\mathscr{G}))_r). \tag{4.3.11}$$

In particular, a resolution of $G(\mathscr{G})$ defines a resolution of $\mathscr{G}$. The argument we have used relies on Lemma 2.14, which was proved over fields of characteristic zero. However in [**24**], Giraud extends this result to the case of smooth schemes over arbitrary fields (see also [**30**]).

REMARK 4.14. The sheaf of differentials attached to E_s in W_s in Remark 4.10 is a subsheaf of the sheaf of differentials on W_s. Furthermore, it is shown there that there is a natural inclusion of $\operatorname{Diff}^c_{E_s} \subset \operatorname{Diff}^c_{E_s}$ for each positive integer c. As in Definition 4.9, a notion of Diff_E-algebras can be defined. One can also mimic the definition of operator G in Theorem 4.11, defined in terms of differentials, and define a new operator, say G_E, in terms of differentials with logarithmic poles.

Theorem 4.11 extends with no change for, say $\mathscr{G} \subset G_E(\mathscr{G})$, simply by replacing operators Γ^{α} by Γ^{α}_E. In this way we get a G_E-saturation of Rees algebras, and

$$\mathscr{G} \subset G_E(\mathscr{G}) \subset G(\mathscr{G}).$$

From this inclusion and formula (4.3.7) it follows that

$$\operatorname{Sing}(\mathscr{G}) = \operatorname{Sing}(G_E(\mathscr{G})) = \operatorname{Sing}(G(\mathscr{G})). \tag{4.3.12}$$

In this way we have two operators: G and G_E, both of which define enlargements of Rees algebras. An advantage of the G_E operator is that it is adapted to our notion of functions w-ord$_s$ in 4.6. In fact, consider the sequence of transformations, say (4.3.2), and recall that for a suitable index b we can attach a sequence of transformations of basic objects, say (4.3.3), together with a factorization

$$J_s = I(H_1)^{b_1} I(H_2)^{b_2} \cdots I(H_s)^{b_s} \cdot \overline{J}_s. \tag{4.3.13}$$

It follows now from the discussion at the end of Remark 4.10 that the functions $\text{w-ord}^s_{\mathscr{G}_s}$ and $\text{w-ord}^s_{G_{E_s}}(\mathscr{G}_s)$ coincide along $\text{Sing}(\mathscr{G}_s) = \text{Sing}(G_{E_s}(\mathscr{G}_s))$.

4.4. Restrictions of differential algebras

Let $\mathscr{G} = \bigoplus I_k \cdot T^k$ be a Rees algebra on a smooth scheme W, defined by ideals $I_k \subset \mathscr{O}_W$, and fix a smooth subscheme $W' \subset W$. Define the *restriction* of $\mathscr{G}$ to W' to be $\mathscr{G}' = \bigoplus I_k \mathscr{O}_{W'} \cdot T^k \subset \mathscr{O}_{W'}[T]$.

If we restrict W to an affine open set, and if $\mathscr{F} = \{f_{N_1}T^{N_1}, \dots, f_{N_s}T^{N_s}\}$, $(f_{N_i} \in \Gamma(\mathscr{O}_W))$, then the restriction of $\mathscr{G}$ is generated by $\{\overline{f}_{N_1}T^{N_1}, \dots, \overline{f}_{N_s}T^{N_1}\}$, where each $\overline{f}_{N_i}$ is the restriction of f_{N_i} to W'.

EXAMPLE 4.15. $W = \text{Spec}(k[X,Y,Z])$, $W' = V(X) = \text{Spec}(k[Y,Z])$. Let $\mathscr{G}$ be the Rees algebra generated by fT^2 $(\subset k[X,Y,Z][T])$, where $f = Z^2 + XZ + XY^4$.

In this case $\mathscr{G}'(\subset k[Y,Z][T])$ is generated by $(Z^2)T^2$. Note that $\text{Sing}(\mathscr{G})$ is a closed point, whereas $\text{Sing}(\mathscr{G}')$ is a line containing it.

The previous example illustrates a general fact: that if $\mathscr{G}'$ denotes the restriction of $\mathscr{G}$ to a smooth subscheme W', then

$$\text{Sing}(\mathscr{G}') \supset \text{Sing}(\mathscr{G}) \cap W'.$$

Now let

$$\begin{array}{ccccccc} W & \xleftarrow{\pi_1} & W_1 & \xleftarrow{\pi_2} & \dots & \xleftarrow{\pi_r} & W_r \\ \mathscr{G} \quad W' & & (\mathscr{G})_1 \quad W'_1 & & & & (\mathscr{G})_r \quad W'_r \end{array} \tag{4.4.1}$$

be a sequence defined by blowing up regular centers $Y_i \subset \text{Sing}(\mathscr{G}_i) \cap W'_i$. So the sequence is permissible for $\mathscr{G}$, and all centers Y_i are also included in W'_i (strict transform of W'). In particular, all W'_i are smooth, and hence (4.4.1) induces a sequence of monoidal transformations

$$W' \xleftarrow{\pi_1} W'_1 \xleftarrow{\pi_2} \dots \xleftarrow{\pi_r} W'_r. \tag{4.4.2}$$

PROPOSITION 4.16. *Let $\mathscr{G}$ be a Diff-algebra on W, W' a smooth hypersurface, and let $\mathscr{G}'$ be the restriction to W'. Then:*

i) $\text{Sing}(\mathscr{G}') = \text{Sing}(\mathscr{G}) \cap W'$.

ii) For any sequence (4.4.1), the induced sequence (4.4.2) is permissible for $\mathscr{G}'$, defining

$$\begin{array}{ccccccc} W' & \xleftarrow{\pi_1} & W'_1 & \xleftarrow{\pi_2} & \dots & \xleftarrow{\pi_r} & W'_r \\ \mathscr{G}' & & (\mathscr{G}')_1 & & & & (\mathscr{G}')_r \end{array} \tag{4.4.3}$$

and

$$\text{Sing}((\mathscr{G}')_r) = W'_r \cap \text{Sing}((\mathscr{G})_r).$$

PROOF. Fix a closed point $x \in \text{Sing}(\mathscr{G}) \subset W$ and a regular system of parameters $\{x_1, \dots, x_n\}$ at $\mathscr{O}_{W,x}$ so that W' is locally defined by $I(W') = \langle x_1 \rangle$. Assume that $\mathscr{G}$ is locally generated by

$$\mathscr{F} = \{f_{n_i}T^{n_i},\ n_i > 0,\ 1 \leq i \leq m\}.$$

As $\mathscr{G} = G(\mathscr{G})$ (i.e. as $\mathscr{G}$ is a Diff-algebra), then
(4.4.4)
$$\mathscr{F}' = \{\Gamma^\alpha(f_{n_i})T^{n_i'-\alpha}/f_{n_i}T^{n_i} \in \mathscr{F}, \alpha = (\alpha_1,\alpha_2,\ldots,\alpha_n) \in (\mathbb{N})^n, \text{ and } 0 \leq |\alpha| < n_i' \leq n_i\}$$
also generates $\mathscr{G}$ locally at x (4.3.5), where each Γ is defined in terms of partial derivatives in the chosen coordinates. Note, in particular, that if, say $f_bT^b \in \mathscr{F}'$, then $\alpha\frac{\partial^r f}{\partial^r x_1} \in \mathscr{F}'$, where α is a non-zero constant. Recall that, locally at x, $\mathrm{Sing}(\mathscr{G}) = \cap_{fT^b \in \mathscr{F}'} \mathrm{Sing}(\langle f \rangle, b)$ (4.3).

Fix an open neighborhood U of x in W, in the setting of 2.5. Note that each $g_{i,r} = \frac{1}{r!}\frac{\partial^r}{\partial^r x_1} f_i$ in (2.2.8) is, up to multiplication by a non-zero constant, an element of $\mathscr{F}'$.

The proof of this proposition follows now from 2.4. In fact, part (i) follows from formula (2.2.3) and part (ii) from formula (2.2.7).

4.5. The inductive function revisited

4.17. We recall here a main property of our *inductive function* discussed in (3.8). Basically there is a simple couple (J'', b'') attached to the biggest value $\max t_s$, and any resolution (3.3.3) induces a sequence (3.3.4) so that $\max t_s > \max t_{s+k}$. In other words, a resolution of the simple couple (J'', b'') forces the maximum value of t to drop. This leads to resolution of basic objects, or at least to reduction to the monomial case, from which it is easy to extend to a resolution of the basic object.

If a sequence of transformations defines a resolution of (J'', b''), it also defines a resolution of the Rees algebra $\mathscr{G}_{(J'',b'')}$ (4.2), and the converse also holds. Similarly, a sequence of transformations defines resolution of the Rees algebra $\mathscr{G}_{(J'',b'')}$ if and only if it defines a resolution of the induced Diff-algebra $G(\mathscr{G}_{(J'',b'')})$ (4.13). The following is a characteristic-free proof of Corollary 2.30.

PROPOSITION 4.18. *Let (J,b) be a simple couple, $J \subset \mathscr{O}_W$, and assume that* $\mathrm{Sing}(J,b)$ *is of pure codimension one in the smooth scheme W. Then:*
i) $\mathrm{Sing}(J,b)$ *is a regular hypersurface, say $\overline{W}$.*
ii) $J = I(\overline{W})^b$.
iii) A resolution of (J,b) is obtained with center $Y = \overline{W}$.

PROOF. Recall that $\mathrm{Sing}(J,b)$ consists of points where the ideal J has order at least b. Since (J,b) is simple, $\mathrm{Sing}(J,b)$ are the points of order exactly b.

Fix a closed point $x \in \mathrm{Sing}(J,b)$ and argue locally at $\mathscr{O}_{W,x}$. Let $\overline{W}$ denote the reduced hypersurface defined by $\mathrm{Sing}(J,b)$ locally at the point. The local ring $\mathscr{O}_{W,x}$ is regular and hence a unique factorization domain.

Since the generic point of every irreducible component of $\overline{W}$ is in $\mathrm{Sing}(J,b)$, it follows that $J = I(\overline{W})^b \cdot \mathscr{A}$ for some ideal $\mathscr{A} \subset \mathscr{O}_{W,x}$. Finally, since the order of J is b, it follows that $\mathscr{A} = \mathscr{O}_{W,x}$, that $\overline{W}$ is irreducible locally at x, and that $\overline{W}$ is also regular at a neighborhood of the point. This proves i) and ii).

As for the proof of (iii), it is worth reiterating a property of our notion of transformations. In fact, the blow-up at a hypersurface is the identity map. However we *declare* $\overline{W}$ to be the exceptional locus of this transformation, in which case the transform is (J_1, b), $J_1 = \mathscr{O}_{W_1}$, and $W_1 = W$.

REMARK 4.19. The local analysis at a closed point $x \in \mathrm{Sing}(J,b)$ in the previous proof shows that if (J,b) is simple and the *local* codimension is one, then $\mathrm{Sing}(J,b)$ is smooth and hence of pure codimension one (locally). This shows that components of codimension one, say $R(1)(\mathrm{Sing}(J,b))$, are open and closed in $\mathrm{Sing}(J,b)$ as well as a union of smooth hypersurfaces. Note also that if $\overline{W}$ denotes as before the hypersurface $\mathrm{Sing}(J,b)$, then $J = I(\overline{W})$ and (J,b) is equivalent, in the sense of idealistic exponents, to $(I(\overline{W}),1)$.

The statement in Prop 4.18 also holds for $\mathscr{G}_{(J,b)}$, since $\mathrm{Sing}(\mathscr{G}_{(J,b)}) = \mathrm{Sing}(J,b)$. Note that $\mathscr{G}_{(J,b)}$ has the same integral closure as $\mathscr{G}_{(I(\overline{W}),1)}$.

4.20. In what follows, for any Rees algebra $\mathscr{G}$, let $[\mathscr{G}]_i$ denote the component of degree i. Note that $[\mathscr{G}_{(J,b)}]_b = J$ (4.2.1) and that $[G(\mathscr{G}_{(J,b)})]_1 = \mathrm{Diff}^{b-1}(J)$ (4.3.5).

For a closed point $x \in \mathrm{Sing}(\mathscr{G})$, let $\mathscr{G}_x$ denote the Rees algebra induced by localization (i.e. $[\mathscr{G}_x]_i = [\mathscr{G}]_i \mathscr{O}_{W,x}$), and let $\hat{\mathscr{G}}$ denote the Rees algebra defined by $\mathscr{G}_x$ at $\hat{\mathscr{O}}_{W,x}$.

If $\mathscr{G}$ is a Rees algebra and if $\overline{W}(\subset W)$ is a smooth hypersurface, then $\mathscr{G}_{\overline{W}}$ will denote the restriction (4.4).

We say that a differential algebra $\mathscr{G}$ is *simple* if $\mathrm{ord}_{\mathscr{G}} = 1$, in the case of fields of characteristic zero, which is the assumption in the following theorem. This is equivalent to the condition that the ideal $[\mathscr{G}]_1$ be order one at the local ring of any point of $\mathrm{Sing}(\mathscr{G})$. Note that if (J,b) is a simple couple, then $G(\mathscr{G}_{(J,b)})$ and $(G(\hat{\mathscr{G}}_{(J,b)}))$ are simple differential algebras.

THEOREM 4.21. (J. Wlodarczyk, **[47]**) *Fix a simple differential algebra $\mathscr{G}$ and a closed point $x \in \mathrm{Sing}(\mathscr{G})$. Let $\overline{W}_1$ and $\overline{W}_2$ be two smooth hypersurfaces in some neighborhood of x, so that $I(\overline{W}_i) \subset [\mathscr{G}]_1$ at $\mathscr{O}_{W,x}$ for $i = 1,2$. Set $\mathscr{H}_i = (\mathscr{G})_{\overline{W}_i}$, $i = 1,2$ (the restriction of $\mathscr{G}$ to $\overline{W}_i$).*

There is a formal isomorphism $\Theta : \hat{\mathscr{O}}_{\overline{W}_1,x} \to \hat{\mathscr{O}}_{\overline{W}_2,x}$ such that $\Theta(\hat{\mathscr{H}}_1) = \hat{\mathscr{H}}_2$ (i.e. such that $\Theta([\hat{\mathscr{H}}_1]_s) = [\hat{\mathscr{H}}_2]_s$, for each $s \geq 1$).

PROOF. As $\mathscr{G}$ is simple, $[\mathscr{G}]_1$ has order one at $\mathscr{O}_{W,x}$. One can choose elements of order one at $\mathscr{O}_{W,x}$, say $x_1, x_1', x_2, x_3, \ldots, x_d$, so that:

i) $I(\overline{W}_1)_x = \langle x_1 \rangle$.

ii) $I(\overline{W}_2)_x = \langle x_1' \rangle$.

iii) Both $\{x_1, x_2, x_3, \ldots, x_d\}$ and $\{x_1', x_2, x_3, \ldots, x_d\}$ are regular systems of parameters at $\mathscr{O}_{W,x}$.

It suffices to show that there is an isomorphism $\Theta : \hat{\mathscr{O}}_{W,x} \to \hat{\mathscr{O}}_{W,x}$ so that $\Theta(x_1) = x_1'$ and $\Theta(\hat{\mathscr{G}}) = \hat{\mathscr{G}}$. Note that $\hat{\mathscr{O}}_{W,x} = k[[x_1, x_2, x_3, \ldots, x_d]]$ and that $\hat{\mathscr{G}}$ is a differential algebra at the ring of formal power series. Let Θ be defined by $\Theta(x_1) = x_1'$, and $\Theta(x_i) = x_i$ for $i = 2, \ldots, d$.

Given $f_s(x_1, x_2, x_3, \ldots, x_d) \in [\hat{\mathscr{G}}]_s$, then $\Theta(f_s(x_1, x_2, x_3, \ldots, x_d)) = f_s(x_1', x_2, x_3, \ldots, x_d)$. So

$$\Theta(f_s(x_1, x_2, x_3, \ldots, x_d)) = f_s(x_1 + (x_1' - x_1), x_2, x_3, \ldots, x_d)$$

and

$$f_s(x_1 + (x_1' - x_1), x_2, x_3, \ldots, x_d) = \sum_{r \geq 0} \frac{1}{r!} \frac{\partial^r f_s}{\partial^r x_1} (x_1' - x_1)^r.$$

In our setting $(x_1' - x_1) \in [\hat{\mathscr{G}}]_1$, so $(x_1' - x_1)^r \in [\hat{\mathscr{G}}]_r$.

For $r \geq s$, $(x_1' - x_1)^r \in [\hat{\mathscr{G}}]_s$ (4.9,i). For $0 \leq r < s$, $\frac{1}{r!} \frac{\partial^r f_s}{\partial^r x_1} \in [\hat{\mathscr{G}}]_{s-r}$, and $\frac{1}{r!} \frac{\partial^r f_s}{\partial^r x_1} (x_1' - x_1)^r \in [\hat{\mathscr{G}}]_s$.

So $\Theta([\hat{\mathscr{G}}]_s) \subset [\hat{\mathscr{G}}]_s$. We now claim that $\Theta([\hat{\mathscr{G}}]_s) = [\hat{\mathscr{G}}]_s$; so let (R,M) denote the (local) ring $\hat{\mathscr{O}}_{W,x} = k[[x_1, x_2, x_3, \ldots, x_d]]$ and let $I(\subset R)$ be the ideal $[\hat{\mathscr{G}}]_s$. Clearly $\Theta(M^s) = M^s$ for any positive integer s, and Θ induces an isomorphism on the finite-dimensional vector space R/M^s. As $\Theta(I) \subset I$, $\Theta(I + M^s) = I + M^s$. Finally, $I = \cap_s (I + M^s)$ by Krull's intersection theorem (c.f. **[5]**, 10.17), which ensures that $\Theta(I) = I$. The theorem follows from the fact that

$$\Theta(\hat{\mathscr{G}}) = \hat{\mathscr{G}} \tag{4.5.1}$$

and $\Theta(x_1) = x'_1$. □

4.22. Recall that if (J,b) is a simple couple, then $G(\mathscr{G}_{(J,b)})$ and $(G(\hat{\mathscr{G}}_{(J,b)}))$ are simple differential algebras. Note that Theorem 4.21 applies to the case $\mathscr{G} = G(\mathscr{G}_{(J,b)})$ and to $I(\overline{W}_1)$ and $I(\overline{W}_2)$ included in $\Delta^{b-1}(J)$ (4.20).

As indicated in 4.17, the objective is to define a resolution of the simple couple (J'',b'') attached to the value $\max t$. This is equivalent to a resolution of $G(\mathscr{G}_{(J'',b'')})$ (4.13).

We now prove resolution for any simple differential algebra and then show why this proves both Prop 2.31, the Main Claim 2.33, and hence induction in resolution theorems.

Consider transformations of Rees algebras, namely, sequences of the form

$$\begin{array}{ccccccc} (W,E) & \stackrel{\pi_1}{\longleftarrow} & (W_1,E_1) & \stackrel{\pi_2}{\longleftarrow} & \cdots & \stackrel{\pi_k}{\longleftarrow} & (W_s,E_s) \\ \mathscr{G} & & (\mathscr{G})_1 & & & & (\mathscr{G})_s. \end{array} \tag{4.5.2}$$

PROPOSITION 4.23. *Let $\mathscr{G}$ be a simple differential algebra over W (4.20). Assume that* $\operatorname{Sing}(\mathscr{G})$ *is of codimension one. Then:*

i) The union of irreducible components of codimension one is a smooth hypersurface, say Y, which is open and closed in $\operatorname{Sing}(\mathscr{G})$.

ii) After blowing up at the permissible center Y, we may assume that $\operatorname{Sing}(\mathscr{G})$ *has codimension at least two in W.*

PROOF. This is a corollary of 4.18 and the local expression of $\operatorname{Sing}(\mathscr{G})$ in 4.3.

PROPOSITION 4.24. *Let $\mathscr{G}$ be a simple differential algebra over W(4.20). Assume that:*

i) $\operatorname{Sing}(\mathscr{G})$ *is of codimension at least two in W.*

ii) There is a smooth hypersurface $\overline{W} \subset W$ such that $I(\overline{W}) \subset [\mathscr{G}]_1$.

Set $\mathscr{H} = \mathscr{G}_{\overline{W}}$ (the restriction). Then:

a) $\operatorname{Sing}(\mathscr{G}) = \operatorname{Sing}(\mathscr{H})(\subset \overline{W})$.

b) Any sequence of transformations, say

$$\begin{array}{ccccccc} (\overline{W},\emptyset) & \stackrel{\pi_1}{\longleftarrow} & (\overline{W}_1,\overline{E}_1) & \stackrel{\pi_2}{\longleftarrow} & \cdots & \stackrel{\pi_r}{\longleftarrow} & (\overline{W}_r,\overline{E}_r) \\ \mathscr{H} & & (\mathscr{H})_1 & & & & (\mathscr{H})_r \end{array} \tag{4.5.3}$$

induces a sequence of transformations of $\mathscr{G}$, say

$$\begin{array}{ccccccc} (W,\emptyset) & \stackrel{\pi_1}{\longleftarrow} & (W_1,E_1) & \stackrel{\pi_2}{\longleftarrow} & \cdots & \stackrel{\pi_r}{\longleftarrow} & (W_r,E_r) \\ \mathscr{G} & & (\mathscr{G})_1 & & & & (\mathscr{G})_r \end{array} \tag{4.5.4}$$

and $\operatorname{Sing}((\mathscr{G})_r) = \operatorname{Sing}(\mathscr{H}_r)$.

In particular, (4.5.4) is a resolution if (4.5.3) is a resolution.

PROOF. The inclusion $\operatorname{Sing}(\mathscr{G}) \subset \overline{W}$ follows from the hypothesis $I(\overline{W}) \subset [\mathscr{G}]_1$. Part (a) follows from (i) of Prop 4.16. Furthermore, induction on r and Prop 4.16 allow us to assume that $\operatorname{Sing}((\mathscr{G})_s) = \operatorname{Sing}(\mathscr{H}_s)$ for $1 \leq s < r$, and that $\operatorname{Sing}(\mathscr{H}_r) = \operatorname{Sing}((\mathscr{G})_r) \cap \overline{W}_r$. So it suffices to show that $\operatorname{Sing}((\mathscr{G})_r) \subset \overline{W}_r$.

The definition of transformation of Rees algebras in 4.4 shows that 4.5.4 induces a sequence of transformations of couples

$$\begin{array}{ccccccc} W & \stackrel{\pi_1}{\longleftarrow} & W_1 & \stackrel{\pi_2}{\longleftarrow} & \cdots & \stackrel{\pi_r}{\longleftarrow} & W_r \\ ([\mathscr{G}]_1,1) & & (([\mathscr{G}]_1)_1,1) & & & & (([\mathscr{G}]_1)_r,1) \end{array} \tag{4.5.5}$$

and $([\mathscr{G}]_1)_r \subset [(\mathscr{G})_r]_1$. In particular, (4.5.5) induces

$$\begin{array}{ccccccc} W & \xleftarrow{\pi_1} & W_1 & \xleftarrow{\pi_2} & \cdots & \xleftarrow{\pi_r} & W_r \\ (I(\overline{W}),1) & & ((I(\overline{W}))_1,1) & & & & ((I(\overline{W}))_r,1) \end{array} \tag{4.5.6}$$

and $(I(\overline{W}))_r \subset [(\mathscr{G})_r]_1$. Finally, the argument in Remark 2.27 shows that $(I(\overline{W}))_r = I(\overline{W}_r)$, and hence that $\mathrm{Sing}((\mathscr{G})_r) \subset \overline{W}_r$. □

PROPOSITION 4.25. *Fix W smooth, or formally smooth, of dimension d. Let $\mathscr{G}$ be a simple differential algebra over W. Assume that $\mathrm{Sing}(\mathscr{G})$ has no component of codimension one, and that:*

i) There is a totally ordered set I^{d-1} and resolution functions as in 1.25, for $d-1$-dimensional basic objects.

ii) There are smooth hypersurfaces $\overline{W}_i$, $i = 1,2$, such that $I(\overline{W}_i) \subset [\mathscr{G}]_1$.

Set $\mathscr{H}_i = (\mathscr{G})_{\overline{W}_i}$, $i = 1,2$ (the restriction of $\mathscr{G}$ to $\overline{W}_i$). Then the resolution of $\mathscr{G}$ defined by the constructive resolution of $\mathscr{H}_1$ in Proposition 4.24 is the same as that defined by the constructive resolution of $\mathscr{H}_2$.

PROOF. In 4.7 we have extended constructive resolutions to Rees algebras. The claim is a corollary of Theorem 4.21, together with Properties **P6)** and **P7)** in 1.28, which we may assume by induction on the dimension of the ambient space.

COROLLARY 4.26. *Resolution functions in dimension $d-1$ provide algorithmic resolution of simple differential algebras in dimension d.*

PROOF. The claim follows from the previous proposition applied at a suitable covering by affine open sets of W, say $\{U_\lambda | \lambda \in \Lambda\}$, so there is a closed and smooth hypersurface $\overline{W}_\lambda$ in U_λ, and $I(\overline{W}_\lambda) \subset [\mathscr{G}]_1(U_\lambda)$ (restriction of the sheaf $[\mathscr{G}]_1$ to U_λ).

4.27. **Proof of Proposition 2.31 and the Main Claim 2.33.**

Set $\mathscr{G} = G(\mathscr{G}_{(J,b)})$ defined on a smooth d-dimensional scheme W. This scheme can be covered by finitely many affine open sets, so following Kollár's notion of tuned ideals we first choose an index N so that $\mathscr{G}$ is a finite extension of the Rees ring $\mathscr{O}_W[[\mathscr{G}]_N T^N]$. As in the proof of the previous corollary, consider a suitable refinement $\{U_\lambda | \lambda \in \Lambda\}$, so there is a closed and smooth hypersurface $\overline{W}_\lambda$ in U_λ, and $I(\overline{W}_\lambda) \subset [\mathscr{G}]_1(U_\lambda)$ (restriction of the sheaf $[\mathscr{G}]_1$ to U_λ). Let $\mathscr{A}_\lambda$ be the restriction of the sheaf of ideals $[\mathscr{G}]_N$ to the affine open set U_λ. Finally set the couple (K_λ, d_λ) in Proposition 2.31 by taking K_λ as the restriction of the ideal $\mathscr{A}_\lambda$ to $\overline{W}_\lambda$, and $d_\lambda = N$.

Set, by restriction, $\mathscr{H}_\lambda = (\mathscr{G})_{\overline{W}_\lambda}$, and note that $\mathscr{H}_\lambda$ is a finite extension of the Rees subring $\mathscr{O}_{\overline{W}_\lambda}[K_\lambda T^N]$.

Recall that $\mathrm{Sing}(J,b) = \mathrm{Sing}(G(\mathscr{G}_{(J,b)}))$ and that $\Delta^{b-1}(J) \subset [\mathscr{G}]_1$, so we may also assume that $I(\overline{W}_\lambda) \subset \Delta^{b-1}(J)(U_\lambda)$.

At a closed $x \in \mathrm{Sing}(J,b) \cap U_\lambda \cap U_\beta$ there is a formal isomorphism (see (4.5.1)) mapping $\mathscr{A}_\lambda$ to $\mathscr{A}_\beta$, $I(\overline{W}_\lambda)$ to $I(\overline{W}_\beta)$, and hence $\mathscr{H}_\lambda$ to $\mathscr{H}_\beta$.

Proposition 2.31 and the Main Claim 2.33 follow now from the previous propositions.

4.28. The previous discussion already justifies the definition of the set I^d in 1.25, proposed in Chapter 3 as well as the properties of the resolution function. We refer to **[13]** for details.

4.6. Elimination and resolution functions

The objective of these final sections is to discuss an alternative approach to resolution of simple basic objects, avoiding the notion of restriction to smooth hypersurfaces as it appears in Proposition 2.31, without modifying the main invariants and the totally ordered set I^d in 1.25.

Starting from this point we will not impose restrictions on the characteristic of the underlying field. Recall that if W is smooth over a field k, for each positive index c, the sheaf of differential operators of degree c is locally free.

In Section 4.3 we presented a local set of generators of the module of differential operators on a smooth scheme defined in terms of a regular system of parameters. Fix a closed point $x \in W$ with residue field k' and a regular system of parameters $\{x_1, \dots, x_n\}$ at $\mathscr{O}_{W,x}$. Let M_x denote the maximal ideal of $\mathscr{O}_{W,x}$. The graded ring associated to this local ring, say $gr_{M_x}(\mathscr{O}_{W,x})$, is a polynomial ring, say $k'[X_1, \dots, X_n]$, where X_i denotes the initial form of x_i. Recall that there is a natural identification of $\mathrm{Spec}(gr_{M_x}(\mathscr{O}_{W,x}))$ with the tangent space of W at the point x. There is, on the one hand, a Taylor morphism, say

$$Tay : \mathscr{O}_{W,x} \to \mathscr{O}_{W,x}[[T_1, \dots, T_n]]$$

that maps x_i to $x_i + T_i$, $1 \leq i \leq n$ (see 4.8); on the other hand, there is a Taylor morphism, say

$$\overline{Tay} : k'[X_1, \dots, X_n] \to k'[X_1, \dots, X_n][T_1, \dots, T_n],$$

that maps X_i to $X_i + T_i$, $1 \leq i \leq n$.

Both are closely related, and some important invariants in desingularization arise from the link between them. In both cases we define operators, for each multi-index Γ, say Γ^α and $\overline{\Gamma}^\alpha$, so that $Tay(f) = \sum_{\alpha \in \mathbb{N}^n} \Gamma^\alpha(f) T^\alpha$ and $\overline{Tay}(F) = \sum_{\alpha \in \mathbb{N}^n} \overline{\Gamma}^\alpha(F) T^\alpha$.

Note here that if F is a homogeneous polynomial of degree N and if $|\alpha| \leq N$, then $\overline{\Gamma}^\alpha(F)$ is either zero or homogeneous of degree $N - |\alpha|$.

Let $\mathscr{G} = \oplus I_s T^s$ be a Rees algebra on the smooth scheme W. If $x \in \mathrm{Sing}(\mathscr{G})$ and if $(\mathscr{G})_x$ denotes the localization at the point, then for each index s, $I_s \subset M_x^s$. We attach a homogeneous ideal to $\mathscr{G}$ at a singular point x, included in $gr_{M_x}(\mathscr{O}_{W,x})$, namely that generated by the class of I_s at the quotient M_x^s / M_x^{s+1}, for all s. This homogeneous ideal defines a cone at $\mathrm{Spec}(k'[X_1, \dots, X_n])$ (the tangent space of W at x).

Recall that $\mathrm{Sing}(\mathscr{G}) = \mathrm{Sing}(G(\mathscr{G}))$. The relation between the two Taylor morphisms discussed above show how the homogeneous ideals at x attached to $G(\mathscr{G})$ and to $\mathscr{G}$ are related. In fact, the graded ideal of $G(\mathscr{G})$ at the point is the smallest homogeneous ideal spanned by the graded ideal of $\mathscr{G}$ and is closed by the action of the differential operators $\overline{\Gamma}^\alpha$; namely, with the property that if F is a homogeneous polynomial of degree N in the ideal and if $|\alpha| \leq N$, then $\overline{\Gamma}^\alpha(F)$ also is in the homogeneous ideal.

Homogeneous ideals in the polynomial ring with this property define a linear subspace at the tangent space of W at x. When k' is a field of characteristic zero, ideals with this property are generated by linear equations (homogeneous of degree one); so after linear change of coordinates we may assume that it is generated by, say $X_1, \dots, X_\tau$ (the first τ coordinates). If k' is a field of positive characteristic, and if we assume for simplicity that the field is perfect, then ideals with this property are those generated by elements of the form $X_i^{p^{e_i}}$ for some positive integer e_i and $1 \leq i \leq \tau$. So both in the case of fields of characteristic zero and the case of positive characteristic there is a natural linear subspace of codimension τ naturally attached to $G(\mathscr{G})$ (and also to $\mathscr{G}$) at the singular point x, or rather , at the tangent space of W at the given point. The codimension of this subspace is called the τ-invariant of the Diff-algebra at the singular point x. The τ-invariant is zero when the homogeneous ideal is zero.

The concept of τ-invariant was introduced by Hironaka and largely studied in desingularization theory ([**30**], [**40**]).

4.29. Let $\mathscr{G}$ be a Rees algebra and fix a closed point $x \in \mathrm{Sing}(\mathscr{G})$. Assume that at an affine open neighborhood of the point, say $U \subset W$, there is a finite set $\mathscr{F} = \{f_1 T^{n_1}, \ldots, f_s T^{n_s}\}$, $n_i \geq 1$ and $f_i \in \mathscr{O}_W(U)$, so that the restriction of $\mathscr{G}$ to U is $\mathscr{O}_W(U)[f_1 T^{n_1}, \ldots, f_s T^{n_s}] (\subset \mathscr{O}_W(U)[T])$. Let

$$\mathscr{G}_x = \bigoplus I_k \cdot T^k (\subset \mathscr{O}_{W,x}[T])$$

be the *localization* of $\mathscr{G}$ at x. As $x \in \mathrm{Sing}(\mathscr{G})$, the order of I_k at $\mathscr{O}_{W,x}$ is at least k. We say that $\mathscr{G}$ is *simple* at the singular point x if for some positive index k, I_k has order k. This amounts to saying that $ord_x(\mathscr{G}) = 1$; or equivalently, that for some $f_c T^{n_c} \in \mathscr{F}$, the element f_c has order n_c at the local regular ring $\mathscr{O}_{W,x}$. Note also that $ord_x(\mathscr{G}) = 1$ if and only if the τ-invariant at the point is non-zero.

The definition of singular locus ensures that $V(\mathrm{Diff}^{n_c - 1}(\langle f_c \rangle)) \supset \mathrm{Sing}(\mathscr{G})$ locally at x.

We may choose the system of parameters $\{x_1, \ldots, x_n\}$ at x, so that at the completion $\hat{\mathscr{O}}_{W,x}$, say $R = k'[[x_1, .., x_n]]$, there is a unit, say u, so that

$$u \cdot f_c = Z^{n_c} + a_1 Z^{n_c - 1} + \cdots + a_{n_c} \in S[Z],$$

$S = k[[x_1, \ldots, x_{n-1}]]$ and $Z = x_n$. A similar result holds at a suitable étale neighborhood of x. We may therefore assume that f_c is a monic polynomial of degree c in $S[Z]$, and of order c in $S[Z]_{\langle M_S, Z \rangle} \subset R$, where S is regular.

Let $\pi : W \longrightarrow W'$ be a smooth morphism defined at an étale neighborhood of x, where W' is smooth, dim W' = dim W-1. We say that π is *transversal* to $\mathscr{G}$ at x if the previous setting holds for $R = \hat{\mathscr{O}}_{W,x}$, $S = \hat{\mathscr{O}}_{W', \pi(x)}$ and for some $f_c T^{n_c} \in \mathscr{F}$, where f_c has order n_c at $\mathscr{O}_{W,x}$.

In these conditions, a transversal morphism π induces a finite morphism

$$\overline{\pi} : \mathrm{Spec}(S[Z]/\langle f_c(Z) \rangle) \longrightarrow \mathrm{Spec}(S).$$

Here we view $H = \mathrm{Spec}(S[Z]/\langle f_c(Z) \rangle)$ as a hypersurface in W. Let $\{ord\ f_c \geq n_c\}$ denote the closed set of points of multiplicity n_{c_1}, so

$$\{ord\ f_c \geq n_c\} = \mathrm{Sing}(\langle f_c \rangle, b) := V(\mathrm{Diff}^{c-1}(\langle f_c(Z) \rangle)) \subset H.$$

In this setting the finite morphism π is one-to-one over a closed subset of $\mathrm{Spec}(S)$, namely on the image of the c-fold points. In fact, as $f_c(Z)$ is monic of degree c, Zariski's multiplicity formula applies to $S \subset B = S[Z]/\langle f_c(Z) \rangle$ (see [**49**], Corollary 1, page 299). It ensures that if Q is a b-fold point in this hypersurface (i.e. a prime of B of multiplicity c) and Q maps to, say P, in S, then Q is the only prime dominating P. In other words, there is a diagram, say

$$\begin{array}{ccc} \mathrm{Spec}(S[Z]/\langle f_c(Z) \rangle) & \xrightarrow{\overline{\pi}} & \mathrm{Spec}(S) \\ \cup & & \cup \\ \{ord\ f_c \geq n_c\} & \xrightarrow{1-\mathrm{to}-1} & \pi(\{ord\ f_c \geq n_c\}). \end{array} \tag{4.6.1}$$

Since $\mathrm{Sing}(\mathscr{G}) \subset \{ord\ f_{c_1} \geq n_{c_1}\}$, π induces a one-to-one map, say

$$\mathrm{Sing}(\mathscr{G}) \xrightarrow{1-\mathrm{to}-1} \pi(\mathrm{Sing}(\mathscr{G})),$$

and hence a set theoretical identification for any transversal $\pi : \mathrm{Spec}(S[Z]) \longrightarrow \mathrm{Spec}(S)$.

REMARK 4.30. Given a Diff-algebra and a closed point $x \in \mathrm{Sing}(\mathscr{G})$, we have defined a linear subspace of codimension τ at the tangent space $\mathrm{Spec}(gr_{M_x}(\mathscr{O}_{W,x}))$. If d denotes the dimension of W, a local projection on a smooth $d-1$-dimensional scheme defines a line at the tangent space at the point (the tangent line to the fiber through the point). The local projection is said to be transversal if the line is not included in the linear subspace. In fact, in such a case one can always find f_cT^c in the Diff-algebra as above.

4.31. Assume that a Diff-algebra $\mathscr{G}$ is simple at any singular point, or equivalently, that the τ-invariant is at least 1 at any closed point $x \in \mathrm{Sing}(\mathscr{G})$. Proposition 4.23 shows that the components of codimension one on the closed set $\mathrm{Sing}(\mathscr{G})$ are open and closed in that set, and that they form a natural center for a monoidal transformation.

THEOREM 4.32. *Let $\mathscr{G}$ be a Diff-algebra over a smooth scheme W, and fix $x \in \mathrm{Sing}(\mathscr{G})$ a closed point that we assume to be simple, and that the local codimension of $\mathrm{Sing}(\mathscr{G})$ at the point is at least two. Let $\pi : W \to W'$ be a smooth morphism defined at an étale neighborhood of x, where W' is smooth, dim W' = dim W-1. Assume that π is transversal at x. Then:*

1) At a suitable neighborhood of $\pi(x)$, there is a Rees algebra $\mathscr{R}_{\mathscr{G},\pi}$ over the smooth scheme W', defined only in terms of $\mathscr{G}$ and π, so that $\pi(\mathrm{Sing}(\mathscr{G})) = \mathrm{Sing}(\mathscr{R}_{\mathscr{G},\pi})$.

2) The morphism π induces a one-to-one map from $\mathrm{Sing}(\mathscr{G})$ to $\mathrm{Sing}(\mathscr{R}_{\mathscr{G},\pi})$. Furthermore, if we set $S = \mathscr{O}_{W',\pi(x)}$ and $S[Z]$ as before, then the one-to-one map is that described above.

3) $\mathscr{R}_{\mathscr{G},\pi} \subset \mathscr{G}$ via the inclusion $\mathscr{O}_{W',\pi(x)}[T] \subset \mathscr{O}_{W,x}[T]$.

4) If $e \geq 1$ denotes the τ-invariant of $\mathscr{G}$ at x, then the τ-invariant of $\mathscr{R}_{\mathscr{G},\pi}$ at $\pi(x)$ is $e-1$.

(See [**46**], Theorem 4.11, and Proposition 5.12.)

$\mathscr{R}_{\mathscr{G}}$ is also a Diff-algebra, and it is called the *elimination algebra* of $\mathscr{G}$ defined by π.

4.33. **Elimination and local presentation.** Let d denote the dimension of W. Fix, as before, an element $f_nT^n \in \mathscr{G}$ so that f_n has order n at the local regular ring $\mathscr{O}_{W,x}$. Up to multiplication by a unit we may assume that $f_n = f_n(Z)$ is a monic polynomial of degree n in $S[Z]$, where $S = \mathscr{O}_{W',\pi(x)}$. Set

$$Tay : S[Z] \to S[Z,T],$$

the S-algebra morphism defined by setting $Tay(Z) = Z + T$. Define operators

$$\Delta^e : S[Z] \to S[Z]$$

so that $Tay(g(Z)) = \sum_{e\geq 0} \Delta^e(g(Z))T^e$ as polynomial in $S[Z][T]$.

Rees algebras on a smooth scheme W are graded subalgebras of $\mathscr{O}_W[T]$. Given two such subalgebras, say $\mathscr{G}_1$ and $\mathscr{G}_2$, let $\mathscr{G}_1 \odot \mathscr{G}_2$ be the smallest algebra spanned by both.

PROPOSITION 4.34. *(Local presentation) Fix a Diff-algebra $\mathscr{G}$ and a closed point $x \in \mathrm{Sing}(\mathscr{G})$, where the local codimension of this closed set in W is at least two, and assume $ord_x(\mathscr{G}) = 1$. Choose a transversal projection $\pi : W \to W'$ and an element $f_nT^n \in \mathscr{G}$ so that f_n has order n at the local regular ring $\mathscr{O}_{W,x}$ and $f_n = f_n(Z)$ is a monic polynomial of degree n in $S[Z]$, where $S = \mathscr{O}_{W',\pi(x)}$. Then, at a suitable neighborhood of x:*

1) $\mathscr{G}$ and $\mathscr{O}_W[\Delta^e(f_n(Z))T^{n-e}, 0 \leq e \leq n-1] \odot \pi^*(\mathscr{R}_{\mathscr{G},\pi})$ are the same, up to integral closure.

2) An inclusion of Diff-algebras $\mathscr{G} \subset \mathscr{G}'$ is finite if and only if $\mathscr{R}_{\mathscr{G},\pi} \subset \mathscr{R}_{\mathscr{G}',\pi}$ is finite.

4.35. **Elimination and transformations.** We analyze the behavior of local presentation when applying monoidal transformations that are permissible for the Rees algebra. Consider again a localization, say $\mathscr{G}_x = \bigoplus I_k \cdot T^k (\subset \mathscr{O}_{W,x}[T])$, of a Diff-algebra $\mathscr{G}$ at a simple closed point $x \in \mathrm{Sing}(\mathscr{G})$. As indicated before, there is an index n and an element $f_n \in I_n$ of order n at $\mathscr{O}_{W,x}$, which we take to be a monic polynomial of degree n in $S[Z]$ at a suitable étale neighborhood, or at the completion $\hat{\mathscr{O}}_{W,x}$, and of order n in $R = S[Z]_{\langle M_S, Z\rangle}$, where S is regular, $R \subset \mathscr{O}_{W,x}$, and $\hat{R} = \hat{\mathscr{O}}_{W,x}$. This defines a finite morphism

$$\pi : \mathrm{Spec}(S[Z]/\langle f_n(Z)\rangle) \to \mathrm{Spec}(S).$$

Here we view $X = \mathrm{Spec}(S[Z]/\langle f_n(Z)\rangle)$ as a hypersurface in W. As indicated before, locally at x, $\mathrm{Sing}(\mathscr{G})$ is included in the n-fold points of this hypersurface (i.e. in $V(\mathrm{Diff}^{n-1}(\langle f_n(Z)\rangle))$.

Under these conditions the multiplicity formula asserts π is one-to-one on $V(\mathrm{Diff}^{n-1}(\langle f_n(Z)\rangle))$. Furthermore, if Y is closed and smooth in $V(\mathrm{Diff}^{n-1}(\langle f_n(Z)\rangle))$, then Y is isomorphic to $\pi(Y)(\subset \mathrm{Spec}(S))$ (see [**49**], Corollary 1, page 299). So both Y in W and $\pi(Y)$ in $\mathrm{Spec}(S)$ are regular centers.

Now let $W \leftarrow W'$ and $\mathrm{Spec}(S) \leftarrow U$ denote the monoidal transformations at Y and $\pi(Y)$, respectively, and let X' denote the strict transform of X. The hypersurface X' has at most points of multiplicity n. Let $F(\subset X')$ denote the closed set of points of multiplicity n. Locally at any point in F, there is a unique finite morphism, say $X' \to U$, compatible with π (the natural diagram commutes).

The regular center Y was chosen in $\mathrm{Sing}(\mathscr{G})$, so a transform, say

$$\mathscr{G}_1 = \bigoplus I_k^{(1)} \cdot T^k (\subset \mathscr{O}_{W'}[T]),$$

is defined. As $I(X) \subset I_n$, it follows that $I(X') \subset I_n^{(1)}$ (where $I(X)$ and $I(X')$ are the sheaves of ideals defining X and X', respectively). In particular, $\mathrm{Sing}(\mathscr{G}_1) \subset F$ (points of multiplicity n). Fix a closed point $y \in \mathrm{Sing}(\mathscr{G}_1)$ mapping to x. Locally at y there is a finite morphism

$$\pi' : \mathrm{Spec}(S'[Z]/\langle f'_n(Z)\rangle) \to U,$$

where f'_n is a strict transform of f_n.

PROPOSITION 4.36. *Set $E = \{H_1\}$, the exceptional hypersurface in W' defined by the monoidal transformation, and let $\overline{E} = \{\overline{H}_1\}$ be the exceptional hypersurface in U defined by the monoidal transformation on* $\mathrm{Spec}(S)$. *Then:*

1) π' is transversal to $\mathscr{G}_1$ at y, and an elimination algebra $\mathscr{R}_{\mathscr{G}_1} \subset \mathscr{O}_{U,\pi'(y)}[T]$ is defined.

2) If $\mathscr{R}_{\mathscr{G}_1} \subset \mathscr{O}_{U,\pi'(y)}[T]$ denotes, as above, the elimination algebra of $\mathscr{G}_1$, and if $\mathscr{R}_{\mathscr{G}} \subset \mathscr{O}_{\mathrm{Spec}(S),\pi(x)}[T]$ is the elimination algebra of $\mathscr{G}$, then $\mathscr{R}_{\mathscr{G}_1} \subset \mathscr{O}_{U,\pi'(y)}[T]$ is the transform of $\mathscr{R}_{\mathscr{G}}$.

3) $\pi'^{-1}(\overline{H}_1) = H_1$.

The proof of this statement is a consequence of a simple property of *stability* of local presentation. In fact, following Proposition 4.34, $\mathscr{G} = \mathscr{O}_W[\Delta^e(f_n(Z))T^{n-e}, 0 \leq e \leq n-1] \odot \pi^*(\mathscr{R}_{\mathscr{G},\pi})$ up to integral closure. Let f'_n denote the strict transform of f_n locally at $y \in W'$, and let Δ_1^e be defined in terms of the local projection π' at y (see 4.33). It is easy to check that $\Delta_1^e(f'_n(Z))T^{n-e}$ is the transform of $\Delta^e(f_n(Z))T^{n-e}$ for each $0 \leq e \leq n-1$. See [**9**].

We have formulated the proposition for one monoidal transformation. A similar result holds for a sequence of monoidal transformations that introduces an exceptional locus E having only normal crossings.

4.37. The following is a corollary of Theorem 4.32 and generalizes Proposition 4.23.

COROLLARY 4.38. *Fix an integer $e \geq 1$. Let $\mathscr{G}$ be a Diff algebra, and assume that the τ-invariant is $\geq e$ at any closed point $x \in \mathrm{Sing}(\mathscr{G})$. Then:*

i) The codimension of $\mathrm{Sing}(\mathscr{G})$ in W is at least e, and the union of components of codimension e, say Y, is open and closed in $\mathrm{Sing}(\mathscr{G})$ and smooth in W.

ii) After blowing up at Y we may assume that $\mathrm{Sing}(\mathscr{G})$ has codimension at least $e+1$ at each closed point of the singular locus.

4.7. Resolution functions via elimination

REMARK 4.39. We have indicated that, up to integral closure, any Rees algebra is of the form $\mathscr{G} = \mathscr{G}_{(J,b)}$. The analog of a basic object is a Rees algebra $\mathscr{G}$ at a smooth scheme W together with a set of closed and smooth hypersurfaces E having only normal crossings. We have defined permissible transformations of such data:

$$\begin{array}{ccccccc} (W,E) & \xleftarrow{\pi_1} & (W_1,E_1) & \xleftarrow{\pi_2} & \cdots & \xleftarrow{\pi_k} & (W_r,E_r). \\ \mathscr{G} & & (\mathscr{G})_1 & & & & (\mathscr{G})_r \end{array} \tag{4.7.1}$$

This sequence of transformations induces transformations of basic objects:

$$(W,(J,b),E) \leftarrow (W_1,(J_1,b),E_1) \leftarrow \cdots \leftarrow (W_s,(J_r,b),E_r), \tag{4.7.2}$$

which we consider together with the natural factorization

$$J_r = I(H_1)^{b_1} I(H_2)^{b_2} \cdots I(H_r)^{b_r} \cdot \overline{J}_r \tag{4.7.3}$$

so that $\overline{J}_r$ does not vanish along H_i, $1 \leq i \leq r$.

Furthermore, $\mathrm{Sing}(\mathscr{G}_r) = \mathrm{Sing}(J_r,b)$, and the function w-ord$_{\mathscr{G}_r}$ is the same as w-ord(J_r,b) (defined in terms of the expression (4.7.3)).

So we fix a dimension d and consider the same inductive function t^d with values in $\mathbb{Q} \times \mathbb{Z}$. We will consider the same totally ordered set I_d and the same resolution function f as for the case of basic objects. As discussed in 3.17, there is a first step to achieve monomialization and then a second step of resolution of the monomial case. Let (W,E) be a pair, where W is a d-dimensional smooth scheme over a field, and consider a Rees algebra $\mathscr{G}$ as above.

One property of the inductive function t^d, when applied to basic objects in 3.3, was that it reduced resolution of d-dimensional basic objects to resolution of simple d-dimensional basic objects, namely those of the form $(W,(J,b),E)$, where (J,b) is a simple couple and $E = \emptyset$ (3.8).

So if we assume resolution of simple basic objects, then $\max t^d$ was forced to drop. And, as the second coordinate of $\max t^d$ can drop only finitely many times, at some point the first coordinate must drop, and ultimately we come to the case where $\max$ w-ord $= 0$ (monomial case), which was simple to resolve (3.3).

The analog of a simple basic object in the class of Rees algebras will be a simple Rees algebra $\mathscr{G}$ (simple at each point of $\mathrm{Sing}(\mathscr{G})$) and a pair $(W, E = \emptyset)$, with W smooth of dimension d.

We assume now that the sequence (4.3.2) is defined so that

$$\max t^d \geq \max t_1^d \cdots \geq \max t_r^d.$$

Recall now the notion of birth in 3.15, defined first for the inductive function t^d. This will lead us to a unique definition of the sequence (4.7.1) for each r.

Our assumption implies that $\max$ w-ord $\geq \max$ w-ord$_1 \geq \cdots \geq \max$ w-ord$_r$.

If $\max \text{w-ord}_r = 0$, then the *birth* was the smallest index $r' \leq r$ for which $\max \text{w-ord}_{r'} = 0$. In such a case, we assume that

$$\begin{array}{ccccccc} (W_{r'}, E_{r'}) & \xleftarrow{\pi_1} & (W_{r'+1}, E_{r'+1}) & \xleftarrow{\pi_2} & \cdots & \xleftarrow{\pi_k} & (W_r, E_r) \\ (\mathscr{G})_{r'} & & (\mathscr{G})_{r'+1} & & & & (\mathscr{G})_r \end{array}$$

is defined by blowing up at $\underline{\text{Max}} h_j$, $r' \leq j \leq r$, and we will extend (4.7.1) to a constructive resolution by successively blowing up at $\underline{\text{Max}} h_r$.

If $\max \text{w-ord}_r > 0$, then we may assume, by induction on r, that $\max t \geq \max t_1 \geq \cdots \geq \max t_r$, and define the *birth* to be the smallest index $r' \leq r$ for which

$$\max t_{r'} = \max t_{r'+1} = \cdots = \max t_r.$$

At level r' the simple basic object $(W_{r'}, (J'', b''), E = \emptyset)$ used in 3.15 will be replaced now by a simple Diff-algebra, say $(\mathscr{G}'')_{r'}$, and a pair $(W_{r'}, E = \emptyset)$. Proposition 4.23 asserts that the irreducible components of $\text{Sing}((\mathscr{G}'')_{r'})$ of codimension one define a permissible center, and that is the choice of center if such components do exist, and after such transformation all components of $\text{Sing}((\mathscr{G}'')_{r'})$ are of codimension at least two. So we assume that

$$(4.7.4) \qquad \begin{array}{ccccccc} (W_{r'}, E_{r'}) & \xleftarrow{\pi_1} & (W_{r'+1}, E_{r'+1}) & \xleftarrow{\pi_2} & \cdots & \xleftarrow{\pi_k} & (W_r, E_r) \\ (\mathscr{G})_{r'} & & (\mathscr{G})_{r'+1} & & & & (\mathscr{G})_r \end{array}$$

is defined by blowing first at $R(1)(\text{Sing}(\mathscr{G}))$, if not empty, and then successively at $\underline{\text{Max}} \overline{f}_r^{d-1}$, where $\overline{f}_r^{d-1}(x)$ is defined by the $d-1$-dimensional Diff-algebra attached to $\mathscr{G}''$ in Theorem 4.32. In other words, there is a sequence

$$(4.7.5) \qquad \begin{array}{ccccccc} (W_{r'}, E = \emptyset) & \xleftarrow{\pi_1} & (W_{r'+1}, E_1) & \xleftarrow{\pi_2} & \cdots & \xleftarrow{\pi_k} & (W_r, E_{r-r'}) \\ (\mathscr{G}'')_{r'} & & (\mathscr{G}'')_{r'+1} & & & & (\mathscr{G}'')_r \end{array}$$

inducing (4.7.4) with the property that $\text{Sing}((\mathscr{G}'')_j) = \underline{\text{Max}} t_j^d$, $r' \leq j \leq r$.

If we fix a closed point $x \in \text{Sing}((\mathscr{G}'')_j)$ and a local transversal projection, say $\pi : W_{r'} \longrightarrow W'$ (smooth morphism defined at an étale neighborhood of x), where W' is smooth, $\dim W' = \dim W$-1, and if we set the elimination algebra $\mathscr{R}_{\mathscr{G},\pi} \subset \mathscr{G}''_{r'}$ and $(W', \overline{E} = \emptyset)$, then there is a sequence

$$(4.7.6) \qquad \begin{array}{ccccccc} (W', \overline{E} = \emptyset) & \xleftarrow{\pi_1} & (W'_1, \overline{E}_1) & \xleftarrow{\pi_2} & \cdots & \xleftarrow{\pi_k} & (W'_{r-r'}, \overline{E}_{r-r'}) \\ \mathscr{R}_{\mathscr{G},\pi} & & (\mathscr{R}_{\mathscr{G},\pi})_1 & & & & (\mathscr{R}_{\mathscr{G},\pi})_{r-r'} \end{array}$$

defined in terms of the resolution functions $\overline{f}_j^{d-1}$. Furthermore, for each index $j \leq r - r'$, $(\mathscr{R}_{\mathscr{G},\pi})_j$ is the elimination algebra of $(\mathscr{G}'')_{r'+j}$ (see Theorem 4.32 and Proposition 4.36).

Note that $\mathscr{G}''_{r'}$ is simple, which can also be formulated by saying that the τ-invariant at any closed point $x \in \text{Sing}(\mathscr{G}''_{r'})$ is at least 1. In the same way in which the simple Diff-algebra $\mathscr{G}''_{r'}$ was attached to $(\mathscr{G})_{r'}$, a new simple Diff-algebra will also be attached to $\mathscr{R}_{\mathscr{G},\pi}$ in the definition of 4.7.6.

A difference of this method, based on elimination algebras, with the approach using Proposition 2.31, is that here there is no notion of restriction. We make use of the inclusion of elimination algebras on the original algebra (Theorem 4.32, 3), and a simple Diff-algebra attached to $\mathscr{R}_{\mathscr{G},\pi}$ (in dimension $d-1$) will give rise to a Diff-algebra, in dimension d, with τ-invariant at least two at any singular point. And so on.

4.40. **Canonical choice of centers**

Let e be a positive integer such that the first e coordinates of $\max f_r^d$, say $(\max f_r^d)_e$, are all in $\mathbb{Q}\times\mathbb{Z}$ (i.e. that $(\max f_r^d)_e \in (\mathbb{Q}\times\mathbb{Z})^e$). Recall that we define the e-birth of $\max f_r^d$ as the smallest index $r'(\leq r)$ for which

$$(\max f_{r'}^d)_e = (\max f_r^d)_e.$$

The closed set $F(r,e)$ was defined as the set of points where the first e coordinates of f_r^d take this highest value $(\max f_r^d)_e$. $F(r,e)$ is closed in $\mathrm{Sing}(\mathcal{G})$, and $(\max f_{r'}^d)_e = (\max f_{r'+1}^d)_e = \cdots = (\max f_r^d)_e$. It follows from our construction that:

1) There is a Rees algebra $\mathcal{G}^e$ at $W_{r'}$ so that $\mathrm{Sing}(\mathcal{G}^e) = F(r',e)$, and $\mathcal{G}^e$ has τ-invariant at least e at any closed point of the singular locus.

2) All centers of monoidal transformations, between the levels r' and r, define

$$\begin{array}{ccccccc} (W_{r'},\emptyset) & \xleftarrow{\pi_1} & (W_{r'+1},E'_1) & \xleftarrow{\pi_2} & \cdots & \xleftarrow{\pi_k} & (W_r,E'_r). \\ \mathcal{G}^e & & (\mathcal{G}^e)_1 & & & & (\mathcal{G}^e)_r \end{array}$$

3) $F(r+i,e) = \mathrm{Sing}((\mathcal{G}^e)_i)$ for $0\leq i\leq r-r'$.

4) The previous sequence is defined so that the first transformation has center at the set of points where $\mathrm{Sing}(\mathcal{G}^e)$ has codimension e at $W_{r'}$, if not empty, and then in accordance to the resolution of an auxiliary Diff algebra $\mathcal{G}^{e+1}$, with τ-invariant at least $e+1$ at any singular point.

As for the canonical choice of centers discussed in 3.19, recall that there are two possibilities for the biggest value of the resolution function at a level r:

A) $\max f_r^d = (\alpha_1,\alpha_2,\ldots,\alpha_e,\infty,\infty,\ldots,\infty)$, $\alpha_i \in (\mathbb{Q}\times\mathbb{Z})$.

B) $\max f_r^d = (\alpha_1,\alpha_2,\ldots,\alpha_e,\gamma,\infty,\infty,\ldots,\infty)$, $\alpha_i \in (\mathbb{Q}\times\mathbb{Z})$, $\gamma\in\Gamma$.

We consider in both cases the first e-coordinates, say $(\max f_r^d)_e = (\alpha_1,\alpha_2,\ldots,\alpha_e)$, and let r' denote the e-birth of $(\max f_r^d)_e$.

In case A), $r'=r$; namely, the e-birth is at the same level r. This case occurs when $\mathcal{G}^e$ in 2) is such that $\mathrm{Sing}(\mathcal{G}^e)$ is a union of components of codimension e.

In case B), $r'<r$, and this case occurs when all auxiliary Diff algebra $\mathcal{G}^{e+1}$ in 4) (with τ-invariant at least $e+1$ at any singular point) have been resolved. This is the case in which $(\mathcal{G}^e)_r$ is e codimensional and "monomial in codimension e".

The canonical choice of center in case (A) is characteristic-free. In the case of characteristic zero the canonical choice of center in Case B) is given by $\max h$, where h is the resolution function for the monomial case, applied here in "codimension e". Constructive resolution of Case (B) remains open over arbitrary fields.

Bibliography

[1] S. S. Abhyankar, 'Good points of a Hypersurface', *Adv. in Math. 68 (1988) 87-256.*

[2] D. Abramovich and A. J. de Jong, 'Smoothness, semistability and toroidal geometry', *Journal of Algebraic Geometry* 6 (1997) 789-801.

[3] D. Abramovich and J. Wang, 'Equivariant resolution of singularities in characteristic 0', *Mathematical Research Letters* 4 (1997) 427-433.

[4] J. M. Aroca, H. Hironaka, and J. L. Vicente, 'The theory of maximal contact', *Mem. Mat. Ins. Jorge Juan (Madrid)* 29 (1975).

[5] M. F. Atiyah, I. G. Macdonald, 'Introduction to commutative algebra', Addison-Wesley Publishing Co., Reading, MA, 1969.

[6] A. Benito and O. Villamayor, 'Stratification of Hypersurface singularities and monoidal transformations'.

[7] E. Bierstone and P. Milman, 'Canonical desingularization in characteristic zero by blowing-up the maxima strata of a local invariant', *Inv. Math.* 128 (1997) 207-302.

[8] E. Bierstone and P. Milman, 'Desingularization algorithms I. Role of exceptional divisors'. *Mosc. Math. J.* 3 (2003), no 3, 751-805, 1197. MR2078560.

[9] A. Bravo and O. Villamayor, 'Stratification of Hypersurface singularities'.

[10] G. Bodnár and J. Schicho, 'A Computer Program for the Resolution of Singularities', *Resolution of singularities. A research book in tribute of Oscar Zariski* (eds H. Hauser, J. Lipman, F. Oort, A. Quirós), Progr. Math. 181 (Birkhäuser, Basel, 2000) pp. 231-238.

[11] G. Bodnár and J. Schicho, 'Automated resolution of singularities for hypersurfaces', *J. Symbolic Comput.* (4) 30 (2000) 401-428.

[12] F. Bogomolov and T. Pantev, 'Weak Hironaka Theorem', *Mathematical Research Letters* 3 (1996) 299-307.

[13] A. Bravo, S. Encinas, and O. Villamayor, 'A simplified proof of desingularization and applications', *Rev. Mat. Iberoamericana* (2005) Vol. 21, no. 2.

[14] A. Bravo and O. Villamayor, 'Strengthening a Theorem of Embedded Desingularization,' *Math. Res. Letters* 8 (2001) 1-11.

[15] A. Bravo and O. Villamayor, 'A Strengthening of resolution of singularities in characteristic zero'. *Proc. London Math. Soc.* (3) 86 (2003) 327-357.

[16] S. D. Cutkosky, 'Resolution of singularities.' *Graduate Studies in Mathematics.* Vol 63, AMS (2004).

[17] S. Encinas and H. Hauser, 'Strong Resolution of Singularities'. *Comment. Math. Helv.* 77 2002, no. 4, pp. 821-845.

[18] S. Encinas, A. Nobile, and O. Villamayor, 'On algorithmic Equiresolution and stratification of Hilbert schemes'. *Proc. London Math. Soc.* (3) 86 (2003), 607-648.

[19] S. Encinas and O. Villamayor, 'Good points and constructive resolution of singularities', *Acta Math.* 181:1 (1998) 109-158.

[20] S. Encinas and O. Villamayor, 'A Course on Constructive Desingularization and Equivariance', *Resolution of Singularities. A research textbook in tribute to Oscar Zariski* (eds H. Hauser, J. Lipman, F. Oort, A. Quirós), Progr. in Math. 181 (Birkhäuser, Basel, 2000), pp. 147-227.

[21] S. Encinas and O. Villamayor, 'A new proof of desingularization over fields of characteristic zero'. Rev. Mat. Iberoamericana 19 (2003), no. 2, 339-353.

[22] A. Fruehbis-Krueger and G. Pfister, Singular library for Resolution of Singularities.

[23] J. Giraud, 'Sur la theorie du contact maximal', *Math. Zeit.*, 137 (1972), 285-310.

[24] J. Giraud, 'Contact maximal en caractéristique positive', *Ann. Scien. de l'Ec. Norm. Sup. 4ème série*, 8 (1975), 201-234.

[25] A. Grothendieck, 'Éléments de géométrie algébrique. IV. Étude locale des schémas et des morphismos de schémas IV' *Inst. Hautes Études Sci. Publ. Math.* (32): 361, 1967 (1972), 285-310.

[26] H. Hauser, 'Excellent surfaces and Their Taut Resolution' in *Resolution of Singularities. A research textbook in tribute to Oscar Zariski*, Eds. H. Hauser, J. Lipman, F. Oort, A. Quirós. Progress in Math. vol **181**, Birkhäuser 2000.

[27] H. Hauser, 'The Hironaka Theorem on Resolution of Singularities (or: A proof we always wanted to understand)', *Bull. Amer. Math. Soc.* (N.S.) 40 (2003), no. 3, 323-403.

[28] H. Hironaka, 'Resolution of singularities of an algebraic variety over a field of characteristic zero I-II', *Ann. Math.*, 79 (1964) 109–326.

[29] H. Hironaka, 'Idealistic exponent of a singularity', *Algebraic Geometry*, The Johns Hopkins centennial lectures, Baltimore, Johns Hopkins University Press (1977), 52-125.

[30] H. Hironaka, 'Theory of infinitely near singular points', *Journal Korean Math. Soc.* 40 (2003), No. 5, pp. 901-920.

[31] H. Hironaka, 'Additive groups associated with points of a projective space'. *Ann. of Math.* (2), 92:327-334, 1970.

[32] H. Hironaka, 'A program for resolution of singularities, in all characteristic $p > 0$, and in all dimension'. preprint of series of lectures in "Summer School on Resolution of Singularities" at International Center of Theoretical Physics. Trieste, June 2006.

[33] H. Kawanoue, 'Toward resolution of singularities over a field of positive characteristic' arXiv: math.AG/ 0607009 3 Aug 2006.

[34] J. Kollár, 'Resolution of Singularities- Seattle Lecture' arXiv: math.AG/ 0508332 v1 17 Aug 2005.

[35] J. Lipman, 'Introduction to Resolution of Singularities', *Proc. Symp. in Pure Math.* 29 (1975) pp. 187-230.

[36] J. Lipman, 'Equisingularity and Simultaneous Resolution of Singularities', *Resolution of Singularities. A research textbook in tribute to Oscar Zariski* (eds H. Hauser, J. Lipman, F. Oort and A. Quirós), Progr. in Math. 181, (Birkhäuser, Basel, 2000) pp. 485-505.

[37] K. Matsuki, 'Notes on the inductive algorithm of resolution of singularities'. Preprint. arXiv: math.AG/0103120.

[38] K. Matsuki, H. Kawanoue. Toward resolution of singularities over a field of positive characteristic (The Kawanoue program) Part II. Basic invariants associated to the idealistic filtration and their properties. Preprint: arXiv:math/0612008v2.

[39] H. Matsumura. 'Commutative algebra', *Mathematics Lecture Note Series, 56*, Benjamin/Cummings Publishing Company, Inc., 2nd edition, 1980.

[40] T. Oda, 'Infinitely very near singular points', *Complex analytic singularities*, Adv. Studies in Pure Math. 8 (North-Holland, 1987) pp. 363–404.

[41] O. Villamayor, 'Constructiveness of Hironaka's resolution', *Ann. Scient. Ec. Norm. Sup.* 4^e serie 22 (1989) 1-32.

[42] O. Villamayor, 'Patching local uniformizations', *Ann. Scient. Ec. Norm. Sup.*, 25 (1992), 629-677.

[43] O. Villamayor, 'Rees algebras on smooth schemes: integral closure and higher differential operators. ArXiv:math. AG/0606795.

[44] O. Villamayor, 'Singularity of the embedded hypersurface and Tschirnhausen revisited.' Academia Nacional de Ciencias en Córdoba (Argentina). Boletín núm. 65.

[45] O. Villamayor, 'On constructive desingularization.' Journal of Symbolic Computation. 39, 2005, pp 465-491.

[46] O. Villamayor, 'Hypersurface singularities in positive characteristic', *Adv. in Math.* 213 (2007), no. 2, pp. 687–733.

[47] J. Wlodarczyk, 'Simple Hironaka resolution in characteristic zero', *J. Amer. Math. Soc.* 18 (2005), no. 4, pp. 779–822.

[48] B. Youssin, 'Newton Polyhedra without coordinates', *Mem. Amer. Math. Soc.* 433 (1990), 1-74, 75-99.

[49] O. Zariski, P. Samuel, 'Commutative Algebra (Vol II)' *University Series in Higher Mathematics*. D. Van Nostrand Company (1960).

Titles in This Series

42 **Holger Brenner, Jürgen Herzog, and Orlando Villamayor (Gemma Colomé-Nin, Teresa Cortadellas Benítez, Juan Elias, and Santiago Zarzuela, Editors),** Three lectures on commutative algebra, 2008

41 **James Haglund,** The q,t-Catalan numbers and the space of diagonal harmonics (with an appendix on the combinatorics of Macdonald polynomials), 2008

40 **Vladimir Pestov,** Dynamics of infinite-dimensional groups. The Ramsey–Dvoretzky–Milman phenomenon, 2006

39 **Oscar Zariski,** The moduli problem for plane branches (with an appendix by Bernard Teissier), 2006

38 **Lars V. Ahlfors,** Lectures on quasiconformal mappings, Second Edition, 2006

37 **Alexander Polishchuk and Leonid Positselski,** Quadratic algebras, 2005

36 **Matilde Marcolli,** Arithmetic noncommutative geometry, 2005

35 **Luca Capogna, Carlos E. Kenig, and Loredana Lanzani,** Harmonic measure: Geometric and analytic points of view, 2005

34 **E. B. Dynkin,** Superdiffusions and positive solutions of nonlinear partial differential equations, 2004

33 **Kristian Seip,** Interpolation and sampling in spaces of analytic functions, 2004

32 **Paul B. Larson,** The stationary tower: Notes on a course by W. Hugh Woodin, 2004

31 **John Roe,** Lectures on coarse geometry, 2003

30 **Anatole Katok,** Combinatorial constructions in ergodic theory and dynamics, 2003

29 **Thomas H. Wolff (Izabella Łaba and Carol Shubin, editors),** Lectures on harmonic analysis, 2003

28 **Skip Garibaldi, Alexander Merkurjev, and Jean-Pierre Serre,** Cohomological invariants in Galois cohomology, 2003

27 **Sun-Yung A. Chang, Paul C. Yang, Karsten Grove, and Jon G. Wolfson,** Conformal, Riemannian and Lagrangian geometry, The 2000 Barrett Lectures, 2002

26 **Susumu Ariki,** Representations of quantum algebras and combinatorics of Young tableaux, 2002

25 **William T. Ross and Harold S. Shapiro,** Generalized analytic continuation, 2002

24 **Victor M. Buchstaber and Taras E. Panov,** Torus actions and their applications in topology and combinatorics, 2002

23 **Luis Barreira and Yakov B. Pesin,** Lyapunov exponents and smooth ergodic theory, 2002

22 **Yves Meyer,** Oscillating patterns in image processing and nonlinear evolution equations, 2001

21 **Bojko Bakalov and Alexander Kirillov, Jr.,** Lectures on tensor categories and modular functors, 2001

20 **Alison M. Etheridge,** An introduction to superprocesses, 2000

19 **R. A. Minlos,** Introduction to mathematical statistical physics, 2000

18 **Hiraku Nakajima,** Lectures on Hilbert schemes of points on surfaces, 1999

17 **Marcel Berger,** Riemannian geometry during the second half of the twentieth century, 2000

16 **Harish-Chandra,** Admissible invariant distributions on reductive p-adic groups (with notes by Stephen DeBacker and Paul J. Sally, Jr.), 1999

15 **Andrew Mathas,** Iwahori-Hecke algebras and Schur algebras of the symmetric group, 1999

14 **Lars Kadison,** New examples of Frobenius extensions, 1999

13 **Yakov M. Eliashberg and William P. Thurston,** Confoliations, 1998

12 **I. G. Macdonald,** Symmetric functions and orthogonal polynomials, 1998

11 **Lars Gårding,** Some points of analysis and their history, 1997

TITLES IN THIS SERIES

10 **Victor Kac,** Vertex algebras for beginners, Second Edition, 1998

9 **Stephen Gelbart,** Lectures on the Arthur-Selberg trace formula, 1996

8 **Bernd Sturmfels,** Gröbner bases and convex polytopes, 1996

7 **Andy R. Magid,** Lectures on differential Galois theory, 1994

6 **Dusa McDuff and Dietmar Salamon,** J-holomorphic curves and quantum cohomology, 1994

5 **V. I. Arnold,** Topological invariants of plane curves and caustics, 1994

4 **David M. Goldschmidt,** Group characters, symmetric functions, and the Hecke algebra, 1993

3 **A. N. Varchenko and P. I. Etingof,** Why the boundary of a round drop becomes a curve of order four, 1992

2 **Fritz John,** Nonlinear wave equations, formation of singularities, 1990

1 **Michael H. Freedman and Feng Luo,** Selected applications of geometry to low-dimensional topology, 1989